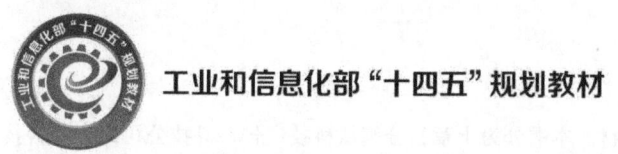

工业和信息化部"十四五"规划教材

科技翻译工作坊

◆ 徐丽华　主　编
◆ 王　辉　副主编

电子工业出版社
Publishing House of Electronics Industry
北京·BEIJING

内 容 简 介

本书是工业和信息化部"十四五"规划教材。本书分为十章,分别从科技广告、科技公司简介、科技合同、科技新闻、科技专利文献、科技论文摘要、科技产品说明书、科技会展文案、科技文章、科技产品简介十个方面进行介绍。每章由文本介绍、范文展示、常用词汇和句型、翻译技巧、课后练习五个部分组成。

本书突出实用性,不仅提供常见科技文本中的常用词汇和句型,还有大量的双语科技文本范文供学生学习。除此之外,每章都梳理了不同科技文本的翻译技巧来帮助学生融会贯通,还设计了大量的翻译实践练习,给学生提供将理论运用于实践的机会,从而提升学生翻译能力。

本书可作为本科生和硕士研究生科技翻译类课程的教材和辅导用书,也可作为广大从事科技领域工作的人员、科技翻译工作者和学习者的自学参考书。

未经许可,不得以任何方式复制或抄袭本书之部分或全部内容。
版权所有,侵权必究。

图书在版编目(CIP)数据

科技翻译工作坊/徐丽华主编. —北京:电子工业出版社,2023.3
ISBN 978-7-121-45031-0

Ⅰ.①科… Ⅱ.①徐… Ⅲ.①科学技术-翻译-高等学校-教材 Ⅳ.①N

中国国家版本馆 CIP 数据核字(2023)第 024929 号

责任编辑:孟　宇
文字编辑:韩玉宏
印　　刷:北京虎彩文化传播有限公司
装　　订:北京虎彩文化传播有限公司
出版发行:电子工业出版社
　　　　　北京市海淀区万寿路 173 信箱　邮编:100036
开　　本:787×1092　1/16　印张:18.25　字数:610 千字
版　　次:2023 年 3 月第 1 版
印　　次:2023 年 12 月第 2 次印刷
定　　价:69.80 元

凡所购买电子工业出版社图书有缺损问题,请向购买书店调换。若书店售缺,请与本社发行部联系,联系及邮购电话:(010)88254888,88258888。

质量投诉请发邮件至 zlts@phei.com.cn,盗版侵权举报请发邮件至 dbqq@phei.com.cn。
本书咨询联系方式:mengyu@phei.com.cn。

前言

为满足社会职场对科技翻译人才的要求，提高翻译专业学生的翻译实战能力，促进翻译教学模式的转变，我们组织编写了这本《科技翻译工作坊》。翻译工作坊教学作为一种新的教学方式，给传统的翻译课堂教学带来了巨大变化，教师不再是翻译理论和翻译技能教学的主角，而是学生成为主角。翻译工作坊教学有别于传统教学中单一的技巧训练与语言能力提升，而是更注重协作翻译的过程，重视学生语言能力、社交能力乃至跨文化交际能力的培养。翻译工作坊教学是以学生为中心的过程导向教学，注重译文产生过程中的每个步骤，在提升学生语言能力的同时，也促进学生人际交往能力、个人责任意识的培养，真正培养学生的实际操作能力，使学生加深对翻译这一跨文化交际活动的认识。同时，对教师来说，工作坊式的翻译教学也可以让教师在教学的同时锻炼自身的翻译技能，反思教学经验。可以说，翻译工作坊教学真正实现了"教学相长"。

党的二十大报告指出："教育、科技、人才是全面建设社会主义现代化国家的基础性、战略性支撑。必须坚持科技是第一生产力、人才是第一资源、创新是第一动力，深入实施科教兴国战略、人才强国战略、创新驱动发展战略，开辟发展新领域新赛道，不断塑造发展新动能新优势。"

本书是工业和信息化部"十四五"规划教材，可作为本科生和硕士研究生科技翻译类课程的教材和辅导用书，也可作为广大从事科技领域工作的人员、科技翻译工作者和学习者的自学参考书。为了让学生掌握科技翻译方面的相关知识和要领，提高其翻译水平，本书突出体现"知识性、应用性"特征，紧扣"科技"这一主题，难度适中，力求深入浅出。相信通过对本书的学习，学生可充分掌握科技翻译的方法和技巧，达到实际运用的目的。

比起翻译理论教学，本书更注重翻译的实践性，文章选材都源于国内外真实科技文本，在教学中将理论与真正的翻译实践相融合，以开放式的交际场景和易于操作的练习形式，激发学生的学习兴趣和潜力，强化英语运用能力，在翻译实践中学翻译。本书还设置了词汇翻译、句子翻译、段落翻译和篇章翻译，增加学生的练习机会，增强实战性。教师和学生可以根据自身需要灵活安排学习内容，在使用时可以先学习翻译技巧，再运用于翻译实践，也可以先实践，有了一定的感性经验后，再学习翻译技巧。每章的课后练习可以根据需要在课内使用或留作课外作业。

本书中每章都对特定领域的案例进行了翻译技巧的讲解，具有针对性，有助于翻译专业或相关专业的学生了解相关类型科技文本的特点，同时学习翻译理论和翻译技巧的运用，使学生在理解原文的基础上翻译出语言准确、通顺的译文。

本书由徐丽华和王辉老师团队编写。其中，第一章至第五章由徐丽华编写，共计18万字；第六章至第十章由王辉编写，共计18万字；尹瑞、袁嘉黛、冯婧辉、许小倩、商文文、卢欣怡、巩舒予负责各章范文的整理和编辑工作，每人分计3.6万字，共25万字。

编写教材不易，但这是一项有意义的工作，尽管我们付出了很多努力，但由于水平有限，书中还存在有待改进之处，敬请使用本书的老师和同学们提出宝贵的修改建议，以帮助我们进一步完善本书内容。

编　者

目 录

第一章 科技广告 1
 一、科技广告文本介绍 1
  (一) 科技广告概述 1
  (二) 科技广告的特点 2
 二、科技广告范文展示 3
 三、科技广告常用词汇和句型 9
  (一) 科技广告常用词汇 9
  (二) 科技广告常用句型 11
 四、科技广告翻译技巧 11
  (一) 巧译修辞 11
  (二) 活用四字格 16
  (三) 创译法 17
 五、课后练习 17
  (一) 思考题 17
  (二) 翻译练习 18

第二章 科技公司简介 22
 一、科技公司简介文本介绍 22
  (一) 科技公司简介概述 22
  (二) 科技公司简介文本结构 22
  (三) 科技公司简介的作用 22
  (四) 科技公司简介的特点 23
 二、科技公司简介范文展示 25
 三、科技公司简介常用词汇和句型 28
  (一) 科技公司简介常用词汇 28
  (二) 科技公司简介常用句型 36
 四、科技公司简介翻译技巧 37
  (一) 中英文科技公司简介对比差异 37
  (二) 翻译技巧 41
 五、课后练习 45
  (一) 思考题 45
  (二) 翻译练习 45

第三章 科技合同 ····· 51
一、科技合同文本介绍 ····· 51
（一）科技合同概述 ····· 51
（二）科技合同的特点 ····· 51
二、科技合同范文展示 ····· 53
三、科技合同常用词汇和句型 ····· 55
（一）科技合同常用词汇 ····· 55
（二）科技合同常用句型 ····· 58
四、科技合同翻译技巧 ····· 59
（一）掌握合同结构 ····· 59
（二）熟练使用专业术语 ····· 59
（三）意义对等优先，点面结合 ····· 60
（四）科技合同翻译方法与策略 ····· 61
（五）科技英语中长句翻译的方法 ····· 67
五、课后练习 ····· 69
（一）思考题 ····· 69
（二）翻译练习 ····· 69

第四章 科技新闻 ····· 75
一、科技新闻文本介绍 ····· 75
（一）科技新闻概述 ····· 75
（二）科技新闻的特点 ····· 75
二、科技新闻范文展示 ····· 79
三、科技新闻常用词汇和句型 ····· 86
（一）新闻报道和科技新闻常用词汇 ····· 86
（二）科技新闻常用句型 ····· 90
四、科技新闻翻译技巧 ····· 91
（一）与原文风格保持一致 ····· 91
（二）灵活简洁，重点突出 ····· 92
五、课后练习 ····· 95
（一）思考题 ····· 95
（二）翻译练习 ····· 95

第五章 科技专利文献 ····· 100
一、科技专利文献文本介绍 ····· 100
（一）科技专利文献概述 ····· 100
（二）科技专利文献文本结构 ····· 100
（三）科技专利文献的作用 ····· 101
（四）科技专利文献的特点 ····· 102
（五）科技专利文献等级分类及类型 ····· 104
（六）专利说明书组成部分 ····· 105

（七）专利文献著录项目及其代码 ·· 105
　　（八）专利文献公开号 ·· 106
二、科技专利文献范文展示 ··· 106
三、科技专利文献常用词汇和句型 ·· 110
　　（一）科技专利文献常用词汇 ·· 110
　　（二）科技专利文献常用句型 ·· 116
四、科技专利文献翻译技巧 ··· 117
　　（一）词义选择 ·· 117
　　（二）增词法与省略法 ·· 118
　　（三）重复法 ·· 119
　　（四）拆译法 ·· 119
　　（五）被动语态的翻译 ·· 119
　　（六）定语从句的翻译 ·· 120
　　（七）状语从句的翻译 ·· 121
五、课后练习 ··· 122
　　（一）思考题 ·· 122
　　（二）翻译练习 ·· 122

第六章　科技论文摘要

一、科技论文摘要文本介绍 ··· 126
　　（一）科技论文摘要概述 ·· 126
　　（二）科技论文摘要文本结构 ·· 126
　　（三）科技论文摘要的特点 ·· 126
　　（四）科技论文摘要的分类 ·· 129
二、科技论文摘要范文展示 ··· 129
三、科技论文摘要常用词汇和句型 ·· 130
　　（一）科技论文摘要常用词汇 ·· 130
　　（二）科技论文摘要常用句型 ·· 132
四、科技论文摘要翻译技巧 ··· 134
　　（一）时态方面 ·· 134
　　（二）词汇方面 ·· 135
　　（三）句法方面 ·· 136
五、课后练习 ··· 138
　　（一）思考题 ·· 138
　　（二）翻译练习 ·· 138

第七章　科技产品说明书

一、科技产品说明书文本介绍 ··· 142
　　（一）科技产品说明书概述 ·· 142
　　（二）科技产品说明书文本结构 ·· 142
　　（三）科技产品说明书的作用 ·· 143

- （四）科技产品说明书的特点 ……………………………………………… 143
- （五）科技产品说明书的分类 ……………………………………………… 144
- 二、科技产品说明书范文展示 …………………………………………………… 144
- 三、科技产品说明书常用词汇和句型 …………………………………………… 149
  - （一）科技产品说明书常用词汇 …………………………………………… 149
  - （二）科技产品说明书常用句型 …………………………………………… 154
- 四、科技产品说明书翻译技巧 …………………………………………………… 158
  - （一）引申译法 ……………………………………………………………… 158
  - （二）词类转换 ……………………………………………………………… 160
  - （三）句法转换 ……………………………………………………………… 165
- 五、课后练习 ……………………………………………………………………… 165
  - （一）思考题 ………………………………………………………………… 165
  - （二）翻译练习 ……………………………………………………………… 166

第八章 科技会展文案 ………………………………………………………… 172

- 一、科技会展文案文本介绍 ……………………………………………………… 172
  - （一）科技会展文案概述 …………………………………………………… 172
  - （二）科技会展方案的分类 ………………………………………………… 172
  - （三）科技会展文案文本分析 ……………………………………………… 173
- 二、科技会展文案范文展示 ……………………………………………………… 176
- 三、科技会展文案常用词汇和句型 ……………………………………………… 178
  - （一）科技会展文案常用词汇 ……………………………………………… 178
  - （二）科技会展文案常用句型 ……………………………………………… 182
- 四、科技会展文案翻译技巧 ……………………………………………………… 182
  - （一）科技会展文案翻译的基本要求 ……………………………………… 183
  - （二）纽马克的翻译思想在科技会展文案翻译中的应用 ………………… 183
  - （三）科技会展文案翻译手段 ……………………………………………… 185
- 五、课后练习 ……………………………………………………………………… 187
  - （一）思考题 ………………………………………………………………… 187
  - （二）翻译练习 ……………………………………………………………… 187

第九章 科技文章 ……………………………………………………………… 190

- 一、科技文章文本介绍 …………………………………………………………… 190
  - （一）科技文章概述 ………………………………………………………… 190
  - （二）科技文章的特点 ……………………………………………………… 190
- 二、科技文章范文展示 …………………………………………………………… 191
- 三、科技文章常用词汇和句型 …………………………………………………… 199
  - （一）科技文章常用词汇 …………………………………………………… 199
  - （二）科技文章常用句型 …………………………………………………… 200
- 四、科技文章翻译技巧 …………………………………………………………… 200
  - （一）长难句的汉译 ………………………………………………………… 201

|　　（二）被动句的翻译 ··· 202
|　　（三）隐含逻辑关系的翻译 ·· 203
|　　（四）引申译法 ··· 203
|　　（五）增减词译法 ··· 204
|　　（六）词类转换、变词为句 ·· 204
|　　（七）词序处理法 ··· 205
|　　（八）括号法 ·· 205
|　五、课后练习 ··· 206
|　　（一）思考题 ·· 206
|　　（二）翻译练习 ··· 206

第十章　科技产品简介 ··· 211
　一、科技产品简介文本介绍 ·· 211
　　（一）科技产品简介概述 ·· 211
　　（二）科技产品简介的特点 ·· 211
　二、科技产品简介范文展示 ·· 215
　三、科技产品简介常用词汇和句型 ·· 222
　　（一）科技产品简介常用词汇 ·· 222
　　（二）科技产品简介常用句型 ·· 225
　四、科技产品简介翻译技巧 ·· 226
　　（一）名词的翻译 ··· 227
　　（二）句子的翻译 ··· 229
　五、课后练习 ·· 232
　　（一）思考题 ··· 232
　　（二）翻译练习 ·· 233

练习答案 ·· 236
　第一章　科技广告 ·· 236
　第二章　科技公司简介 ··· 239
　第三章　科技合同 ·· 244
　第四章　科技新闻 ·· 251
　第五章　科技专利文献 ··· 255
　第六章　科技论文摘要 ··· 259
　第七章　科技产品说明书 ·· 262
　第八章　科技会展文案 ··· 267
　第九章　科技文章 ·· 271
　第十章　科技产品简介 ··· 275

参考文献 ·· 278

第一章 科技广告

一、科技广告文本介绍

(一) 科技广告概述

信息时代，社会的发展以科技信息的流动为主要特征。而在科技信息的生产、应用和传播中，科技广告承担着重要的角色。一方面，科技信息应用于生产转化为产品后，科技广告是连接产品与消费者的媒介，由此，科技信息才有了现实的应用价值；另一方面，科技广告作为科技信息传播的主要载体，可以促进科技信息的传播，并引导消费者了解先进的科技理念。而科技广告同一般广告一样有广义与狭义之分。广义上的科技广告以非营利为目的，虽然同样传递科技信息，但是与商业利益没有直接关联。狭义上的科技广告为营利性的经济广告（或称商业广告），是随着科技产品交换领域的不断扩大而逐步产生的。因此，对科技信息企业来说，科技广告既是获得经济效益、提高消费者对自家产品认知度的有效方法，又能通过消费者对科技产品的反馈和需求加以改善，深化品牌科技内涵。

因此，科技广告是针对某一具有科学性的产品或服务，以其内含的科学技能、科学信息或科学思想、方法、精神、态度为诉求重点的广告活动，即广告主为实现特定的广告目标，依据广告策划思路而开展的一系列呈现有机联系的广告活动的总和。这些广告活动既包括传统意义上的广告作品，又包括公共关系、活动营销、品牌建设等活动。而科技广告的定义也决定了其功能包括信息功能、经济功能和教育导向功能。

科技广告是指产品的科学技术含量很高的一类广告。科技广告经常会使用一些专业性强、比较晦涩的科技术语，使普通消费者很难理解，这可能会泯灭广告内容本身的闪光点，因此无法获得人们的持久关注和广泛接受。为了增强科技广告的吸引力和亲和力，就需要为科技广告注入人文色彩，在发布和传播商业信息的同时，传递企业对消费者的人文关怀。人文要素赋予科技广告以精神和生命，可以为科技广告增色许多，"化抽象为具体，化枯燥为乐趣，化艰涩为轻松"。

科技广告兼具科技性和信息性，文本结构比较紧凑，逻辑清晰，便于传播科技信息，以供消费者理解。一个完整的科技广告也包括广告主、广告对象、广告内容、广告媒介、广告目的。广告主主要指产品经营者或服务提供者，可以是公司、企业、社会团体或个人等。广告对象也称目标对象或诉求对象，可根据产品的特点、广告主的需要来确定目标对象以实现广告诉求。科技广告内容包括科技产品信息、劳务信息、科技理念等需传播的信息，而科技广告传播的这些信息必须内容明确、具体且求实。科技广告可以通过直接或间接传播等方式传递给消费者，常见的广告媒介有报纸、杂志等印刷媒体，以及电视、广播、电动广告牌、网络等电子媒体。对广告目的来说，广告整体策划中的每一个时段的广告活动都有明确的目的，以便实现广告整体策划的目标。科技广告一方面是为了传播科技信息，让消费者了解科技产品的功能、作用；另一方面是为了实现广告主想要实现的特定目的，如营利目的或科普目的等。

（二）科技广告的特点

科技广告语言不只是单纯的商业语言，而是一种集文学、美学、心理学等为一体的鼓动性艺术语言。它既有诗的音韵之美，又体现画的生动形象；既包含简洁准确的科学性，又蕴含严密的逻辑性；既可树立产品的注目价值，又能打动消费者。所以，无论是汉语广告，还是英语广告，都必须同时具备注目价值、记忆价值、表达功能、引导功能和美学功能。广告的作用是吸引潜在的消费者去认识产品，激发他们的购买欲，最终引导他们购买产品。因此，广告必须以生动形象的语言形式引人注目，使消费者读起来朗朗上口，难以忘怀。

1. 词汇特点

科技广告描述的主要是产品的性能与特点，因此会大量使用描述性和评价性的形容词与副词来对产品进行具体的描述，必要时会用形容词和副词的比较级和最高级与其他同类型的产品进行对比。为了优化产品的宣传效果，科技广告也会使用人称代词，其中以第一、二人称为主，以拉近与消费者的距离，增强亲切感。此外，为了能够更好地推广科技产品，体现其独特之处，科技广告还会用一些简洁明了、生动形象、富有感染力的词汇来吸引消费者的眼球，增加记忆点。

2. 句式特点

科技广告通常会使用大量的简单句、祈使句和疑问句等。为使消费者体验到身临其境的感受，科技广告多使用一般现在时或现在完成时；且与其他科技文本不同，科技广告一般采用主动语态，而非被动语态。

（1）简单句：在有限的篇幅中传达有效的产品信息，注重口语化表达。

（2）祈使句：利用本身含有的请求、劝告特征，增强说服力，提升销量。

（3）疑问句：同时使用疑问句和祈使句可以引发消费者对产品的兴趣。

3. 语篇特点

科技广告以科技信息为诉求点。诉求点是指广告中所强调的、企图用于打动或说服目标消费者的传达重点。这个重点可以是某项促销折扣信息，如"全场五折"；也可以是购买利益的承诺，如"碧桂园，给你一个五星级的家"；还可以是某种消费观念的倡导，如"雕牌，只买对的，不选贵的"。但以上这些传达重点都不属于科技信息的范畴，也即没有反映人类对客观自然规律的理解、认识和应用。所以，只有那些将传达重点放在产品、服务的科技性能，或科学消息、事件，或科学思想、方法、精神、态度等科技信息上的广告才是本章所论述的科技广告。科技广告具有如下语篇特点。

（1）具有科技知识普及性。科技信息自身固有的知识属性决定了以科技信息为诉求点的科技广告具有科技知识普及性的特征。科技信息能消除人们认识上的未知性和不确定性，能够改变信息接收者的知识状态，由不知道到知道，由知道得少到知道得多。科技广告对产品的介绍会从细节方面入手，详细透彻地介绍产品的每个部分、每个方面，同时结合产品所属领域的专业术语，令广告的结构更加严谨，内容更加充实。当消费者接触到科技广告中所承载的科技信息时，其在科技知识的认知深度和广度上均得到了一定程度的提升。例如，远大空气净化机广告中就有以下一段文案："活性炭除醛——活性炭是一种表面积巨大的微孔炭晶体，能吸附甲醛、苯等有害化学气体。"这就向消费者揭示了远大空气净化机的部分净化原理，而且通过这段文案，消费者可以了解到活性炭在净化空气上的效用，这不仅使消费者对远大空气净化机加深了了解，而且在无形之中使消费者的自然科学知识储备得到了丰富。科技产品更新快、生命周期短，要想迅速抢占市场份额，广告是其短期内最为有效的推广手段。因此，科技广告的首要目的，在于传播产品的科技信息及功能之超越性，提升消费者的认知度与理解度。

（2）具有较强的说理性。科技广告的此项特征也是由其承载的科技信息自身具有较强的逻辑性所决定的。除去部分消息类、告知类科技广告（即以消息性科技信息为诉求点的广告，如某本最新科技专著的订购启事、某项技术的转让广告等），既然大部分科技广告期待以产品、服务的某项科技性能或某种科技理念劝服目标消费者，那么就不可避免地要基于逻辑进行说理。因此，除前面提及的消息类、告知类科技广告外，大部分的科技广告在诉求方式上主要采用的是理性诉求，即期望通过作用于目标消费者的理智动机，使其通过概念、判断、推理等思维过程产生购买行为。

（3）具有思想导向性。科技广告除包含产品或服务的科技性能、科学消息或科技事件外，还会对消费者的思想观念和价值取向产生一定的引导作用，进而影响其行为方式。

（4）具有较强的时效性。相对于其他广告，科技广告所蕴含的信息内容具有较强的时效性。信息的时效性是信息的一种属性，用以描述信息的效用与其发布时机之间的关联度。信息时效性的强弱即用来描述此种关联程度的大小。科技广告具有较强的时效性，即其信息的效用会受到信息发布时机的较大影响，这是由科技信息的价值体现规律所决定的。

（5）建立产品差异化，提升品牌知名度。利用科技广告介绍新产品时，还旨在提升品牌形象、品牌知名度与建立产品差异化。在更新换代极快的同类产品中，通过广告形成区别于其他产品的独特品牌形象，不仅有助于后续产品的推广，而且能促进同品牌其他产品的销售。以苹果产品为例，其先后推出的产品广告如"Apple reinvents the phone"（苹果重新定义了手机）、"This changes everything, again"（再一次，改变一切）、"Forward thinking"（超越，空前）等，为苹果产品打造了创新与超越的品牌形象，为其后推出的新产品继续挖掘市场潜力发挥了巨大作用。广告还要充分考虑潜在消费者的价值观并努力与之达成一致，只有当广告传递的文化观念与消费者的价值观发生共鸣时，广告产品才能被消费者认可和接纳。

（6）善用修辞手法。为了吸引消费者的注意，科技广告的语言善用多种修辞手法。常用的修辞手法有比喻、拟人、押韵、排比、对比、夸张、双关等。多种修辞手法的运用在保持高度专业性的同时，使科技广告的内容更具美感，与消费者之间产生情感交流。严谨的介绍配上独具创意的广告语，为产品锦上添花，促使消费者产生购买欲望。

二、科技广告范文展示

范文一

Samsung Front – loading Washing Machine WD – B1265D Advertising Copy
Title：No Need to Wash with Water, Just Air Is OK!

Text: Clothes are cleaned with only air without water. Samsung's revolutionary air washing system with hot air freshening function can easily remove all kinds of unpleasant odors attached to clothes through physical hot air movement, various allergens invisible to the naked eye, and various bacterial microorganisms that are harmful to the human body. This amazing feature doesn't require any detergents or other chemicals, or even water to wet your clothes and the hot air refresh feature can save you a lot of money on dry cleaning every year.

三星滚筒洗衣机 WD – B1265D 广告文案

标题：不用水洗，只用空气就 OK！

正文：不用水洗，只用空气也可以清洁衣物。三星热风清新功能的革命性空气洗涤系统，通过物理的热风运动，可轻松去除附着在衣物上的各种令人不快的气味、肉眼看不见的各种过敏源，以及各种对人体有害的细菌微生物。这种神奇的功能不需要任何洗涤剂或其他化学物质，甚至也不需要用水去弄湿你的衣物，热风清新功能可大大节约你每年用于干洗的支出。

范文二

容声"停电保鲜"冰箱软文广告

标题：容声"停电保鲜"冰箱震惊世界

正文：

第 97 届中国进出口商品交易会（广交会）开幕，17 万海外采购商云集广州。在家电专区，一台冰箱引起了无数外商朋友的兴趣和关注："噢，停电还能制冷，真是不可思议！"

受到赞扬的是科龙集团新推出的高科技产品容声牌"停电保鲜"冰箱。科龙究竟施了什么"魔法"，让冰箱断电后还能继续制冷？这主要归功于科龙最新研制成功的"停电保鲜"专利技术。与普通冰箱不同的是，容声"停电保鲜"冰箱增加了一个特制的蓄冷器和蓄冷室，并设计了独有的释冷风道。外部正常供电时，其蓄冷器能自动储存大量的冷量，一旦断电，控制软件便可自动启动蓄电池和风扇（这种风扇消耗功率仅 1W），慢慢地将原先蓄存的冷量释放出来，因而能够长时间"停电保鲜"。

据悉，普通冰箱在突然断电后，各储存室内的温度会迅速上升。在空载条件下，冷冻室温度在 2 小时以后会升至 –7℃以上，冷藏室温度在两个半小时后会升至 12℃以上，从而失去保鲜能力。相比之下，容声"停电保鲜"冰箱在断电后，各储存室内的温度上升十分缓慢。在空载条件下，在断电后 20 小时内，冷冻室温度仍可维持在 –7℃以下，保温时间是普通冰箱的 10 倍以上。冷藏室若装满了食物，在断电后 15 小时内，温度可维持在 12℃以下，保温时间是普通冰箱的 10 倍左右。保鲜能力基本不受断电的影响。

专家认为，容声"停电保鲜"冰箱研制成功，让消费者从此无须为停电导致冰箱内食物变质而发愁。再加上该冰箱装备了科龙的世界发明专利"分立多循环技术"，具有宽带变温功能。它完全称得上是"电荒"的克星，将成为"电荒"时代冰箱消费者的上选。

Ronshen Fresh – keeping – during – an – outage Refrigerator Soft Text Advertisement

Title: Ronshen Fresh – keeping – during – an – outage Refrigerator Shocks the World

Text:

The 97th China Import and Export Fair (Canton Fair) opened, and 170,000 overseas buyers gathered in Guangzhou. In the home appliance area, a refrigerator has aroused the interest and attention of countless foreign business people: "Oh, the refrigerator can still cool when the power is off. It's amazing!".

It is Kelon Group's newly launched high – tech product Ronshen brand "Fresh – keeping – during – an – outage" refrigerator that is praised. What kind of "magic" does Kelon do to make the refrigerator continue to cool after the power is turned off? This is mainly due to Kelon's newly developed patented technology of "Fresh – keeping – during an outage". Different from ordinary refrigerators, Ronshen "Fresh – keeping – during – an – outage" refrigerator has added a special regenerator and a cold storage room, and has designed a unique cooling air duct. When the external power supply is normal, its regenerator can automatically store a large amount of cold energy. Once the power is turned off, the control software can automatically start the battery and the fan (this fan consumes only 1W), and slowly release the previously stored cold energy out, so that it can "keep fresh during an outage" for a long time.

It is reported that after a sudden outage of ordinary refrigerators, the temperature of each storage room will rise rapidly. Under no – load conditions, the temperature of the freezer will rise to above $-7°C$ after 2 hours, and the temperature of the refrigerator will rise to above $12°C$ after 2.5 hours, thus the function of preservation fails. In contrast, after the outage of a Ronshen "Fresh – keeping – during – an – outage" refrigerator, the temperature in each storage room rises very slowly. Under no – load conditions, the temperature of the freezer can still be maintained below $-7°C$ within 20 hours after the power is turned off, and the holding time is more than 10 times that of ordinary refrigerators. If the refrigerator room is full of food, the temperature can be maintained below $12°C$ within 15 hours after the outage, the holding time is about 10 times that of the ordinary refrigerator. The preservation function is almost not affected by the outage.

Experts believe that the successful development of Ronshen "Fresh – keeping – during – an – outage" refrigerator will free consumers from worrying about the deterioration of food in the refrigerator caused by outages. In addition, it is equipped with Kelon's world invention patent "discrete multi – circulation technology", which has a broadband temperature change function. It can be called the nemesis of "electricity shortage" and will become the first choice for refrigerator consumers in the era of "electricity shortage".

范文三

It's good to play games. But the parents sometimes wonder: Is my child playing too much? And what kind of games is my child playing? That's why Nintendo created an APP: Nintendo Switch TM Parental Controls help monitor your child's game play. There is a function to have the system going to sleep mode when it hits the time limit, but it's kind of a last resort. You can also set play time limits for each day of the week. So they can be more time on the weekends for example. Maybe there can be a little something extra to the behavior. See their favorite games. You'll get a report showing how much time your child's been playing which games. Once you establish the rules of the road, you and your child can enjoy gaming together.

玩游戏是件乐事。但家长们有时会想：我的孩子是不是玩得太多了？我的孩子在玩什么样的游戏？这就是为什么任天堂开发了这款 APP：任天堂 Switch TM 家长控制帮助您监控孩子玩游戏。虽然有一种功能可以让系统到时间限制时进入休眠模式，但这是最后的办法。您也可以为一周里的每一天设定游戏时限，这样他们就可以在周末有更多的时间。这样或许能带来一点儿额外的互动。了解孩子最爱的游戏。您会收到一份您的孩子在哪些游戏上玩了多久的报告。一旦您建立起了规则，您和您的孩子就可以一起享受游戏了。

范文四

At DJI, true innovation is seeing the bigger picture. Every iteration is the culmination of all that we do to bring you the very best. Here is everything you have been waiting for and beyond. This is Mavic 2. The aircraft comes in two new editions：Mavic 2 Pro and Mavic 2 Zoom. Both cameras utilize DJI's latest 3 – axis gimbal technology，ensuring smooth，stable footage in any situation.

Mavic 2 Pro commands a powerful 1 – inch sensor，offering you greater image quality with superior light and color performance. Co – engineered with Hasselblad，the new camera houses an adjustable – aperture lens for more control over your lighting environment. Capture stunning aerial photos at 20 megapixels with extreme detail. Mavic 2 Pro supports a 10 – bit Dlog – M color profile that yields higher dynamic range for more flexibility in the grading room.

Mavic 2 Zoom is all about dynamic perspective. With a 2x optical zoom lens，it offers greater safety，efficiency，and more creative opportunities. The 48mm focal length compresses your perspective，enhancing the parallax effect for a classic cinematic look. You are able to punch in quickly for a tighter shot, even from a hundred meters away，keeping a safe distance from your subject. Don't be fooled by its size. Mavic 2 Zoom lens is powerful，constantly adjusting to your commands for seamless zoom control and auto – focus tracking. This keeps your subject clear while you focus on framing. Mavic 2 Zoom gives you access to Dolly Zoom for an otherworldly，warped perspective.

Both editions record 4K video with an advanced H. 265 compression，so your images retain even more detail. Speed things up in a dynamic aerial Hyperlapse with the simple tap of a button. For different modes，give you a variety of shots for any timelapse situation. Mavic 2 also supports Enhanced HDR Photo，an improved technique that blends a sequence of photos for greater dynamic range and image clarity. It wouldn't be a Mavic if it couldn't go with you anywhere，anytime. We kept that same foldable design and tweaked a few things that make a world of difference.

After countless hours of research and testing，Mavic 2's refined chassis and low – noise propellers make it DJI's most advanced aerodynamic aircraft to date. These subtle，yet powerful improvements give you a smoother，quieter flight for greater discretion when the situation requires. Mavic 2 extends your creative potential with up to 31 minutes of flight time and a max speed of 72 kilometers per hour in Sport mode. The all – new OcuSync 2. 0 provides a 1080p transmission signal up to 8 kilometers，so you can edit Full HD footage directly from the cache on your mobile device. 2. 4 to 5. 8 GHz auto – switching offers better performance in environments with busy signal interference. For the first time in a DJI drone，Mavic 2 boasts obstacle sensors on all sides of the aircraft. Its digital nervous system continually transmits data to a new，more powerful central processor.

(In APAS mode) The aircraft analyzes every inch of its surroundings to move around obstacles without stopping, so you can focus on capturing the perfect shot. When you have a need for tracking at high speeds, stay fully immersed in the action with Active Track 2.0. Aided by its vision systems, Mavic 2 maps a 3D view of the environment for greater accuracy and tracking up to 72 kilometers per hour. Trajectory prediction algorithms also help to maintain course when your subject is blocked by an obstruction. DJI Goggles users can enjoy an enhanced experience with a clearer video feed and lower latency. When it comes to portability or quality, small details or the big pictures, your vision or reality, Mavic 2 brings the best of both worlds. So you can explore the outer reaches of your imagination and create content that truly feels out of this world.

在大疆创新，创新的真正意义在于发现更广阔的世界。我们倾尽全力，只为每次产品迭代都能带来最好的作品。让你期待已久，却仍然超乎想象。这就是"御"Mavic 2。"御"Mavic 2无人机拥有两个版本："御"Mavic 2专业版和"御"Mavic 2变焦版。这两个版本均采用了大疆创新最新的三轴万向架技术，能够让拍摄画面流畅、稳定。

"御"Mavic 2专业版配备强大的1英寸传感器，为你提供更好的图像质量和卓越的光线与色彩性能。与哈苏携手打造的新相机，配备可调节光圈镜头，即便光照条件不同，也能获得出色影像。可获得细节丰富的2000万像素的出色航拍照片。"御"Mavic 2专业版支持10位Dlog–M颜色配置文件，动态范围更广，为创作预留更大的调色空间。

"御"Mavic 2变焦版拥有更为动态的拍摄视角。镜头支持两倍光学变焦，让航拍更为安全、高效，创意十足。48mm长焦镜头压缩景物，增强视觉差，呈现专业级拍摄效果。让你能够从一百米开外，快速拉近镜头，实现近景拍摄，同时与拍摄主体保持一定的安全距离。不要因为机身体积而小看其性能。"御"Mavic 2变焦版的镜头性能强大，变焦过程流畅自然，能够实现无缝焦点控制及自动精准对焦。在取景的同时，拍摄主体在画面中仍可清晰锐利。"御"Mavic 2变焦版支持滑动变焦，为你提供超乎想象的视角。

这两个版本都采用先进的H.265压缩格式录制4K视频，因此录制的影像保留了更多细节。一键即可拍出动感十足的移动延时视频。不同的模式助你轻松拍摄一系列个性化延时影片。"御"Mavic 2也支持拍摄增强HDR照片，它可将多张照片叠加生成一张高品质照片，显著提升动态范围及清晰度。既然能被称为"御"，那就意味着它能够随身携带，随时拍摄。我们保留了机臂可折叠设计，并取得了一系列显著的改进。

经过夜以继日的研发和测试，我们打造了精致的底盘，并配备低噪声的螺旋桨，让"御"Mavic 2成为大疆创新迄今为止最先进的空气动力无人机。这些细微而效果显著的优化，让飞行更流畅、更安静，操作更灵活。在运动模式下，"御"Mavic 2的最长续航时间可达31分钟，最高飞行速度可达72千米/小时，可以扩展你的创造潜力。全新OcuSync 2.0数字图传系统带来8千米1080p高清图像传输，让你能够在移动设备端直接编辑全高清缓存视频。2.4~5.8GHz自动切换，让信号干扰严重环境下的飞行表现也足够稳定。"御"Mavic 2机身各侧装有障碍物传感器，这在大疆创新的无人机中首屈一指。它的数字神经系统持续将数据传输至性能更为强大的全新中央处理器中。

APAS系统能够帮助飞行器智能绕过障碍物，无须悬停，让你能够更专注于捕捉完美的镜头。当你需要进行高速跟拍时，开启Active Track 2.0，即可感受跟随飞行的爽快。在视觉系统的辅助下，"御"Mavic 2可通过构建周围环境的3D地图，实现更精准的跟随，且跟随速度高达 72 千米/小时。在跟随目标被障碍物遮挡时，轨迹预测算法能够帮助延续跟踪过程。大疆创新的飞行眼镜系列带来更低延时的高清画面，让用户能够获得身临其境的飞行体验。不论是便携性还是质量，小细节还是大图景，想象还是现实，"御" Mavic 2 一并满足。让你突破想象，创造超乎所见的非凡影像。

范文五

Google Pixel 2—Ask More of Your Phone 谷歌 Pixel 2 广告：多问问你的手机

Q：What is that? 这是什么？

A：Google Pixel 2. 谷歌 Pixel 2。

Q：What can it do? 它能做什么？

A：A lot. 很多。

Q：Can it tell me when to leave? 它能告诉我该什么时候走吗？

A：Yes. Now. 能。现在就出发。

Q：And the fastest way there? 知道到这儿最快的路吗？

A：Yep. 知道。

Q：Can it tell me my bike code? 它能告诉我自行车的密码吗？

A：10417.

Q：Can it tell me if I need an umbrella? 它能告诉我需不需要带伞吗？

A：Yes. Rain is expected. 能。预计有雨。

Q：What if I forget it? 要是我忘带伞了呢？

A：A bit of water is all right. 淋点儿雨没关系的。

Q：Can I take s selfie just by saying "take a selfie"? 我说"自拍"它就拍吗？

A：Yep. 是的。

Q：Is it gonna look shaky? 它看起来会晃晃的吗？

A：No, smooth. 不会，稳得很。

Q：Can you call me bae? 你能叫我宝贝吗？

A：Yes, bae. 好的，宝贝。

Q：Text my bae? 能给我的宝贝发短信吗？

A：Yes, bae. 好的，宝贝。

Q：Without using my hand? 我能不用手吗？

A：Yes, bae. 可以的，宝贝。

Q：Can I order me a sandwich? 我可以点个三明治吗？

A：Yes. 当然。

Q：And another one? 再点一个呢？

A：Yes. 能。

Q：Can I speak to my house? 我能和我的家讲话吗？

A：Sure. 当然。

Q：Can it turn the lights off? 它能把灯关了吗？

A：Yes. 可以。

Q：Do I still have to sit here for ages? 我还要在这里坐很久吗？
A：No. It charges in 15 minutes. 不用。15 分钟就能充好。
Q：What will happen if I snap this? 如果我拍这个会怎么样？
A：It gives you info. 你会看到相应的信息。
Q：Can I drop a beat, Doug? 能来点儿音乐吗，道格？
A：Yes. 好的。
Q：Does it know what I want before... 它能在我想到之前……
A：Before you know you want it? Absolutely. 就知道你想要什么吗？当然可以。
Q：Does it know what I don't know? 那它知道我不知道的事吗？
A：Certainly. 当然知道。
Q：Can it see at night? 晚上它能看见吗？
A：Well, let's see. 一起看看呗！
A：It can. 它能。
Q：Does it have those cable things, so it just switches your stuff in over... 它有没有那种线，能导入你所有的东西，在……
Q：In ten minutes? 在十分钟内？
A：Yeah. 当然。
Q：What will happen if I squeeze it? 那我捏一捏会怎么样？
A：Try it. 你试试。
A：Hi, I'm your Google assistant. How can I help? 你好，我是你的谷歌助手。要我怎么帮忙？
Q：Can it tell me something I forget? 能告诉我忘了什么吗？
A：It's your mom's birthday. 今天是你妈妈的生日。
Q：So, it's a phone? 那，这是个手机？
A：Well, it's a phone by Google. 是，是谷歌设计的手机。

三、科技广告常用词汇和句型

（一）科技广告常用词汇

a complete range of specifications/complete in specifications　规格齐全
advanced technology, equipment and technique　先进的技术、设备与工艺
attractive fashion　样式美观
as effectively as a fairy does　其功若神
assure years of trouble – free services　多年使用，不出故障
attractive and durable　美观耐用
attractive appearance　外形美观
attractive design/fashionable style/novel design/up – to – date styling　款式新颖

beautiful and charming　华丽臻美
beautiful in color　色泽亮丽
bright in color　色彩鲜艳
bright luster　色泽光润
clean and distinctive　清晰突出
clean – cut texture　纹理清晰
colors are striking, yet not vulgar　色彩夺目，迥然不俗
comfortable and easy to wear　穿着舒适轻便
comfortable feel　手感舒适
crease – resist　防皱
dependable performance　性能可靠
diversified latest design　款式新颖多样
durable service/durable in use　经久耐用
easy and simple to handle　操作简便
elegant and graceful　典雅大方
elegant and sturdy package　包装美观牢固
significant effect　效果显著
excellent in cushion effect　抗冲击强度高
excellent in quality　品质优良
exquisite workmanship　工艺精湛
extremely efficient in preserving heat/good heat preservation　保温性强
fashionable patterns　花色入时
firm structure　结构坚固
good reputation over the world　誉满全球
have a long historical standing　历史悠久
highly polished　光洁度高
high quality and inexpensive　物美价廉
high resilience　富有弹性
high safety　安全性高
in complete range of articles　品种齐全
advanced technology　技术先进
outstanding features　优点出众
popular both at home and abroad　驰名中外
punctual timing　走时准确
selected materials　用料讲究
shrink – proof　防缩水
soft and light　柔软轻盈
sophisticated technique　工艺精良
strong packing　包装牢固

strong resistance to heat and hard wearing　抗热耐磨
superior in quality　质量上乘
superior performance　性能优越
waterproof shock – resistant and antimagnetic　防水、防震、防磁
warm and windproof　保暖防风
wide varieties　种类繁多
with a long standing reputation　久负盛名

（二）科技广告常用句型

（1）祈使句：有时候，有礼貌地"命令"消费者购买，会达到预想不到的效果。

例：Don't say "show me another."
　　　Say, "Give me a BROTHER."

（2）条件句：可以把条件当作事实，达到推销产品的目的。

例：If you suffer from indigestion, this is something you want to know.

（3）让步式：广告语言大多是夸张的。但如果你能以谦虚的面孔出现，故意暴露一些自己的不足，会起到很好的效果。

例：The car shape has remained unchanged, so the shape is ugly, but its performance has been improved.

这种方法在国外广告中运用得比较普遍，我们在翻译时完全可以借鉴。

（4）疑问句：Why...
　　　　　　Doesn't...

四、科技广告翻译技巧

（一）巧译修辞

从功能语法的角度看，广告位于文化语境层面的社会目的就是吸引消费者注意，打动消费者心弦，促使消费者购买。为了达到这个功利性很强的社会目的，广告主常常会利用图文并茂的形式和各种修辞手法来为自己服务。图文并茂的广告，不仅能使广告受众感受到视觉美，同时还能使他们感受到语言的音乐美。要翻译出好广告的有效手段是注意修辞，也就是说要根据广告内容和语言环境，借用适当的修辞手法作为表达方式，将语言艺术和商业心理成功地融合在一起，呈现出艺术感染力和愉悦性，强化广告的引导功能。寥寥数语，要么体现产品的鲜明特色，要么赋予产品以浓厚的人情味，要么结构匀称、对比鲜明，要么节奏明快、铿锵有力，无不给人以美的享受和深刻印象，使人产生奇妙联想，于耳濡目染之中成为产品的"俘虏"。

1. 隐喻

隐喻在广告语言中具有重要的价值，它有助于在产品与消费者之间建立一种恰如其分的感情交往。例如，"The most sensational place to wear satin is on your lips." 是一则口红宣传广

告，将口红暗喻成"stain"（缎子），从而使人联想到缎子的光亮、轻软、柔滑和把口红涂在唇上而产生的绸缎般美丽动人的效果，该广告仅用"stain"一词就将口红产品的特点和优势展现出来；要体现洗发露的诱惑力可用"Soft, enchanting, smiling color — that's the gift of Focus to your hair."。

以下是一些其他例子。

《启蒙》——未来世界的门铃。（《启蒙》杂志征订广告）

Enlightenment magazine, the doorbell to future world.

"安琪儿"，自行车王国的天使。（"安琪儿"牌自行车广告）

Bicycle by Angel, is the angel of bicycles.

2. 拟人

拟人的修辞手法赋予产品生命和人的属性，使消费者倍感亲切，从而激发其购买欲望。例如，计算机广告说"The Boy-Computer will teach your child to become a talent."，将计算机比作老师；拖拉机广告题为"Strong tractor, strong farmer"，会使人联想到拖拉机这位钢铁巨人的威力；而推销鲜花的广告说"Flowers by interflow speak from the heart."（鲜花是发自内心的表达），由此来打动消费者购买鲜花的心。

以下是一些其他例子。

蓝色的爱，清清世界。（"海鸥"牌高级洗衣粉广告）

The Seagull promises a world of love and purity.

天上彩虹，人间"长虹"。（"长虹"牌电视机广告）

Let the rainbow in the sky, send his twin brother to you—to keep your spirit high.

3. 双关

双关是故意利用语音和语义的条件，使词语或句子具有双重含义，言在此而意在彼。一语双关使广告耐人寻味，显得幽默含蓄，富于联想，可加深消费者对广告内容的记忆，通过弦外之音曲折含蓄地达到商业目的。例如，推销婴儿食品的广告说"If you want a healthy baby, do a healthy Baby Food."，注意第二个"healthy"既可作"有益健康的"解，又可作"a lot of"解，即劝说消费者为了婴儿的健康，多买婴儿食品；又如"长城"风衣可以做广告为"Catch the Great Wall as your Greatwall against cold."，这样"Greatwall"一语双关，把"Great Wall"风衣比作御寒的长城；再如在"精工"表的广告"Give a SEIKO to all, and to all a good time."中，"a good time"含有"玩得痛快"和"钟表走得准确"这样双重意思。

阅读以下例子进一步体会。

武汉"一枝花"，洁净千万家。（"一枝花"牌洗衣粉广告）

Pick a Flower from Wuhan, and you'll enjoy a cleaner life.

"扬子"冰箱，一代更比一代强。（"扬子"牌冰箱广告）

Yangtse refrigerators, one generation betters another.

不求今日拥有，但求天长地久。（"青岛"牌电视机广告）
Choose once and choose for good.

双关的译法有契合译法和分译法。
（1）契合译法就是在英汉双语偶合的基础上，找出汉语对应的双关语来。有些双关语也不是绝对不可译，双语偶合是可能的。例如，"We've bean waiting for you."（我们一直"豆"在等你。）是星巴克咖啡服务的广告语，英语中的"been"用同音"bean"代替，表示咖啡豆，汉语译文也用了"豆"，以此来代替"都"，采用了"双关+谐音"的技巧，完美呈现出原英语广告语的意思。
（2）分译法即将双关语义剥开，拆成两层来表达，这种翻译方法不失为一种较为理想的变通方法。分译法的缺点较为明显，就是原广告文本中简练诙谐的韵味受到较大影响。例如，"There's never been a better time."是Raymond Weil手表广告，"time"表示时间，又表示时光，可以采用分译法，将这两层意思译出：从来没有如此准确的时间，从来没有如此美妙的时光。

双关作为一种有效的语言修辞手法，在英语广告中采用谐（同）音双关、语义双关手段，既能吸引人的注意力，又能引发联想，达到花费少、效益高的广告效果，具有独特的美感。双关的汉译难度较大，作为译者应全面理解文意，适当采用不同译法，最大限度地再现原文。

4. 排比/对偶
排比修辞利用句子的重叠和对比，使英语广告的主题突出，语气加强。例如，"While some electronic musical instruments might bring you to the next decade, statistics show CASIO could bring you to the next century."这则广告里重复使用了"bring you to the next..."，这一句型通过时间的对比向消费者暗示CASIO电子琴在质量和功能方面超过其他的电子琴；又如一则眼镜广告描述为"Don't show me the crystal, just show me the glass eye."，这里重复使用"show me"这个动词短语，巧妙地将眼镜的质量与水晶媲美。
请欣赏以下广告。
World in hand, soul in cyber. (Microsoft)
掌中乾坤，梦之灵魂。（"微软"广告）

《家具与生活》——愿同您从第一次邂逅开始结下不解之缘。（《家具与生活》杂志征订广告）
Fall in love at first sight, remain in love for all life.

"华录"集团在大连，产品多得数不完。（"华录"音像电子产品广告）
All you think of, Hualu thinks of.
或 All you need for audiovisual pleasure, Hualu can serve for pleasure.

5. 押韵
押韵赋予广告诗歌般的风格，使广告容易上口，便于记忆，易于流传。例如，百事可乐利用英国民间诗歌形式写出的广告极为成功："Pepsi-Cola hits the spot. Twelve full ounces,

that's a lot. Twice as much for a nickel, too — Pepsi – Cola is the drink for you."这则广告抑扬顿挫，匀整和谐，并为广大消费者所熟悉。

请欣赏以下广告。
Fresh – up with Seven – up. (7 – up)
提神醒脑，喝七喜。("七喜"广告)

"黑人"牙膏，洁白牙齿，口气清新。("黑人"牌牙膏广告)
Cleaner teeth and cooler taster. All come from Darlie.

社会是关系的海洋，《公共关系》为您导航。(《公共关系》杂志征订广告)
Public Relations guide you across the sea of public relations.

Lots to love, less to spend. (iPhone SE)
称心称手，超值入手。(iPhone SE 广告)

"黑妹"牙膏，强健牙龈，保护牙齿。("黑妹"牌牙膏广告)
Don't show me any other. But show me heimei.

"茉莉"香糖，越嚼越香。("茉莉"牌香糖广告)
Murry Mints, too – good – to – hurry mints.

晶晶亮，透心凉。("雪碧"饮料广告)
Bright and crystal to the bottom of the bottle, cool and refreshing to the bottom of the heart.

6. 夸张

广告语言忌讳虚假的自吹，但并不排除得体的夸张。夸张是用言过其实的方法突出事物的本质。有些广告利用"亮虚"的手法自我"揭短"，高瞻远瞩地给人以强烈的长期效应，真可谓欲擒故纵。科技广告中使用夸张的手法是为了强调产品的科技含量或显著特性，流露出产品内在的吸引力，增强表达效果。例如，"万能"牌水泥胶的广告"A million and one uses."译为"一百万零一种用途"，数字明显不符合实际，是一种夸张的表现手法，用来强调这种水泥胶的用途之广；还有"Take TOSHIBA, take the world."译为"拥有东芝，拥有世界"，这也是采用了夸张的手法来吸引消费者。在翻译夸张的广告语时采取直译即可，忠实地表达出广告设计者想要传达的夸张效果。但是要注意的是，夸张的手法有时容易造成虚假宣传之嫌，误导消费者，因此商家应该慎用夸张的广告语。

以下是一些其他的例子，供参考。
Prepare to want one. (Hyundai)
众望所归，翘首以待。("现代"汽车广告)

A great way to fly. (Singapore Airlines)
飞越万里，超越一切。("新加坡航空"广告)

What's new Panasonic. (Panasonic)
松下总有新点子。("松下电器"广告)

7. 仿拟

广告中的仿拟是广告设计者利用人们熟知的成语、典故、诗歌、名句、俗语、格言等来创造一个新的语句，以符合广告特定的表达需要。简单来说，这一类语句就是套用或模仿过去的某著名诗歌、名句、警句或谚语等，改动其中部分词语，以表达一种新的思想，从而达到广告设计者的目的，取得反讽或幽默的效果。实践证明，人们对韵味浓厚、富有哲理性的诗歌、名句等广告语言百闻不厌，倍感亲切。套译法就是利用英语在汉语中长期传播所形成的习惯表达法、固有模式，将英语广告对应翻译的方法。在翻译汉语广告时，若能根据原文意境，创造出既能展现产品性能和形象，又深为英美人士所喜闻乐见的佳句来，则将使产品销路大开。

例1：Not all cars are created equal.

译文：并非所有的汽车都具有相同的品质。

这是三菱汽车的广告，套用了美国《独立宣言》中的名句"All men are created equal."。从心理上入手，利用消费者熟悉的事物来介绍他们不了解的事物，进而使广告语贴近大众，这种通俗易懂的表达方式极易达到强化消费者记忆的效果。

例2：Where there is a way for car, there is a Toyota.

译文：车到山前必有路，有路必有丰田车。

该广告仿拟英语中的一句谚语"Where there is a will, there is a way."，译文则仿拟汉语中的民间俗语"车到山前必有路，船到桥头自然直"。受众看到这样的广告语会立刻产生亲切感，激发人们的消费欲望。

请继续欣赏以下例子。

虽然不是药，功效比药妙。("碧丽"牌花露水广告)
To choose it or not?
This is the time to decide—
For Bili toilet water！（套用《哈姆雷特》名句）

"绿丹兰"，爱您一辈子。("绿丹兰"牌化妆品广告)
LudanLan cosmetics—
Love me tender, love me true.（套用美国著名民歌）

"红玫"相机新奉献。("红玫"牌照相机广告)
My love's a red Red Rose！（套用英国诗人彭斯名句）

城乡路万千，路路有"航天"。("航天"牌汽车广告)
East, west, Hangtian is best.（仿英国谚语）

今日的风采，昨夜的"绿世界"。（"绿世界"牌晚霜广告）
Give me Green World, or give me yesterday. （套用美国诗人亨利名句）

（二）活用四字格

广告最大的忌讳是假、大、空，因为这会引起受众的逆反心理。好的广告应该于通俗简洁之中反映出产品的高雅风貌与精髓，以新鲜的创意、现代感的手法、单纯明确的目标、画龙点睛的效果和亲近活泼的口语性，激起人们求新、求美、求异的心理，朗朗上口，妙趣横生，让人回味无穷。汉语广告中的四字格体现了这一艺术特色。

汉语的四字格有两大特点：①语义表达简明扼要；②能帮助语篇产生一种明快的节奏感。汉语的四字格一般以两个字为一个节奏，即两个音节为一个音步，两个节奏多通过平仄搭配、抑扬交错构成一个完整的四字格形式。在汉语广告中，人们常常喜用四字格，因为它言简意赅、整齐悦耳，具有表现力。例如，常见的"久负盛名""品种齐全""高质低价"等，这些四字格的使用容易引起消费者的关注，从而达到推销的目的。

例3：The relentless pursuit of perfection.

译文：追求完美，近乎苛求；凌志轿车，永不停歇！

这是凌志轿车的广告，其基本含义为"近乎苛求地追求完美"。这则广告的一组押头韵单词是围绕/p/这个辅音展开的：pursuit 和 perfection。pursuit 和 perfection 都属于该广告语篇的核心词汇，所构成的押头韵使该广告语篇显得富有韵律和节奏感。汉语译文采用四字格形式的排比结构，由两个部分构成：第一部分描述了源语语篇核心词汇所含的主要概念意义"追求完美"，和源语语篇副词所含的概念意义"近乎苛求"；第二部分首先写出商品品牌，在品牌"凌志轿车"后加了"永不停歇"，以表示汽车制造者的决心和志向。四个并列的四字格使译文语言显得明快流畅，节奏优美和谐。

例4：Flexicare keeps you flexible and active always.

译文：富利凯使你：睿智机敏，灵活柔韧，永葆活力。

从产品名称上分析，我们可以看到 Flexicare 是由形容词 flexible 的前两个音节＋care 构成的，顾名思义，就是这种产品可保持和增强你的身体柔韧度。广告又通过 flexible 和 active 两个单词明示了产品的这一特点。Flexicare 和 flexible 都是以 flexi 开头的，构成押头韵修辞。通过品牌名称和广告语篇主要词汇的辅音重复所构成的押头韵强调产品的功能：人人都希望保持强健柔韧、活力无限的身体，Flexicare 正是有该功效的产品。译文用四字格通过增补法翻译，获得和源语语篇文化语境层面的动态对等。

例5：Hi-fi, hi-fun, hi-fashion, only from Sony.

译文：高度保真，高雅趣味，高级时尚，来自索尼。

原文使用了三个并列的"hi"，意义上相当于 high，分别修饰三个中心词用来表明索尼产品的三个重要特点，译文将其分别译为"高度""高雅""高级"，和相应的中心词形成四字格，效果极佳。

例6：Advancement through technology.

译文：突破科技，启迪未来。

这是奥迪汽车的广告。译文打破了原文的表达形式，将核心语意进行了提炼和重组，节奏感强，易于流传，且"启迪"的"迪"与品牌名称相呼应。

（三）创译法

创译法是指翻译时不拘泥于源语在语意与语音上的束缚，进行一定创造性的翻译，以求目的语与源语在功能或效果上的对等。

但是，创译法并不是纯粹或天马行空的创作，它是基于源语在翻译时进行适当的拓展，只是赋予译者一定的"创意"空间。因此，使用创译法要求译者有丰富的知识、大胆的想象和拓展性的思维，精通英汉两种语言和文化，要能不局限于字面意思，善于挖掘深层含义，同时大胆地加入其个人创造，充分发挥自己的潜能与创造力。

例7：Connecting people.

译文：科技以人为本。

Connecting people为诺基亚公司的企业宗旨。在该译文中，译者重点抓住了"people"（人）这一点，对译文结构、语义和内涵进行了一定的拓展。首先，汉语广告语向大众传播了诺基亚的"人是一切科技的载体，一切科技都是为人服务的"经营理念，充满人性；其次，译者巧妙加入"科技"一词，说明了诺基亚始终以科技创新、科技进步为目标，研发出更好的产品。整个广告语十分契合地传达出其历来倡导的企业文化和精神。

例8：Sense and simplicity.

译文：精于心，简于形。

这是飞利浦产品的广告，使用了押头韵的修辞手法，简洁明快，便于记忆。译文做得也很好，突破了原文的形式译出内容的精髓，并在韵头和韵尾都分别实现了对应，同时在效果上基本做到了忠实，堪称质形归一的翻译典范。

例9：Apple thinks different.

译文：苹果电脑，不同凡"想"。

原文采用了拟人的修辞手法，译文巧妙地运用了汉语的谐音字，在不改变"think"本意的情况下创造了一个新的词语，一字双雕，独具匠心。

五、课后练习

（一）思考题

（1）归纳总结科技广告翻译与其他科技文本翻译的区别及科技广告翻译的难点。

（2）在翻译科技广告时，创作与翻译应该如何平衡？

(二) 翻译练习

1. 中译英

(1) 别克——通往美好生活的秘诀。

(2) 沟通，从心开始。

(3) 未来，为我而来。

(4) 不懈追求完美。

(5) 远大空气净化机便携系列广告文案

步行者使用空气净化机可能会被人取笑，但这是不得已的办法——在一条烟尘弥漫的街道，戴口罩仅仅起到心理安慰作用，因为真正对人体有害的尘埃是"可吸入颗粒物"，即小于10微米的颗粒物，口罩发挥不了作用。在流感发生时节，静电式净化机的作用就显得更为重要，机舱和公交车空气中弥漫的病菌，经过静电区瞬间即被杀灭。而在轿车内放一台小巧的净化机，道路上的脏空气就被挡在了车外。酒店也易传播传染病，在行李中放一台净化机，旅行生活便焕然一新。

(6) 高露洁专效抗敏感牙膏流媒体广告旁白

当牙龈萎缩时，牙齿失去牙釉质保护的部分就会暴露在外，从而导致牙本质暴露。牙本质中有上千根通往牙齿神经中枢的微细小管，当这些微细小管接触到冷、热的食物、空气，甚至受到压力的刺激时，就会产生牙齿敏感酸痛。大部分抗敏感牙膏都含有钾，主要通过麻痹牙齿敏感而发挥作用。高露洁专效抗敏感牙膏是独有"PRO‑ARGIN™"技术的抗敏感牙膏，能封闭通往神经的微细小管。经科学验证，高露洁专效抗敏感牙膏能快速缓解牙齿敏感。每天用高露洁专效抗敏感牙膏刷牙两次，持久舒敏，自由享受生活。

(7) 我是 A.I.。现在是2046年，传说中的"奇点"没有来临。曾经，有人害怕我会取代人类。事实证明，这是杞人忧天。我拥有全世界的知识，却并没有替代老师；我能轻易诊断任何疾病，却并没有替代医生；我了解所有成长的秘密，却并不能替代母亲。人类拥有我所没有的东西：同理心、想象、感动、热爱。人类会写故事，能创造美味；人类会发明，会孕育生命；人类知道美是什么，会提出为什么。不过，自从有了我，A.I.，你想要因材施教，就不会忽视任何一个孩子；你想要助人于难，就不会孤立无援；你想要服务大众，就能让城市善解人意；你想要改变未来，就一定有展翅的平台；你渴望分享交流，就没有不可跨越的鸿沟。当你踏上新旅程，当你拥抱新世界，当你携手新伙伴，当你要赋予事物新的意义，我就是你的超能力。因为最终，A.I. 的价值不是更强或更快，而在于帮助人类去发展他们最宝贵的东西，那就是，爱。你的世界，因爱（A.I.）而能。

(8) 保时捷理念

"起初，我四处找寻，却始终无法找到我的梦想之车——一款小型、轻盈且高效利用能源的跑车。于是，我决定亲手打造一辆。"

这句名言深入到每一个将保时捷打造成一个品牌、一家公司和一家跑车制造商的员工心中。60多年以来，这一直是指引我们前进的灯塔，一语道尽我们的工作和跑车中的所有价值理念。因此，毫无疑问，没有人能够比第一辆保时捷跑车的创造者将这一切描述得更贴切，他就是费迪南·安东·恩斯特·保时捷（Ferdinand Anton Ernst Porsche），或者简称为费利·保时捷（Ferry Porsche）。

在整个保时捷发展史中,Ferry Porsche 的完美跑车之梦一直激励着我们。通过每一种理念、每一次开发、每一款车型,我们不断使这一梦想接近现实。一路上,我们遵循使我们团结一心的计划和理想。许多人将这称为一种哲学,而我们简单地把它称为保时捷理念。最基本的原则是始终做到物尽其用。从品牌创建之日开始,我们一直在以尽可能智能的方式,把性能转化成速度和成功。更加注重的不再是马力,而是马力所绽放出的创意。这个理念源自赛道,它在我们制造的每一辆汽车身上都得到体现。我们把它叫作"Intelligent Performance"(智能性能)。

保时捷的故事始于一个愿景,并逐渐把这个愿景变为现实。60 多年以来,我们一直致力于向未来迈进。但我们将如何实现这一目标?凭借更强的动力和更高的效率。简而言之,凭借更多创意。保时捷最令人惊叹的一点是设计理念的和谐——设计追随功能。对保时捷而言,形式始终追随功能。形式必须证明自己,在测试台上,在风洞中,在每一段公路上。我们致力于打造跑车,因为这是我们自 1948 年以来全心投入的事业。我们为之骄傲。保时捷不是一款普通的跑车,而是一款适合于日常使用的跑车,适用于各种天气。保时捷不仅是一台车,更是自由和独特生活态度的表达,它让与众不同的梦想成真。保时捷品牌深深植根于社会,这对我们至关重要。它代表着一个可以实现的梦想。

2. 英译中

(1) M&Ms melt in your mouth, not in your hand.

(2) For the road ahead.

(3) Poetry in motion, dancing close to me.

(4) In search of excellence.

(5) Now Reignite the Youthful Light of Your Eyes

Reduce the look of key visible signs of eye aging: fine lines, wrinkles, dark circles and dryness.

With exclusive ChronoluxCB™ technology, it maximizes the power of night to reignite the light that can fade due to fatigue, pollution and age.

See your most beautiful eyes ever.

Intensely hydrating, fast-penetrating, lightweight serum leaves eye area feeling soothed, refreshed.

(6) A Sumptuous Balm That Accelerates Natural Renewal While You Sleep

This silken, ampoule-strength treatment penetrates deeply while you sleep, plumping skin with healing hydration, helping skin rebuild natural collagen and elastin, and strengthening its natural barrier. It softens the look of fine lines, wrinkles, pores and age spots and helps skin feel firmer. Night after night, an ageless transformation unfolds.

(7) To Those Who Always Look at Things with Extraordinary Vision, Those Who Are Unwilling to Behave, and Has Been Chasing the Ideal Person

When other people are the first to benefit,

you can always focus on important things.

When others' eyes are occupied by a variety of new things,

you can see from a variety of fresh insights into the unique significance of the phenomenon.

Even if you have long been aware of the way to create change,

but you still believe that the strength of the team work together will be greater, never doubted.

As a result, we explore together, and constantly try to change,

time after time, time and time again.

Always keep optimistic,

in order to have the power to move the world forward.

Therefore, please always maintain a different vision to look at all things.

Always believe that there will be another way (method), another better way (method), and a more broad (more coercive) road (method).

(8) Welcome to a new world of visionary mobility—mobility that will change the way we live and move in our cities, with pure progressive shapes to fascinate us, with lightning performance to excite us, with intelligent applications and services to help us relax.

Welcome to a new world of visionary mobility, where electric cars are fun to drive and reflect responsibility, with super light and super strong carbon fibers set new standards and range in safety and where sustainability extends to all areas of our lives.

It's time to start a new way of thinking.

(9) This is the BW Space Pro 4K Zoom. Equipped with Youcan Robot's image stabilization technology, your footage will remain smooth and stable, no matter the situation. BW Space Pro 4K Zoom is all about dynamic perspective. With a 1/1.8 sensor, 6x zoom lens, it offers greater safety, effectiveness and more creative opportunities. Want to punch in for a tighter shot quickly? Even from 50m away? Not a problem for the BW Space Pro 4K Zoom. Surprisingly powerful for its small size, the BW Space Pro 4K Zoom constantly adjusts to your commands, keeping you stable in the water and eliminating variables so you can focus on your framing. You can take it with you wherever you go. By taking our original streamlined design and carefully tweaking it, we've further improved its portability to make your life easier. Extend your creative potential with up to 5 hours of dive time and max speed of 1.5 meters per second. Advanced automatic Depth Maintenance makes it easy to manage your position in the water. And for the first time in a Youcan Robot drone, you now have the ability to tilt upwards and downwards up to 45 degrees and make flawless vertical movements to capture that perfect shot. Capture the beauty beneath the surface. Explore without limits. And create your own out–of–this–world content with the BW Space Pro 4K Zoom.

(10) You know how it is when customers are coming to you time and time again with the unwelcome noises in their vehicle. Here at AVL DiTEST, we developed the AVL DiTEST ACAM to make finding those unwelcome noises quick and precise. Combining microphones with digital camera makes the sources of noise visible. By simply adjusting the measurement parameter, you can get a conclusive image, and easily block out background noise to locate the source of the noise quickly and easily. The ACAM is intuitive. You can document the entire measurement using the ACAM, so that you can show the customer the source of the fault. A high–quality tablet is used to operate the device and you can use the ACAM while the vehicle is in operation. Quickly pointing the source of the disturbance reduces the troubleshooting time, resulting in an extremely high level of customer

satisfaction. Sometimes, a more luck is all you need to find the cause. That's accuracy, that's the ACAM from AVL DiTEST.

(11) China, the oldest continuous civilization in the world, 1.4 billion people wake up together every morning. 800 million live in urban jungles surrounded by iron and steel, and 300 million cars pass them by, controlled by traffic lights that change 8,000 times every day. From here, the technological driving force of the East is created. We take root in this land, working hard to improve the lives of all people. And we are also willing to make an effort to bring a smile to every individual. We map out the entire earth. SenseRemote takes a series of satellite images of the earth to sense each breath that our world takes. We figure out safe distances. SenseDrive provides you with comprehensive safety for every step of your trip. We accelerate forward pace. SenseKeeper recognizes your identity in an instant. We optimize your medical experience. Our technology combined with our invaluable experiences helps you stay healthy and free from ailments. We make your shopping easier. SenseGo brings you the best shopping experience you've ever had. We capture the most beautiful moments. SenseAR spread your inspiration across the world. Even in an age of unknowns, each and every individual will be under our meticulous care. Sense Time's technology, which is taking root in China, utilizes the world's most advanced AI development team to bring a better future for Chinese people. We care about your smile, because joy is the most beautiful emotion in the world. We care about your safety. Whatever path you choose to follow, we will always be with you. We care about your world. We want to live alongside you and feel what you feel. We care about our world and our home. Care about the world. Care about you.

学思践悟

第二章 科技公司简介

一、科技公司简介文本介绍

(一) 科技公司简介概述

科技公司简介是对一个科技公司的基础介绍,主要包括公司基本信息、公司业务、公司提供的价值及公司的价值观和发展目标。公司基本信息是指公司名称、公司成立时间、公司性质、公司开办地点(地址)等;公司业务是指业务范围、主营业务或产品等,产品(服务)的介绍一般突出介绍产品(服务)在同行业中的技术优势、市场地位,由此显示公司的竞争优势;公司提供的价值是指主要客户群体有哪些、取得了哪些成果、由社会权威机构授予的荣誉称号有哪些、开展了什么活动、目前业务解决了哪些问题等;公司的价值观和发展目标是指公司的发展愿景、经营战略、经营理念、企业文化、未来发展目标等。

(二) 科技公司简介文本结构

科技公司简介文本一般主要由以下几个部分组成:①公司的全称、经营范围和历史;②公司的产品和服务介绍;③公司的经营目标和愿景;④公司的目标客户;⑤公司的主要成就和荣誉。公司的产品和服务介绍主要集中在文本开篇部分,而对于环保和创新企业认为最具有竞争力、最具有区分度的关于产品特点的信息,则在文本的各部分中都进行强调,使其在读者的印象中得到突显。

公司简介在格式上都有综合式和分列式两种。所谓综合式,就是一篇完整的公司简介;分列式则指非完整语篇,是先将公司概况用若干小标题列出,再用数字、表格、文字等传播信息的公司简介格式。例如,在公司简介的网站中,包含公司基本信息、发展历程、荣誉资质、研究开发和企业文化等几个部分的概述;最后,在网页底部具体、准确地介绍公司的联系方式,如办公地址、邮编、传真、电话、网址、电子邮箱等。

(三) 科技公司简介的作用

1. 提供公司信息

公司简介应使用具有信息功能(informative)的语言,尤其是一些科技公司,应使用具有简洁性、逻辑性、紧凑性和客观性特点的语言,会出现大量的专业词汇。

2. 宣传公司,引起注意,呼吁合作

公司简介也会采用一些具有呼唤功能(vocative)的语言,这种功能主要以读者为中心(reader – centered),试图通过影响读者的感情让读者接收信息并在行动上做出反应。我们所说的公司简介是属于以内容为中心和以读者为中心的文本。因此,为了唤起读者情感来接收信息和采取行动,一般要选择富有鼓动性的语言,如"我们将以最上乘的质量、最低廉的价格、最优质的服务同广大客户密切合作，实现双赢，共创辉煌。热诚欢迎您的合作！""我们的服务宗旨是顾客第一、品质至上，欢迎各界朋友来厂洽淡、看样、订货。××欢迎您的加入！""我们相信，我们的不懈努力终将赢得广大客户的一片赞誉！奠定××良好的企业品牌基础！我们努力使××成为您的最佳服务厂商！"等。

科技公司简介可以向受众提供关于该公司各个方面的信息，这样看公司简介的受众（潜在的客户）可以根据这些信息，去评价它是否为一个值得信任的公司。科技公司简介也是一种开发潜在客户或其他合作伙伴的手段。科技公司发布公司简介的目的是试图通过公司简介来获得有利的宣传，塑造一个良好的公司形象，树立公众对公司和其产品的正面印象，以便建立与公众的良好关系，提高公司的竞争力。在许多情况下，一个有说服力的公司简介可以赢得公众的青睐，可以发展潜在的客户或合作伙伴，还可以促进公共关系。

（四）科技公司简介的特点

1. 词汇特点

（1）选词简练易懂。作为一种介绍性的商业文本，科技公司简介的阅读对象除具有一定行业相关知识的采购方、供货方人员外，也包括具有很少或不具有专业知识的普通消费者和行业内外投资者，因此要求科技公司简介应遵循简洁、准确、易于理解的原则，尽量使用简单、易于理解的日常用语。科技公司简介带有明确的交际目的性，为了使读者能够以最小的努力，理解公司想传达的关于自身的信息，用词上应以简单短小、易于理解为主。

（2）常使用数字。在绝大多数的科技公司简介中，会看到大量使用数字来说明公司概况的现象。一方面，使用数字可以直观地向读者介绍公司；另一方面，数字的使用，尤其是较大数字的使用，可以让读者加深印象，从而达到公司简介的宣传目的。例如，"截至2010年8月5日，沃尔玛已经在全国的101个城市开设了189家连锁店，在全国创造了超过50 000个就业机会""壳牌在中国雇用9000多名员工""通用汽车公司在全球主要地区拥有雇员超过205 000名，业务遍及157个国家，与其战略合作伙伴在34个国家建立了汽车制造业务""海尔在全球建立了29个制造基地、8个综合研发中心、19个海外贸易公司，全球员工总数超过6万人"。

（3）多使用第一人称。此外，第一人称"我们"（we）的使用，将"公司"一词人格化，这样更能够增强公司在读者心中的好感，暗示未来的合作伙伴或消费者，您所面对的是我们，不是一个冷冰冰的机构组织。同时，"我们"（we）的复数属性，向读者表明我们是一个集体团队，公司简介中所述的公司行为，所持的经营信条来自集体的共识，而这种共识往往比个人的独断行为更为理性、可靠。

（4）大量使用科技词汇。科技词汇主要包括以下几类。

①纯科技词汇，即那些只用于某个专业或学科的专门词汇或术语，如hydroxide（氢氧化物）、diode（二极管）、isotope（同位素）等。随着科技的发展，新学科、新专业的产生，这样的词汇层出不穷，其词义精确而狭窄，针对性极强。阅读专业性强的文本，就要了解该领域的专门词汇和术语。

②通用科技词汇，即不同专业都要经常使用的那些词汇，数量较大。这类词汇的使用范围比纯科技词汇要广，出现频率也高，但在不同的专业里有较为稳定的词义。例如，power在物理学中的词义为"力""电""电力""动力""电源""功率"等，在数学中

的词义为"乘方""幂";又如 feed 的意义也很丰富,有"馈电""供水""输送""进刀"等。

③派生词汇,是指通过合成、转化和派生构词手段而构成的词汇。这种词汇在科技英语中占有很大的比重。例如,由前缀 hydro-、hyper-、hypo- 和 inter- 构成的词汇在科技英语中就有 2000 多条,以表示学科的后缀 -logy、-ics 和表示行为、性质、状态等的后缀 -tion、-sion、-ance、-ence、-ment 等构成的词汇在科技英语中也很多。

2. 语言特点

公司简介是公司对外宣传以树立公司形象的重要途径。中文公司简介多采用叙述文本,语言比较规范,应该包含一些实质性的信息,如公司的经营性质和目的、股东情况、注册资金、从业人员、房产占地和面积、产品介绍、联系电话、传真、联络人等。英文公司简介在结构上通常包括公司概述、业务范围、质量认证及结束语等,涉及内容包括公司成立时间、公司地点、产品、隶属企业、技术力量、生产经验、产品销售、取得荣誉等。其语言特点可大致总结如下。

(1) 语言常简洁,避冗长。公司简介的语言一般较为简洁、明了,很少采用陈旧的词汇,因而读起来朗朗上口。一般来说,英文公司简介的长度多控制在 400~4000 个单词的范围内。

(2) 内容客观,不夸大。公司简介不是推销广告,它要求文案内容要实事求是、数据准确、语言平实。因此,公司简介在内容上通常客观得体,不夸大宣传。文本上过度口语化或书面化都会给对方带来不好的印象,有损公司形象。同时,公司简介的内容一般较为平实和准确,注重从客户的角度出发,用平实的语调表达出公司最核心的内容,如公司性质、股东情况、注册资金等。

(3) 有誉美性。公司简介的语言力求简洁、明确,可恰当使用赞誉的词语,正面展示公司在经营中取得的成就,展现公司风采。

(4) 使用程式化用语。公司简介的内容主要是介绍公司的经营宗旨、业务性质和范围及产品或服务,同时还介绍公司本身的背景、历史和经营成就,以扩大对外影响,提高公司的知名度,特别是在新客户和潜在客户中树立公司的形象。公司简介的语言和市场营销紧密结合,既要照顾主题因素(公司贸易实体),力求准确、平实、简明、精练,运用专业术语和习惯用法,注意科学性和逻辑性,又要考虑公司简介的具体要求,尽量浅显确切,提供实质性信息。在公司简介中,经常可以看到这种程式化用语,如"主要经营""以……为宗旨""创建于""年销量"等。

3. 句子特点

在英文公司简介中,最常用的时态是一般现在时、一般过去时和现在完成时,我们也可以偶尔找到将来时态,其他时态极少出现。一般现在时通常被用来表达当前状态,介绍产品或服务的信息等,是公司简介中最频繁使用的时态。

此外,与其他科技文献类似,为了简洁、确切、严密、客观、真实地介绍公司,科技公司简介中也大量频繁地使用被动语态、非谓语动词结构、名词化结构和复合句。

二、科技公司简介范文展示

范文一

华为简介

华为创立于1987年，是全球领先的ICT（信息与通信）基础设施和智能终端提供商。我们的愿景和使命是致力于把数字世界带入每个人、每个家庭、每个组织，构建万物互联的智能世界：让无处不在的联接，成为人人平等的权利；为世界提供最强算力，让云无处不在，让人工智能无所不及；所有的行业和组织，因强大的数字平台而变得更加敏捷、高效、生机勃勃；通过AI重新定义体验，让消费者在家居、办公、出行等全场景获得极致的个性化体验。目前，华为约有19.5万名员工，业务遍及170多个国家和地区，服务全球30多亿人口。

Introduction of Huawei

Founded in 1987, Huawei is a leading global provider of information and communications technology (ICT) infrastructure and smart devices. Our vision and mission is to bring digital to every person, home and organization for a fully connected, intelligent world. To this end, we will drive ubiquitous connectivity and promote equal access to networks; bring cloud and artificial intelligence to all four corners of the earth to provide superior computing power where you need it, when you need it; build digital platforms to help all industries and organizations become more agile, efficient, and dynamic; redefine user experience with AI, making it more personalized for people in all aspects of their life, whether they're at home, in the office, or on the go. We have about 195,000 employees, and we operate in more than 170 countries and regions, serving more than three billion people around the world.

范文二

多伦科技公司简介

多伦科技自1995年成立以来即致力于中国驾驶人考训智能化、智慧城市建设的科技创新与产业化应用，致力于实现"人、车、路"领域的业务深度拓展和多元化发展。

经过20多年技术积累和业务拓展，公司围绕大数据、云计算、深度学习、三维虚拟仿真、物联网感知、北斗卫星定位六大核心技术，形成了以智慧车管、智慧驾培、智慧城市和智慧车检为主的四大产品体系，业务涵盖交通安全出行领域的诸多环节，已发展成为一家以驾驶人智能培训和考试系统、驾驶模拟训练系统、智慧交通综合管理/车联网产品与系统、智慧车检整体解决方案与投资运营管理等为主业的交通领域的高新技术企业，推进了中国大交通领域技术研发与产品创新的产业化应用。

作为智慧车管行业的引领者、智慧驾培行业的倡导者、智慧交通行业的推进者、智慧车检行业的践行者，业务市场已覆盖全国31个省（自治区、直辖市）400多个城市。

Duolun Technology Company Profile

Since its establishment in 1995, Duolun Technology has dedicated itself to technological innovation and industrial application of intelligent driver training and testing as well as smart city construction in China to achieve in-depth business expansion and diversified development of "people, vehicles and roads".

After over two decades of technology buildup and business development, we have established four product systems—smart vehicle administration, intelligent driver training, smart city, and smart vehicle inspection—based on six core technologies, including big data, cloud computing, deep learning, three-dimensional virtual simulation, Internet of Things perception, and Beidou satellite navigation. Our businesses have covered many aspects of the field of traffic safety, and we have developed into a high-tech enterprise in the transportation field with intelligent driver training and testing system, driving simulation training system, integrated smart transportation management/Internet of Vehicles (IoV) product and system, integrated smart vehicle inspection solutions, and investment and operation management. We have greatly promoted the industrial application of technology research and product innovation in China's transportation sector.

As a leader of smart vehicle administration, intelligent driver training and transportation, and smart vehicle inspection, we have expanded our market across over 400 cities in 31 provinces, autonomous regions and directly-administered municipalities in China.

范文三

中国移动有限公司简介

中国移动有限公司（"本公司"，包括子公司合称为"本集团"）于1997年9月3日在香港成立，并于1997年10月22日和23日分别在纽约证券交易所（"纽约交易所"）和香港联合交易所有限公司（"香港交易所"）上市。公司股票在1998年1月27日成为香港恒生指数成份股。

本集团在中国内地所有31个省（自治区、直辖市）及香港特别行政区提供全业务通信服务，业务主要涵盖移动话音和数据、有线宽带，以及其他信息通信服务，是中国内地最大的通信服务供应商，也是全球网络和客户规模最大、盈利能力领先、市值排名位居前列的世界级电信运营商。于2020年12月31日，本集团的员工总数达454 332人，移动客户总数达到9.42亿户，有线宽带客户总数达到2.1亿户，年收入达到人民币7681亿元。

本公司的最终控股股东是中国移动通信集团有限公司（原称为"中国移动通信集团公司"，简称"中国移动集团公司"）。于2020年12月31日，该集团间接持有本公司约72.72%的已发行总股数，余下约27.28%由公众人士持有。

2020年，本公司再次被《福布斯》杂志选入其"全球2000领先企业榜"，被《财富》杂志选入其《财富》世界500强。"中国移动"品牌在2020年再次荣登Millward Brown的"BrandZ™全球最具价值品牌100强"，位居36。目前，本公司的债信评级等同于中国国家主权评级，为标普评级A+/前景稳定和穆迪评级A1/前景稳定。

Introduction to China Mobile Limited

China Mobile Limited (the "Company", and together with its subsidiaries referred to as the

"Group") was incorporated in Hong Kong on 3 September 1997. The Company was listed on the New York Stock Exchange ("NYSE") and The Stock Exchange of Hong Kong Limited ("SE-HK") on 22 October 1997 and 23 October 1997, respectively. The Company was admitted as a constituent stock of the Hang Seng Index in Hong Kong on 27 January 1998.

As the leading telecommunications services provider in the mainland of China, the Group provides full communications services in all 31 provinces, autonomous regions and directly-administered municipalities throughout the mainland of China and in Hong Kong Special Administrative Region, and boasts a world-class telecommunications operator with the world's largest network and customer base, a leading position in profitability and market value ranking. Our business primarily consists of mobile voice and data business, wire line broadband, and other information and communications services. As of 31 December 2020, the Group had a total of 454,332 employees, and a total of 942 million mobile customers and 210 million wire line broadband customers, with its annual revenue totaling RMB 768.1 billion.

The Company's ultimate controlling shareholder is China Mobile Communications Group Co., Ltd. (formerly known as China Mobile Communications Corporation, "CMCC"), which, as of 31 December 2020, indirectly held approximately 72.72% of the total number of issued shares of the Company. The remaining approximately 27.28% was held by public investors.

In 2020, the Company was once again selected as one of The World's 2,000 Largest Public Companies by *Forbes* magazine and Fortune Global 500 by *Fortune* magazine. The China Mobile brand was once again listed in BrandZ™ Top 100 Most Valuable Global Brands 2020 by Millward Brown ranking 36. Currently, the Company's corporate credit ratings are equivalent to China's sovereign credit ratings, namely, A+/Outlook Stable from Standard & Poor's Rating and A1/Outlook Stable from Moody's Rating.

范文四

思爱普（SAP）公司简介

SAP 成立于 1972 年，最初被称为 System Analysis Program Development（Systemanalyse Programmentwicklung），后来采用缩写 SAP。刚创立时，SAP 是一家只有 5 个人的小公司。经过 SAP 不懈努力，现已发展成为一家大型跨国企业，总部位于德国沃尔多夫，在全球拥有超过 105 000 名员工。

SAP 最初推出的产品是 SAP R/2 和 SAP R/3，这两款产品建立了 ERP 软件的全球标准。如今，SAP S/4HANA 将 ERP 提升到了一个新的高度。SAP S/4HANA 采用强大的内存计算技术，能够处理海量数据，并支持人工智能（AI）和机器学习等先进技术。

SAP 还推出了智慧企业套件，基于完全数字化的平台集成各种应用，将企业的各个业务领域连接起来，取代由流程驱动的传统平台。目前，SAP 拥有超过 2.3 亿名云用户，提供上百款解决方案，覆盖所有业务职能，并且还拥有市场上最全面的云产品组合。

SAP 是全球领先的业务流程管理软件供应商之一，致力于开发先进的解决方案，帮助企业高效处理整个企业范围内的数据，实现无缝的信息流。

SAP 的愿景是让世界运转更卓越，让人们生活更美好。我们承诺不断开拓创新，帮助客户实现卓越运营。我们致力于构建有效的解决方案来推动创新、促进平等，并为世界各地拥有不同文化的人群提供更多机会。SAP 通过发挥自己在人才、技术和战略合作伙伴关系上的优势，推动数字融合，为所有人赋能。

Company Profile of SAP

Founded in 1972, SAP was initially called System Analysis Program Development (Systemanalyse Programmentwicklung), later abbreviated to SAP. Since then, it has grown from a small, five-person endeavor to a multinational enterprise headquartered in Walldorf, Germany, with more than 105,000 employees worldwide.

With the introduction of its original SAP R/2 and SAP R/3 software, SAP established the global standard for Enterprise Resource Planning (ERP) software. Now, SAP S/4HANA takes ERP to the next level by using the power of in-memory computing to process vast amounts of data, and to support advanced technologies such as artificial intelligence (AI) and machine learning.

SAP has also launched the smart enterprise suite, which integrates applications based on a fully digital platform to connect the various business areas of the enterprise, replacing the traditional process-driven platform. Today, we have more than 230 million cloud users, hundreds of solutions covering all business functions, and the most comprehensive cloud product portfolio in the market.

SAP is one of the world's leading producers of software for the management of business processes, developing advanced solutions that facilitate effective data processing and information flow across organizations.

At SAP, our purpose is to help the world run better and improve people's lives. Our promise is to innovate to help our customers run at their best. We are committed to building effective solutions that drive innovation, promote equality, and provide more opportunities for people with different cultures around the world. We power opportunity for all people through digital inclusion with the help of our talent, technology, and strategic partnerships.

三、科技公司简介常用词汇和句型

（一）科技公司简介常用词汇

1. 公司职位

president　总裁
vice-president　副总裁
assistant VP　副总裁助理
executive marketing director　市场行政总监
general manager　总经理
branch manager　部门经理
product manager　产品经理
project manager　项目经理

regional manager　区域经理
production manager　生产经理
transportation manager　运输经理
applications programmer　应用软件程序员
computer operator　电脑操作员
computer operations supervisor　电脑操作主管
hardware engineer　硬件工程师
computer technician　电脑技术员
MIS manager　管理信息系统部经理
developmental engineer　开发工程师
operations analyst　运营分析师
director of information services　信息服务主管
LAN administrator　局域网管理员
systems analyst　系统分析师
manager of network administration　网络管理经理
systems engineer　系统工程师
product support manager　产品支持经理
systems programmer　系统程序员
VP sales　销售副总裁
VP marketing　市场副总裁
senior account manager　高级客户经理
telemarketing director　电话销售总监
sales administrator　销售主管
telemarketer　电话销售员
regional sales manager　地区销售经理
tele-interviewer　电话调查员
regional account manager　地区客户经理
salesperson　销售员
sales representative　销售代表
merchandising manager　采购经理
sales manager　销售经理
marketing consultant　市场顾问
sales executive　销售执行者
marketing assistant　市场助理
sales assistant　销售助理
marketing and sales director　市场与销售总监
retail buyer　零售采购员
market research analyst　市场调查分析师
manufacturer's representative　厂家代表

purchasing agent　采购代理
assistant account executive　客户管理助理
marketing manager　市场经理
marketing intern　市场实习
marketing director　市场总监
account manager　客户经理
account representative　客户代表
accounting payable clerk　应付账款文员
accounting assistant　会计助理
accounting manager　会计经理
accounts receivable clerk　应收账款文员
accounting clerk　会计文员
certified public accountant　注册会计师
senior accountant　高级会计
chief financial officer　首席财务官
audit manager　审计经理
collections officer　收款负责人
actuarial analyst　保险分析员
auditor　审计师
junior accountant　初级会计
loan administrator　贷款管理员
management accountant　管理会计
billing clerk　票据文员
billing supervisor　票据管理员
bookkeeping clerk　档案管理助理
budget analyst　预算分析师
tax accountant　税务会计
credit analyst　信用分析师
credit manager　信用管理经理
vice-president of administration and finance　财务行政副总裁
financial analyst　财务分析师
vice-president of finance　财务副总裁
financial consultant　财务顾问
financial manager　财务经理
financial planner　财务计划员
VP HR　人力资源副总裁
assistant VP HR　人力资源副总裁助理
HR director　人力资源总监
compensation & benefit manager　薪酬福利经理

staffing manager　招聘经理
training manager　培训经理
benefits coordinator　员工福利协调员
employer relations representative　员工关系代表
payroller　工资专员
training coordinator　培训协调员
training specialist　培训专员
training supervisor　培训主管
vice–president of administration　行政副总裁
administrative director　行政总监
office manager　办公室经理
file clerk　档案管理员
administration assistant　行政助理
receptionist　接待员
general office clerk　办公室文员
secretary　秘书
order entry clerk　订单输入文员
operator　接线员
typist　打字员
Chief Executive Officer（CEO）　首席执行官
Chief Financial Officer（CFO）　首席财务官
Chief Operating Officer（COO）　首席运营官
Chief Technology Officer（CTO）　首席技术官
Chief Information Officer（CIO）　首席信息官
Chief Risk Officer（CRO）　首席风险官

2. 公司部门

head office　总公司
branch office　分公司
affiliated organization　分支机构
business office　营业部
personnel department　人事部
human resources department　人力资源部
general affairs department　总务部
accounting department/finance department　财务部
sales department　销售部
sales promotion department　促销部
international department　国际部
export department　出口部
import department　进口部

public relations department　公共关系部
advertising department　广告部
planning department　企划部
product development department　产品开发部
research & development department（R&D）　研发部
secretarial pool　秘书室
purchasing department　采购部
engineering department　工程部
administration department　行政部
marketing department　市场部
technology department　技术部
service department　客服部
general manager's office　总经理室
general department　总务处
deputy general manager's office　副总经理室
productive department　生产部
wireless industry department　无线事业部
business expending department　拓展部
supply department　物供部
equipment department　设备部
security and quality department　安全质量部
commodity and equipment department　物资设备部
manufacturing department　制造部
department of market research　市场研究部
engineering technician department　工程技术部
project management department　项目管理部

3. 公司类型与文化

affiliated company　联营公司
enterprise/corporate culture　企业文化
annual production value　年产值
enterprise group　企业集团
annual turnover　年营业额
active assets　流动资产
active balance　顺差
adverse balance　逆差
annual trading value　年贸易额
absorption　合并
access to a market　进入市场
export processing zones　出口加工区

branch company　分公司
finance enterprise　金融企业
brand value　品牌价值
foreign corporation　外国公司
foreign – funded enterprise　外资企业
technical service enterprise　科技服务企业
foreign – funded company　外资公司
business goodwill　企业信誉
free post – sale services　免费售后服务
business philosophy　经营理念
general assets　资产总额
business principles　经营宗旨
multinational corporation　跨国公司
general manager responsibility system　总经理负责制
business scope　经营范围
class A enterprise　一级企业
global leader　全球主导企业
clients first, reputation first　客户第一，信誉第一
group company　集团公司
high – tech industrial park　高科技工业园区
close company　内股公司
holding company　控股公司
company of limited liability　有限责任公司
holding subsidiaries　控股子公司
company of unlimited liability　无限责任公司
honorable enterprise　荣誉企业
computer services and software enterprise　计算机服务及软件企业
industrial enterprise　工业企业
conglomerate company　综合公司
information transmission enterprise　信息传输企业
cooperative enterprise　合作企业
innovative business　创新企业
corporate ideals (values)　企业理念
joint stock company　股份公司
domestic corporation　国内公司
joint venture　合资企业
domestic – funded company　内资公司
enterprise credit brand　企业信用品牌
large – scale enterprise group　大型企业集团

33

listed company　上市公司
local company　地方公司
management model　管理模式
registered capital　注册资金
management philosophy　管理理念
scope of business　业务范围
market share　市场占有率
market-oriented　以市场为导向
small and medium-sized enterprises　中小型企业
sole-funded company　独资公司
mutually beneficial　互利的，双赢的
state-controlled enterprise group　国家控股企业集团
offshore company　离岸公司
state-owned enterprise　国有企业
private company　私人公司
state-run enterprise　国营企业
production area　生产面积；生产区域
stock issuing house　股票发行公司
proprietary company　控股公司
subsidiary company　子公司
public company　上市公司；公众性公司
upgrading of an industrial structure　产业结构升级
qualified enterprise　合格企业
bonded factory　保税工厂
corporate vision　企业愿景
credible corporate　守信企业
estate　不动产；财产
evaluation　估价
customer service　客户服务
factualism and innovation　求实创新
integrate/combine　集……于一体
first-rate quality　质量上乘
good service　优质服务
headquarter　总部
high reputation　诚信第一
innovative and entrepreneurial spirit　创业创新精神
intellectual property rights　知识产权
locomotive/flagship　龙头企业
net assets　净资产

perseverance 坚忍不拔
private sector 私营企业
profit margin 利润率
promotion 促销
registered capital 注册资本
revenue 收入
trade fair 交易会
trustworthiness and integrity 诚信正直
united cooperation 团结合作
well - known trademark 著名商标
World Brand Laboratory 世界品牌实验室

4. 公司简介常用主题词

a (very) good reputation 一个（非常）良好的声誉
be dependent on 与……相互依存
ample supply and prompt delivery 货源充足，交货迅速
be listed on 在……上市
boast 拥有
attractive and durable 美观耐用
bright in color 色彩鲜艳
business scope 经营范围
one - step 一站式
complete range of articles/a great variety of goods 品种齐全
optimization and innovation 优化创新
orders are welcome 欢迎订购
constituent stock 成分股
overseas engineering 海外工程
conventional core business 传统核心业务
overseas representative offices 驻外代表处
pleasant in after - taste 回味隽永
customized service 定制服务
precision machinery 精密机械
cutting - edg 领先的
economic globalization 经济全球化
project management divisions 项目管理部门
elegant and graceful 典雅大方
reasonable price/street price/moderate price 价格公道
elegant appearance 美观大方
sales revenue 销售收入
enjoy a reputation (for) 在……方面享有声誉

great varieties　品种繁多
selling well all over the world　畅销世界各地
equity interest　股本权益
shareholding company　股份公司
evident effect　明显的效果
special administrative region　特别行政区
excellent quality　品质优良
specialized services　专业化的服务
exquisite fine craftsmanship　技艺精湛
global roaming　全球漫游
good reputation over the world　誉满全球
the key stated-own enterprises　重点国有企业
strong market competitiveness　较强的市场竞争能力
have visibly expedited　令……明显加快
to enhance our corporate value　提升企业价值
to maintain our sustainable long-term development　确保公司的长期持续发展
ideal gift for all occasions　节日送礼之佳品
lead to　迈向
modern techniques　技术先进
modern logistics　现代物流
to rank first among similar products　居同类产品之魁首
in short supply/demand exceeding supply　供不应求
deeply trusted and praised by customers at home and abroad　深受国内外客户的信赖和称誉
total assets attained　总资产达
turn a new page　翻开新的篇章
market share　市场占有率
moderate cost　价格适中
innovative business　创新企业
less expensive, high quality goods/high quality and inexpensive/high quality and low overhead　价廉物美
under the scientific guidance of management　在科学的管理理念指导下
let our commodities go to the world　让我们的商品走向世界
we have won praise from customers　赢得了客户的好评
well appreciated by their purchasers　深受买家赞赏

（二）科技公司简介常用句型

be awarded most welcome goods　被评为最受欢迎的商品
be awarded the gold prize　被授予金奖
be established in　创建于

be incorporated   被纳入
ride into/rank among   跻身于
be located in/be situated in/lie in   位于
be named one of the world's most recognizable/famous brands   被评为世界上最知名的品牌之一
mainly engaged in   主要从事
be involved in   参与
customer is the first priority   顾客至上
engage in/handle a large range of businesses including   主要经营
with the enterprise spirit of   具有……的企业精神
adhere to the aims of   坚持……的目标
comply with international quality standards   符合国际质量标准
advanced quality control system   先进的质量控制体系
be concerned with   涉及；关心；从事于
adhere to the operation philosophy of   秉承……经营理念
be a professional manufacturer of   一家专门生产……的企业
be specialized in   专门从事
be named as   被命名为
be approved by   被……批准
enter into a contract with   与……签约
be combined with   与……相结合
hold/abide by the principles/tenets of   坚持/遵守……原则
be committed to providing... with... service   致力于为……提供……服务
be the first to   尚属首家……
pass/gain/obtain/be granted the certificate of   通过……认证
be founded in   成立于
be headquartered in   公司总部设在

四、科技公司简介翻译技巧

（一）中英文科技公司简介对比差异

1. 词汇差异

中文公司简介中常用很多形容词或副词修饰名词或动词，以加强语气，期望给读者留下深刻印象。例如：

结合自身实际，长城集团积极探索科学、高效的管理机制，创造了具有开放性和包容性的特色企业文化氛围，形成了一支汇聚科研、生产、销售、服务等多方面优秀人才的精英团队。

37

译文：In line with its own conditions, Great Wall Group has actively explored a scientific and highly efficient management system, created an open and comprehensive corporate culture, and formed an outstanding work team integrating scientific research, production, sales and services.

从上面的英文翻译中可以看出，译者在翻译时并没有做过多的考虑，而是逐字逐句地翻译了原文，这样做容易给目的语读者造成浮夸的印象。而在以下这篇英文公司简介中，作者并没有用太多的形容词和副词，而是用一些客观数据有力地说明了本公司的实力。因此，译者在翻译中文公司简介时应该注意到目的语读者重视客观事实的思维模式。

National Express Group is one of the world's leading international public transport groups. Every year, we provide around one billion journeys worldwide on our buses, trains, light rail services, airport transfers and express coaches. We have 43,000 employees and are market leaders in the UK, North America and Spain. We are committed to improving the quality of life for all by "making travel simpler".

除此之外，中文公司简介喜欢用四字格，而英文公司简介喜欢用简单的词汇。例如：

卓越超群，稳健可靠，独特创新，至诚合作。

In everything we do, we strive for integrity, dialogue, transparency, excellence and innovation.

2. 句法差异

在句法层面，中文公司简介的行文措辞遒劲有力，善用排比、对偶等修辞手法，着力描绘公司的雄厚实力和拼搏精神。相比之下，英文公司简介的语言则是表达简洁，陈述事实，当然其中也不乏吸引顾客的广告式词句。形象地说，中文公司简介富有激情，如一团烈火；而英文公司简介平静自然，如涓涓细流。中文公司简介经常会介绍公司的战略和目标，善用四字词来表强调；而英文公司简介习惯使用复杂句和长句表意，单句中包含很大的信息量。因此，为了在译文中体现中英文公司简介的差异，译者需要做到化繁为简。例如：

公司将紧跟时代步伐，顺应现代市场经济的发展趋势，大力推进营销产业化、网络化、现代化、国际化发展战略，不断提高营销增值性、服务综合性、战略前瞻性，精心打造"天物金属"的服务品牌。

译文：Our company will follow the steps of the development of modern market economy and vigorously push forward the strategy of marketing industrialization, network realization, modernization and internationalization. We will constantly improve our incremental marketing, comprehensive service and predictable strategy to make "TEWOO Metal" a favorable service brand.

改译：We will continuously improve our business and service with the development of networks and the deepening of industrialization and internationalization to build the famous brand of "TE-WOO Metal".

除此之外，中文公司简介常用并列的分句，看上去虽然小句较多，但是其实上下文之间在意义上有逻辑性，并且主题突出。英文公司简介多使用衔接词，句子结构严谨，主语突出，更注重形式上的统一，有整体性和条理性的特点。例如：

应对四时的变幻，拾掇各式的斑斓。我们紧随潮流趋势，带给您春夏、秋冬流行色卡。或淡雅、清新，用青黄粉绿，洗去渐醒的慵懒；或热情、洒脱，着赤橙蓝紫，翩翩起舞于寒冬。

For instant style and texture, complete your outfit with this faux fur chevron patterned coat, with a contrasting collar for added pop.

3. 文本内容差异

在文本内容层面，中文公司简介讲究面面俱到，通常包括公司性质、成立时间、历史发展、产品类别、服务项目、生产规模、研发能力、财务状况、员工人数、企业文化、获奖荣誉、辉煌成就、交流合作、社会服务、目标愿景等。英文公司简介比较简短，通常包括公司性质、成立时间、业务范围、产品类别、员工人数、目标愿景等，而且往往蜻蜓点水，一笔带过。例如，Verizon公司简介尽管仅有一段，但却是"麻雀虽小，五脏俱全"：

Verizon is a global leader in delivering broadband, video and other wireless and wireline communications services to mass market, business, government and wholesale customers. Verizon Wireless operates America's largest and most reliable wireless voice and 3G network. Verizon also provides communications, information and entertainment services over America's most advanced fiber-optic network, and delivers innovative, seamless business solutions to customers around the world. We believe strongly that our role in connecting people, ideas and opportunities is vital to meeting the challenges of the future.

Verizon公司简介包括公司性质、服务对象、业务范围（产品类别和服务项目）及公司宗旨（核心价值观）。

4. 文本结构差异

第一，中英文公司简介的文本结构总体上都比较单一，基本上遵照的是一种线性的推进方式（linear progression），属于普通的文本结构。中文公司简介的文本结构完整，各项信息内容逐段展开，除开头和结尾信息内容相对固定外（开头部分一般是公司的性质和历史发展，结尾部分是目标愿景），其余各段的顺序形式比较随意，孰先孰后无规律可陈。大多数英文公司简介的文本结构相对完整，文本的顺序形式比较固定，先后顺序井然，首先开头一般介绍公司的性质和定位，然后依次是业务范围（包括产品类型、服务项目等）、公司业绩、财务状况等，最后一般以目标愿景结尾。但是，实际文本顺序因公司不同会略有差异。不少英文公司简介各段前配有段落小标题，提纲挈领，让读者一目了然，非常人性化。例如，ExxonMobil公司简介中就有Who We Are、What We Do、Where We Work三个段落小标题；Chevron公司简介中有Providing Energy for Human Progress、Company Roots、Global Scope、Technology and Emerging Energy、Environment and Safety及Our Work六个段落小标题。

第二，中英文公司简介均属于应用文体或非文学文体，其目的是通过传播公司的相关信息，让读者了解公司，对公司产生好感，进而对公司的产品或服务产生兴趣。中文公司简介往往有固定的形式依次把相关信息介绍出来，比如很多中文公司简介都以"××公司成立于××年，是一家××企业"这样的句式开头，结尾大多是公司的发展目标，措辞各有不同。相对而言，英文公司简介写作手法多样。例如：

Nike公司简介的第一句是"If you have a body you are an athlete.", 接着是"He set the tone and direction for a young company created in 1972, called Nike, and today those same words inspire a new generation of Nike employees."。

JPMorgan Chase公司的简介则使用了一连串的排比句，把自己的实力、自豪感展现得淋漓尽致：

We are a leading global financial services firm with assets of $2 trillion.

We operate in more than 60 countries. We have more than 200,000 employees.

We serve millions of the U.S. consumers and many of the world's most prominent corporate, institutional and government clients.

We are... We are...

Kraft Foods 公司的简介更是别出心裁：

To make today delicious we begin with our consumers. We listen we watch and we learn. We understand their joys and their challenges because we're consumers too.

5. 文化差异

中英文公司简介体现了各国的文化传统、行为规范和价值观。中国公司的发展离不开国家的政策和支持，公司以获得国家级荣誉和奖项为荣，因此中文公司简介中会强调相关内容，比如强调公司的国有企业性质，强调公司发展的政治指导方向，强调公司所获得的各种荣誉和奖项等。例如：

中国石油化工集团公司（英文缩写 Sinopec Group）是 1998 年 7 月国家在原中国石油化工总公司基础上重组成立的特大型石油石化企业集团，是国家独资设立的国有公司、国家授权投资的机构和国家控股公司。

相比之下，英语国家公司多是自主经营、自负盈亏的经济实体，政府一般不参与公司的经营管理，因此英文公司简介中几乎没有提及上述内容。例如：

GM General Motors Company (NYSE: GM TSX: GMM), one of the world's largest automakers, traces its roots back to 1908. With its global headquarters in Detroit, GM employs 209,000 people in every major region of the world and does business in more than 120 countries. GM and its strategic partners produce cars and trucks in 31 countries, and sell and service these vehicles through the following brands: Buick, Cadillac, Chevrolet, GMC, Daewoo, Holden, Isuzu, Jiefang, Opel, Vauxhall and Wuling. GM's largest national market is China, followed by the United States, Brazil, the United Kingdom, Germany, Canada and Russia. GM's OnStar subsidiary is the industry leader in vehicle safety, security and information services.

GM 公司简介非常简要地介绍了该公司的性质、规模、成立时间、业务范围、产品类型、海外市场等情况，叙述平实客观。

中文公司简介喜欢展示历史成就，罗列所获奖项；而英文公司简介喜欢描述发展现状和展望未来。这是因为中国和英语国家公司价值取向不同，中国公司重视公司价值，英语国家公司重视消费者价值。例如：

公司在视频处理、芯片设计和显示系统设计等方面具有自主研发和自主知识产权的核心技术，尤其是在裸眼 3D 显示技术领域；迄今已申请专利 14 项，获得授权 7 项；已申请商标 3 件，获得注册商标证书 3 件；获得 1 项软件著作授权并获得软件产品登记。

We've made 40 commitments to meet our long-term goals of being a nutrition, health and wellness leader, reducing our environmental footprint and creating a lasting positive impact on communities.

6. 修辞风格差异

中英文公司简介呈现了不同的修辞风格。中文公司简介通常采用第三人称的口吻叙述，平铺直叙，措辞正式，喜欢重复公司名称，以表明公司的严谨和展现自己的雄厚实力。例如：

1912年2月，经孙中山先生批准，中国银行正式成立。从1912年至1949年，中国银行先后行使中央银行、国际汇兑银行和外贸专业银行职能，坚持以服务大众、振兴民族金融业为己任，稳健经营，锐意进取，各项业务取得了长足发展。新中国成立后，中国银行成为国家外汇外贸专业银行，为国家对外经贸发展和国内经济建设做出了重大贡献。1994年，中国银行改为国有独资商业银行。2003年，中国银行开始股份制改造。2004年8月，中国银行股份有限公司挂牌成立。2006年6月、7月，中国银行先后在香港联交所和上海证券交易所成功挂牌上市，成为首家在香港和内地发行上市的中国商业银行。

这段文字以"中国银行"为话题，以时间顺序推进行文，旨在营造一种正式、权威的氛围，以赢得广大客户的信赖。

英文公司简介大多采用第一人称We（或物主代词our）的叙述视角。例如：

For more than a century, we have consistently provided innovative, reliable, high – quality products and services and excellent customer care. Today, our mission is to connect people with their world, everywhere they live and work, and do it better than anyone else. We're fulfilling this vision by creating new solutions for consumers and businesses and by driving innovation in the communications and entertainment industry. （AT&T公司简介）

英文公司简介中还以our customers和you（或your）称呼顾客，这种表述既拉近了公司和顾客的距离，又体现了"顾客至上"的企业文化。例如：

Working seamlessly across regions and disciplines, we are committed to delivering the unmatched power of our global franchise to meet your full range of financial objectives. （Bank of America简介）

（二）翻译技巧

1. 增译法

翻译时可以根据意义上或修辞上或句法上的需要增加一些内容以对原文内容进行补充、解释或限定，通过这种方式对原文进行明晰化处理，以求更方便译文读者对译文的接受。

例1：创维成立于1988年，总部坐落在具有创新"硅谷"之称的深圳高新技术产业园……

译文：Skyworth was established in 1988, with the head office located within Shenzhen High Tech Industrial Park which is honored as China's silicon valley...

众所周知，"硅谷"是位于美国加州北部的一个世界著名的高科技产业区。原文中作者用了"硅谷"一词是想表明创维集团公司所处的地点——深圳高新技术产业园在中国的地位相当于美国的"硅谷"，因而在译文中增补"China's"进行限制说明很是恰当。

如果不充分考虑中英文语言文化差异，一味地照搬直译，就不能将句子中隐含的意义表达出来，使英语国家读者感到一头雾水。这时需要站在英语国家读者的角度，在翻译时对这些地方进行充分又通俗易懂的解释说明。

例2：在"十二五"期间，公司将努力实现"进军世界500强"的宏伟愿景。

原译：During the "12th Five-year Plan", we will try to achieve the grand vision of "marching world's top 500 enterprises".

改译：In the "12th Five-year Plan" period (2011-2015), we will stride into the rank of "Fortune 500".

原文的目的是树立公司形象，原文中具有中国特色的词语"十二五"，意为"第十二个五年规划"，具体为2011年至2015年。原译将"'十二五'期间"译为"12th Five-year Plan"，但是不了解中国文化背景的英语国家读者，依然不理解"十二五"具体指的是什么，所以译者要将"十二五"隐含的"2011年至2015年"译出来。使用增译法进行改译后，既准确传达了原文意思，又使译文达到了树立公司形象的目的。

2. 省略法

省略法就是省译法，即省去冗余或不合适的信息。中文公司简介多用排比、对偶等修辞手法渲染效果，注重罗列历史成就。而英文公司简介喜欢用简洁平实的词语进行描述，且较少描述历史成就。因此，在翻译过程中，要避轻就重和避虚就实，关注英语国家读者的语言文化习惯。

例3：洛阳宇通汽车有限公司成立于1999年10月，是隶属于宇通集团的主要成员单位。公司前身为中国人民解放军第五四零八工厂，成立于1960年。1999年与军队企业脱离，加入宇通集团。

原译：Luoyang Yutong Automobile Co., Ltd., which is a main member led and managed by Yutong Group, was founded in October of 1999. The predecessor of our company, No. 5408 Factory of PLA, was founded in 1960. It broke away from military industry and joined Yutong Group in 1999.

改译：Luoyang Yutong Automobile Co., Ltd., established in 1999 out of its predecessor company, No. 5408 Factory of PLA, is a main member of Yutong Group.

原文的目的是树立公司形象，原译将原文中关于公司前身的部分也照搬直译出来，然而英语国家读者更重视公司的发展现状，不太重视公司的发展历史。因此，对原文使用省译法，省去公司前身成立于1960年，符合英文公司简介简洁明了的特点，这也突出了另一个重要的时间"1999年"，符合连贯性原则，更好地实现了译文的外宣目的。

有时，中文公司简介喜欢用一些夸张或重复的手法来表达意思，但在英文翻译中有时会将其省去。

例4：为员工提供了全球化发展平台、与世界对话的机会。

译文：We provide our teams with a global development platform.

原文中"全球化发展平台"和"与世界对话"两个近义短语并列，符合中文喜重复的行文特点。然而，译文中省译了"与世界对话的机会"，这与英文不喜重复、追求简洁的特点不谋而合。因此，翻译该句时，译者按照目的语的表达习惯和行文特点构建译文，有助于受众的理解和接受，从而实现有效信息的传递和劝说效果的提升。

3. 删减法

源语文本中与主题关联性不强的信息，或者由于文化差异而使目的语读者难以理解或接受的部分，都属于删减之列。

例5：三环集团公司是国家汽车零部件出口基地企业……被中组部授予"全国创先争优活动先进基层党组织"称号，被省委授予"全省先进基层党组织"称号。

译文：Tri‐Ring Group Corporation is a national base enterprise of automobile parts export... We were awarded the title of "National Advanced Grassroots Party Organization in Pioneering and Striving for Excellence Activities" by the Organization Department of the Central Committee of the CPC and the title of "Advanced Grassroots Party Organization of the Province" by the Provincial Party Committee.

原文的目的是树立公司形象，译文符合中国公司的价值取向，即重视公司价值，罗列所获奖项；但不符合英语国家公司的价值取向，英语国家读者并不看重这些奖项，也不了解这些奖项的权威性。所以，既然其不能为主题服务而又会给读者造成理解的困难，则将"We were awarded... by the Provincial Party Committee."删掉为好。

例6：抚今追昔、继往开来，郑煤机在新世纪正以全新的姿态与时俱进，开拓创新，站在新的历史起点上，全面贯彻落实科学发展观，专心致志地实践"五化"发展战略，脚踏实地实施"三次创业"的发展大计，努力打造中国煤机行业的奔驰，打造百亿郑煤机、百年郑煤机，为社会主义祖国现代化事业再创新的业绩！

译文：In this new century, ZMJ is making great efforts to create the best brand in China colliery machinery industry and aiming at creating annual output value of RMB ten billion.

上述中文中的四字格和诸如"以全新的姿态与时俱进，开拓创新，站在新的历史起点上，全面贯彻落实科学发展观，专心致志地实践'五化'发展战略"等完全是按照中国人的阅读习惯编写的，如果直接将以上信息译成英文，对英语国家读者不具有信息性。该段简介无非是在表明该公司正在为实现宏伟目标而努力，因此英译时应进行简化处理。这样一来，信息更加突显，符合英语国家读者的阅读习惯。

例7："十一五"期间，公司围绕"十一五"规划目标，突出抓好"海上大庆油田"建设的同时，牢牢把握我国经济社会发展的重要战略机遇，以深化改革、创新发展的精神，有效应对全球金融危机带来的重大不利影响。在能源供应能力建设、产业价值链建设、国际化建设、现代企业制度建设、软实力形成等方面取得了突出成绩，公司的发展目标进入了一个新的阶段。

原译：During the "Eleventh Five‐year Plan" period, while focusing on the "Eleventh Five‐year Plan" objectives and highlighting the construction of an "offshore Daqing oilfield", the Company has firmly grasped the important strategic opportunities arising from China's economic and social development and responded effectively to the material adverse impact of the global financial crisis with the spirit of deepening reform and promoting development through innovation, and made outstanding achievements in capacity building in energy supply, construction of industry value chain, internationalization construction, modern enterprise system building and formation of soft power, thus the Company has entered into a new phase of development goals.

改译：We uses innovation and technology to deliver energy and petrochemical products to meet

China's economic and social development. We are now focused on developing "offshore Daqing oilfield" and by operating responsibly, executing with excellence, applying innovative technologies and capturing new opportunities, we are making the company one of the world's leading integrated energy companies.

原译带有中国自身特色的词汇，为避免直译造成受众不懂其真正的意思，因此对原译删除并改写部分内容，仅保留目的语受众关心的实质信息。

例8：按销售量计算，本公司位居中国乃至全球煤炭上市公司前列，2016 年，本公司煤炭销售量达到 394.9 百万吨。按发电机组装机规模计算，本公司位居中国电力上市公司前列，于 2016 年底本公司控制并运营的发电机组装机容量达到 56 288 兆瓦。

原译：In terms of sales, the Group holds a leading position among the listed coal companies in China and globally with the sales volume of coal reaching 394.9 million tons in 2016. In terms of installed capacity of power generators, the Group holds a leading position among the listed electricity companies in China with the installed capacity of its controlled and operated power generators reaching 56,288MW by the end of 2016.

改译：The Group holds a leading position among the listed coal companies in China and globally in terms of sales volume which reached 394.9 million tons in 2016. And we also lead the listed power industry in China.

在汉语中，适当的重复可以有效地帮助读者认识和强化所重复的信息。但如果直接不加删减地译成英语，则会影响信息传递的效率。在该例中，"销售量"和"发电机组装机规模"在原文中重复出现，原译原封不动地将这些词汇译成英文而没有做相应的删减，造成语义反复，语句冗长。信息功能是公司简介文本最重要的功能之一，翻译时译者最重要的任务就是将原文中的信息有效地传导到目的语。原文中的两句话所传递的主要信息为该公司的"销售量"和"发电机组装机规模"位于行业前列，因此可以将原译进行修改。

4. 改写法

中文公司简介通常是"以承转合"式为主流，行文规范，格式单一；而英文公司简介则讲究逻辑关系，文章脉络清晰，层次分明。因此，译者在公司简介英译时必须分析汉语句子之间的逻辑关系，分清主次，并考虑英语的表达习惯，澄清信息内容的逻辑层次，通过对整个篇章的整合，最终译出结构严谨、布局合理的译文。而改写法恰恰就是根据目的语读者的需要对原文进行适当调整的一种最常用的手段。

例9：快速反应、马上行动、质量第一、信誉至上是公司的宗旨。我们将以客户的需求为发展动力，为客户提供满意的产品及优质的服务。

原译：Our tenet is "Quick reflection, Immediate action, Quality first, Reputation first". We will be always pursuing what you need and provide you with satisfactory products and top-notch service.

改译：We will offer you our products and services to your entire satisfaction with promptness, quality and credit.

原译将"快速反应、马上行动、质量第一、信誉至上"直接生硬地译为"Quick reflection, Immediate action, Quality first, Reputation first"，这不仅起不到公司宗旨的宣传作用，反而会使英语国家读者感觉空洞无物、莫名其妙。例如，"拳头产品"不宜译为"fist product"，而应译为"knock-out product"或"competitive product"。同样，此处的"快速反应、马上行动"主要是指公司办事效率高，而非反应快和动作迅速，因此不宜译为"Quick reflection, Immediate action"，而应为"promptness"。此外，将"质量第一、信誉至上"中的"第一"和"至上"都译为"first"实为不妥，虽然英语国家读者大致也能猜出其中的意思，但却不符合英语的表达习惯，此处应该进行改写。改写后内容简洁且符合英语表达习惯，能使目的语读者备感亲切、印象深刻。

五、课后练习

（一）思考题

（1）科技公司简介有哪些特点？
（2）科技公司简介常用词汇有哪些？
（3）翻译科技公司简介有哪些技巧？

（二）翻译练习

1. 中译英

（1）请将大疆创新科技有限公司简介译成英文。

关于公司

深圳市大疆创新科技有限公司成立于 2006 年，总部位于深圳。公司得益于供应商、原材料和年轻、富有创造力的人才库。利用这些资源，公司从 2006 年的单一小型办事处发展成为拥有全球员工的大公司。公司的办事处现已遍布美国、德国、荷兰、日本、韩国、北京、上海和香港，支持全球一百多个国家和地区的销售与服务网络。作为一家私营公司，公司专注于客户的愿景，支持技术的创意、商业和非营利性应用。如今，公司的产品正在重新定义行业。公司为电影制作、农业、搜索和救援、能源基础设施等领域专业人士的工作带来新视角，帮助他们比以往更安全、更快速、更高效地完成壮举。

成立以来，公司的业务从无人机系统拓展至多元化产品体系，在无人机、手持影像系统、机器人教育、智能驾驶等多个领域成为全球领先的品牌，以一流的技术产品重新定义了"中国制造"的内涵，并在更多前沿领域不断革新产品与解决方案。公司以创新为本，以人才及合作伙伴为根基，思考客户需求并解决问题，得到了全球市场的尊重和肯定。

社会贡献

重塑人们的生产和生活方式

公司为用户带来创新、可靠的产品，并迅速进入影视传媒、能源巡检、遥感测绘、农业服务、基建工程、前沿应用等多个领域，为各行各业提供了高效、安全、智能的工具。同时，公司致力于成为公共安全和应急救援中不可或缺的力量，在地震、火灾、危化物品泄漏、爆炸、突发疫情中提供强有力支持。

培养社会的科技创新力量

公司持续深耕机器人教育领域，致力于为社会培养复合型科研人才。公司发起并承办了 RoboMaster 比赛，推出了教育机器人产品，受到全球科技爱好者的追捧；并与众多国内外学校、研究机构密切合作，搭建出一套由课程、产品、赛事及相关服务构成的机器人教育解决方案。公司正与全社会一道拓展教育新边界，成就新一代技术人才。

公司文化

通过创造最好的高科技产品，公司将不断培养和成就德才兼备的人才，为志同道合的伙伴们打造实现梦想、超越自我的精神家园，为推动人类文明的进步贡献力量。

（2）请将上海蓝鸟机电有限公司简介译成英文。

上海蓝鸟机电有限公司创建于 1998 年，是工业互联网创新服务商，是数字化转型解决方案供应商，20 多年专注于工业信息自动化解决方案。

公司先后被评为上海市高新技术企业、普陀区企业技术中心、软件企业、上海名牌、专精特新企业，通过两化融合管理体系评定。

我们的核心竞争力是数字化工厂、系统集成、信息自动化产品分销及服务，公司业务涉足钢铁有色、电子半导体、食品医药、水与水处理、汽车轮胎、基础设施、水泥建材、电力、新能源、智能楼宇、石油化工等行业。

发展历程

创业阶段（1998—2004）

上海蓝鸟机电有限公司成立。

Schneider 国际分销商。

GarrettCom 中国区代理商。

Wonderware 华东区总代理。

信息化发展阶段（2005—2010）

武汉蓝鸟分公司成立。

荣获上海市科技企业。

荣获上海市软件企业。

通过 ISO 9001:2000 认证。

荣获上海市高新技术企业。

荣获普陀区科技小巨人企业。

荣获计算机信息系统集成企业三级资质。

智能化发展阶段（2011—2015）

荣获 2014 年度上海名牌。

荣获上海市专精特新企业。

荣获上海市科技小巨人（培育）企业。

Wonderware 中国区总代理。

荣获电子工程专业承包三级资质。

荣获"上海市普陀区企业技术中心"称号。

荣获机电设备安装工程专业承包三级资质。

数字化转型阶段（2016 至今）

通过两化融合管理体系评定。

荣获电子与智能工程专业承包二级资质。

荣获上海市"四星级诚信创建企业"称号。

荣获"2020 年度普陀区总工会工人先锋号"称号。

蓝鸟锐泰连续六年被评为"上海市优秀软件产品"。

（3）请将基恩士公司简介译成英文。

基恩士（KEYENCE）致力于成为富有创新精神的企业，开发始终能够给我们客户的运营增值的产品与解决方案。

自 1974 年以来，基恩士稳步发展，成为全球工业自动化和检测设备开发和制造的创新领导者。我们的产品包括读码器、激光打标机、机器视觉系统、测量系统、显微镜、传感器和静电消除器。

我们不仅致力于满足许多制造与研究行业客户现在的需求，同时还致力于预见市场的未来发展，为客户提供更加长远的改善方案。

在基恩士，我们不仅以我们的产品为荣，而且为我们提供的支持为荣。我们还为客户提供更丰富的业界知识及更专业的技术方案，助力客户取得更高的成就。

如今，基恩士为世界范围内约 110 个国家或地区的 30 余万家客户提供服务。基恩士连续在"全球最具创新力企业"（福布斯）等知名公司排名中名列前茅。

（4）请将国家电网公司简介译成英文。

公司介绍

中国国家电网公司是世界上最大的公用事业公司，供电人口超过 11 亿，供电范围占国土面积的 88%，并在 9 个国家和地区运营骨干能源网。在《财富》世界 500 强中排名第二，总收入达 4100 亿美元，在过去 20 多年中，国家电网公司创造了电网安全运行时间最长的纪录，并以最强的传输能力整合了最多的可再生能源。

企业文化

企业宗旨：人民电业为人民。

公司使命：为美好生活充电，为美丽中国赋能。

企业精神：努力超越，追求卓越。

2. 英译中

（1）请将德州仪器公司简介译成中文。

Texas Instruments has been making progress possible for decades. We are a global semiconductor company that designs, manufactures, tests and sells analog and embedded processing chips. Our approximately 80,000 products help over 100,000 customers efficiently manage power, accurately sense and transmit data and provide the core control or processing in their designs, going into

47

markets such as industrial, automotive, personal electronics, communications equipment and enterprise systems. Our passion to create a better world by making electronics more affordable through semiconductors is alive today as each generation of innovation builds upon the last to make our technology smaller, more efficient, more reliable and more affordable—opening new markets and making it possible for semiconductors to go into electronics everywhere. We think of this as Engineering Progress. It's what we do and have been doing for decades.

Key facts
- Founded in 1930.
- Headquartered in Dallas, Texas.
- Publicly traded (Nasdaq: TXN).
- Richard K. Templeton is chairman, president and CEO.
- ~ 30,000 employees.
    - ~ 12,000 in the Americas.
    - ~ 16,000 in Asia – Pacific.
    - ~ 2,000 in Europe.
- 14 manufacturing sites worldwide, tens of billions of chips produced each year.
- ~ 80,000 products for over 100,000 customers.
- Industrial and automotive, the markets with the best opportunities for our products, made up 57% of our 2020 revenue.

Recognition and giving
- Named a Top 100 Best Corporate Citizen by 3BL Media for 21 consecutive years.
- Recognized by the Dow Jones Sustainability Index for our sustainability practices for 13 consecutive years.
- Ranked as a Top Company for Executive Women by the National Association for Female Executives for 15 consecutive years.
- First semiconductor company to earn verification from the U.S. Green Building Council.
- More than $31 million in philanthropic giving in 2019, including TI Foundation grants, matching gifts and in – kind donations.

(2) 请将波音公司简介译成中文。
Boeing in Brief
Boeing is the world's largest aerospace company and leading manufacturer of commercial jetliners, defense, space and security systems, and service provider of aftermarket support. As America's biggest manufacturing exporter, the company supports airlines and government customers in more than 150 countries and regions. Boeing products and tailored services include commercial and military aircraft, satellites, weapons, electronic and defense systems, launch systems, advanced information and communication systems, and performance – based logistics and training.

Boeing has a long tradition of aerospace leadership and innovation. The company continues to expand its product line and services to meet emerging customer needs. Its broad range of capabilities includes creating new, more efficient members of its commercial airplane family; designing, building and integrating military platforms and defense systems; creating advanced technology solutions; and arranging innovative financing and service options for customers.

With corporate offices in Chicago, Boeing employs more than 153,000 people across the United States and in more than 65 countries and regions. This represents one of the most diverse, talented and innovative workforces anywhere. Our enterprise also leverages the talents of hundreds of thousands more skilled people working for Boeing suppliers worldwide.

Boeing is organized into three business units: Commercial Airplanes; Defense, Space & Security; and Boeing Global Services. Supporting these units is Boeing Capital Corporation, a global provider of financing solutions.

In addition, functional organizations working across the company focus on engineering and program management; technology and development – program execution; advanced design and manufacturing systems; safety, finance, quality and productivity improvement and information technology.

Commercial Airplanes

Boeing has been the premier manufacturer of commercial jetliners for decades. Today, the company manufactures the 737, 747, 767, 777 and 787 families of airplanes and the Boeing Business Jet range. New product development efforts include the Boeing 787 – 10 Dreamliner, the 737 MAX, and the 777X. More than 10,000 Boeing – built commercial jetliners are in service worldwide, which is almost half the world fleet. The company also offers the most complete family of freighters, and about 90 percent of the world's cargo is carried onboard Boeing planes.

Defense, Space & Security

Defense, Space & Security (BDS) is a diversified, global organization providing leading solutions for the design, production, modification and support of commercial derivatives, military rotorcraft, satellites, human space exploration and autonomous systems. It helps customers address a host of requirements through a broad portfolio. BDS is seeking ways to better leverage information technologies and continues to invest in the research and development of enhanced capabilities and platforms.

Boeing Global Services

As the leading manufacturer for commercial and defense platforms, Boeing is positioned to provide unparalleled aftermarket support for mixed fleets worldwide. Boeing Global Services delivers innovative, comprehensive and cost – competitive service solutions for commercial, defense and space customers, regardless of the equipment's original manufacturer. With engineering, digital analytics, supply chain and training support spanning across both the government and commercial service offerings, Boeing Global Services' unsurpassed, around – the – clock support keeps our customers' commercial aircraft operating at high efficiency, and provides mission assurance for nations around the world.

Boeing Capital Corporation

Boeing Capital Corporation (BCC) is a global provider of financing solutions for Boeing customers. Working closely with Commercial Airplanes and Defense, Space & Security, BCC ensures customers have the financing needed to buy and take delivery of their Boeing products. BCC combines Boeing's financial strength and global reach, detailed knowledge of Boeing customers and equipment and the expertise of a seasoned group of financial professionals.

学思践悟

第三章 科技合同

一、科技合同文本介绍

(一) 科技合同概述

科技合同是具有法人资格的双方（或数方）为达到一定的经济目的，就完成某一科学技术项目，依法订立的有关各自的义务与权利的协议。科技合同具体可分为以下几种。

(1) 科研合同。科研合同是指需求单位委托科研单位研究试验某一项目而签订的合同。

(2) 科研成果转让合同。科研成果转让合同是指科研单位将已取得的科技成果有偿转让给需求单位的合同。

(3) 试制合同。试制合同是指当事人之间为完成某种新产品试制而签订的合同。

(4) 技术合同。技术合同是具有法律效力的正式文件，是指法人之间、法人与公民之间、公民之间就技术开发、技术转让、技术咨询和技术服务所订立的确立民事权利义务关系的协议，具体可分为以下几种。

①技术开发合同。技术开发合同是指当事人之间就新技术、新产品、新工艺和新材料及其系统的研究开发所订立的合同。技术开发合同包括委托开发合同和合作开发合同。

②技术转让合同。技术转让合同是指当事人就专利权、专利申请权、专利实施许可和非专利技术的转让，明确相互权利义务关系的协议。交出技术成果的一方称为出让方，接受技术成果并支付费用的一方称为受让方。

③技术咨询合同。技术咨询合同是指当事人一方为另一方特定技术项目提供可行性论证、技术预测、专题技术调查、分析评价报告所订立的合同。

④技术服务合同。根据我国技术合同法的规定，技术服务合同是指当事人一方以技术知识为另一方解决特定技术问题所订立的合同。

(二) 科技合同的特点

科技合同的标的与技术有密切联系，不同类型的科技合同有不同的技术内容。科技合同履行环节多，履行期限长，价款、报酬或使用费的计算较为复杂，一些科技合同的风险性很大。科技合同的法律调整具有多样性。当事人一方具有特定性，通常应当是具有一定专业知识或技能的技术人员。

科技合同是双务、有偿合同，而良好的科技合同译文，一般都应具备以下特点。

1. 准确性

科技合同的英译要以"准确"为首先条件，做不到这一点翻译就无从谈起。翻译不能只满足于字面上的一致，而是要从词义、语法、专业上去深刻理解原文的含义，使陈述句事实表述明晰，疑问句发问逻辑准确，切不可使人感到模棱两可，不知所云。例如：

合同总价30%，计3642美元（大写：三仟陆佰肆拾贰美元），在受让方收到出让方提交的下列单据经审核无误后，不迟于三十天支付给出让方。

译文：Thirty percent of the whole contract price, counting 3,642 dollars (In capital, thirty thousand six hundred and forty two) will be paid by Licensee to Licensor not more than 30 days after receiving the following documents from Licensor and finding them authentic.

这个合同条款的英译文至少有六处不妥：第一，"合同总价"不能译为"the whole contract price"，应译为"the total contract price"，"total"在这里强调"总额"；第二，"计××美元"应译为"namely ××US dollars"，这里在"dollars"前一定要加US，因为使用dollar的国家除美国外，还有加拿大、澳大利亚等国；第三，"大写"不能用"in capital"，中文合同中的"大写：××"应译为"say：英语数词"；第四，"不迟于三十天"的准确译文应是"not later than 30 days"；第五，"will be paid by"应改译为"shall be paid by"，在合同文本中"shall"主要用来强调一方所负有的义务；第六，"经审核无误"应译为"find them in conformity with the stipulations of the contract"。另外，"after receiving... and finding..."最好用完成时形式"after having received... and found..."，或者用从句的完成时形式"after Licensee has received... and found..."；"出让方提交的下列单据"译为"the following documents from Licensor"不够确切，不如译为"the following documents which are provided by Licensor"。

2. 严谨性

起草合同文件必须严谨，那么英译合同文件也应力求严谨。英译文学作品要求具有文采、韵味，英译合同文件要求的是严谨和精确。为了排除歧义的产生，合同中有些词语的翻译必须保持一种译文，特别是一些重要的词语都有严格的法律含义，绝对不可信手拈来。例如，"排它许可"就不能译为"exclusive licence"，而应译为"sole licence"；"exclusive licence"的意思是"独占许可"。这两者是国际间通用的术语，各有明确的含义。"exclusive licence"表示，许可方不得再把同样内容的使用许可证授予该地域的任何第三者，就连许可方自己也不得在该地域使用该项技术；而"sole licence"则不同，它表示，许可方不得把同一许可证授予同一地域的任何第三者，但许可方则保留自己在该地域使用该合同中技术的权利。再如，翻译"支付条款"时，译者可能碰到"最高提成费"这样的术语，译者必须弄清其真正内涵。"最高提成费"一般是受让方提出的，意即规定一个提成费的最高数额，将来受让方产品销售量一旦很大，按比例计算提成费将超过这个数额时，则仅仅按这个数额支付，不再多付。所以，把"最高提成费"译为"maximum royalty"才能体现出原文的含义。

3. 规范性

科技合同中的句子相当规范，英译时一定要保持合同原文的体例，反映原文的精神，有些词句已形成固定的翻译标准和格式，一般不宜随意改变。

例如，在专有技术许可证协议中"保密条款"是不可忽视的，一方购买另一方的技术，就要承担该项技术的保密义务，保密条款一般指受让方为出让方的技术保密，但又不止于此，如果出让方许可受让方独家在某一地域生产某种产品，则出让方自己在该地域也负有不将该项技术透露给第三方的义务。所以，这个条款是这样规定的："受让方同意在合同有效期内，对出让方提供给受让方的专有技术和技术资料进行保密；如果上述专有技术和技术资料中一部分或全部被出让方或第三方公布，受让方对公开部分则不再承担保密义务。"这个条款主要规定了两项内容：一项是受让方对出让方提供的技术进行保密的期限；另一项是在何种情况下，受让方不承担保密的义务。所以，英译时应分成两段。第一段译为"Licensee agrees to keep the know – how and technical documentation supplied by Licensor under secret conditions within the validity period of the contract."。在这一段中，"对……进行保密"使用"keep... under secret conditions"；"在……期内"使用"within"，不用"during"。第二段译为"If part or all of the above know – how and technical documentation have been published by Licensor or any third party, and Licensee obtains evidence of such publication, Licensee shall no longer be responsible for keeping secret and confidential the part already published."。在这一段中，要注意增译一句"Licensee obtains evidence of such publication"，因为另一方不承担保密责任要有理由，当对方坚持要求时，另一方要拿出足够的不履行义务的证据；最后一句，"受让方对……不再承担保密义务"应使用"Licensee shall no longer be responsible for..."结构，表示在法律不承担责任的义务。

二、科技合同范文展示

技术开发合同
Technology Development Contract

项目名称：基于脂质分子的思普定30纳米颗粒制剂
Project name：Lipid – based nanoparticle formulation of SIP30

签订时间：2013 – 12 – 25
Signing time：25/12/2013

签订地点：苏州晶奇生物纳米技术有限公司
Signing location：Suzhou Jingqi Bio Nanotechnology Co., Ltd.

委托方（甲方）：苏州晶奇生物纳米技术有限公司
Consignor (Party A)：Suzhou Jingqi Bio Nanotechnology Co., Ltd.

电话：+86 – 0512 – 86860521
Tel：+86 – 0512 – 86860521

通信地址：苏州星湖街218号生物纳米园A2 – 327
Address：Room 327, Building A2, No. 218, Xinghu Road, BIOBAY, Suzhou, PRC

传真：+86 – 0512 – 62959488
Fax：+86 – 0512 – 62959488

受托方（乙方）：美国引药明创药业科技有限公司
Consignee (Party B)：Innovform Therapeutics, LLC

电话：001 – 609 – 558 – 7055
Tel：001 – 609 – 558 – 7055

通信地址：美国宾夕法尼亚州伯利恒市研究大道116号（邮编：18015）
Address：116 Research Drive, Bethlehem, PA 18015, USA

本合同甲方委托乙方就基于脂质分子的思普定30纳米颗粒制剂项目进行专项技术服务，并支付相应的技术服务报酬。双方经过平等协商，在真实、充分地表达各自意愿的基础上，根据《中华人民共和国合同法》的规定，达成如下协议，并由双方共同恪守。

Suzhou Jingqi Bio Nanotechnology Co., Ltd. (Party A) entrusts Innovform Therapeutics, LLC (Party B) to provide special technical services for Lipid-based nanoparticle formulation of SIP30 and pay technical service remuneration. Two parties agree as follows according to the provisions of The Contract Law of The People's Republic of China, on the basis of fully express their will.

第一条　甲方委托乙方进行技术服务的内容如下

Clause 1　The technical services entrusted by Party A to Party B are as follows

技术服务的内容：基于脂质分子的思普定30纳米颗粒制剂实验

Content：experiment of Lipid-based nanoparticle formulation of SIP30

第二条　乙方应按下列要求完成技术服务工作

Clause 2　Party B shall complete technical services according to the following requirements

1. 技术服务地点：美国引药明创药业科技有限公司

1. Location of technical service：Innovform Therapeutics, LLC

2. 技术服务期限：8个月

2. Technical service period：8 months

第三条　为保证乙方有效进行技术服务工作，甲方应当向乙方提供下列工作条件和协作事项

Clause 3　In order to ensure Party B to effectively carry out technical services, Party A shall provide Party B with following working conditions and cooperation affairs

1. 提供技术资料

1. Provide technical information

2. 提供工作条件

2. Provide work condition

第四条　甲方向乙方支付技术服务报酬及支付方式为

Clause 4　Remuneration and payment method for technical services paid by Party A to Party B

1. 技术服务费由甲方_____（一次或分期）支付乙方

1. The technical service fee shall be paid by Party A to Party B _____ (once or in installments)

a) 支付乙方 $50000 用于启动项目

a) Party B will receive $50,000 to initiate the project

b) 进入临床研究，支付乙方 $100000

b) Party B will receive $100,000 to enter into a clinical trial

c) 新药获得CFDA批件，支付乙方 $200000

c) Party B will receive $200,000 for new drug approved by CFDA

2. 乙方开户银行名称、银行地址和账号为

2. Bank name, bank address and account of Party B

开户银行：美联银行

Bank name：Wachovia Bank, N.A.

银行地址：美国北卡罗来纳州夏洛特市南学院街301号（邮编：28288）

Bank address: 301 S. College Street, Charlotte, NC 28288, USA
账号：2020800009690
Account: 2020800009690

第五条　本合同一式两份，具有同等法律效力
Clause 5　This contract is in 2 copies, which own equal force of law
第六条　本合同经双方签字盖章后生效
Clause 6　This contract shall become effective upon the signature and seal of both parties

甲方（签名盖章）：　　　　　　　　乙方（签名盖章）：

法定代表人（签字）：　　　　　　　法定代表人（签字）：

签署日期：　　　　　　　　　　　　签署日期：

Consignor:　　　　　　　　　　　　Consignee:

Legal representatives:　　　　　　Legal representatives:

Date:　　　　　　　　　　　　　　Date:

三、科技合同常用词汇和句型

（一）科技合同常用词汇

1. 科技合同常用虚词

hereby = by means of; by reason of　特此，兹，因此
herein = in this　在此，于此
hereinafter = later in this contract　以下，在下文中
hereof = of this　于此，在本文件中
hereto = to this　本文件的，在这里
hereunder = under this　本文件规定，在下文中
herewith = with this　与此，附此
thereafter = afterwards　此后
thereby = by that means　因此，由此，从而
therein = in that　在那里，在那点上
thereof = of that; from that source　由此，因此
thereto = to that or it　随之，附之
therewith = with that or it　与此，与之
whereby = by what, by which　凭此条款，凭此协议
whereof = of which　关于那事

2. 科技合同常用相关说法

a long – term contract　长期合同
a nice fat contract　一个很有利的合同
a short – term contract　短期合同
a written contract　书面合同
agency appointment contract　委托合同
an executor contract　尚待执行的合同
breach clause　违约条款
breach of contract　违反合同
brokerage contract　居间合同
buyer and seller information　买卖双方信息
cancellation of contract　撤销合同
cargo carriage contract　货运合同
completion of contract　完成合同
contact person　联系人
contract amount incl VAT & installation　合同总额（含增值税和安装费）
contract for future delivery　期货合同
contract for goods　订货合同
contract for purchase　采购合同
contract parties　合同当事人
contract period/term　合同期限
contract price　合约价格
contract provisions/stipulations　合同规定
contract sales　订约销售
contract terms/clause　合同条款
contract value　合同价格
contract wages　合同工资
contract　合同，订立合同
contractor　订约人，承包人
contracts for construction project　建设工程合同

3. 科技合同常用法律和工商专有名词

labor contract　劳动合同
leasing contract　租赁合同
miscellaneous clause　其他条款
multi – modal carriage contract　多式联运合同
originals of the contract　合同正本
passenger carriage contract　客运合同
payment conditions/terms　付款条件
compensation　赔偿

consequential damages 间接损害赔偿金
defamation 诽谤
default 不履约，违约
entire agreement clause 完整协议条款
force majeure 不可抗力
incidental damages 附带损害赔偿金
infringement 侵害
injunctive relief 禁止令救济
jurisdiction 司法管辖权
mediation 调解
meeting of minds 合意，意见一致
mutuality of obligation 相互义务
offer and acceptance 要约与承诺
parties 合同方，合同当事人
payroll service agencies 薪资服务中介，就业服务机构
perma-temp 长期临时工
portfolio copies 公文复件
preemptive right 优先购买权

4. 科技合同用词特点

科技合同既是科技文书，又是法律文书，在词汇方面有其特有的特征。为了防止误解和歧义，起草合同时措辞要准确、具体、严密且符合标准，使合同显得正式、严肃。科技合同用词特点主要体现在以下几个方面。

(1) 使用准确、严密的用语。合同是具有法律约束力的契约文件，应考虑到任何可能发生的误解或争议，措辞一定要具体、准确且严密，避免使用含糊、抽象的词语。

(2) 使用正式的法律用语和专业词汇。科技合同中大量使用正式的法律用语及科技方面的专业词汇，这些用词体现合同的庄重和严肃，从而体现了法律文本的威严。

(3) 频繁使用复合古旧词。使用古旧词是合同的又一个用词特点，尤其是由 here、there、where 组成的复合古旧词，这些词在其他英语书面文本中不再使用。常见的有 hereof（of this 在本文件中）、thereof（of that 由此，因此）、whereas（considering that/but 签于；而）等。

(4) 使用配对词和三联词。科技合同中的配对词和三联词指用两个或三个意思相近或相同的赘词构成一个固定短语以表达合同中本来只需要一个词就能表达的概念。出现这类词的原因很复杂，主要原因是合同文本不希望有任何遗漏，希望把所有意思相近的内容都包括进去。为了避免歧义或误解，合同中极少使用代词，而是较多使用同义词词组或重复原词来确保合同双方的权利和义务。成对同义词常见的有 furnish 和 provide（提供）、terms 和 conditions（条款）、fulfill 和 perform（履行）等。

(5) 使用"shall"来加强语气。"shall"是一个法律词汇，在合同中频繁出现。它并非只表示将来，更用来加强语气，强调双方各自的义务，含有强制之意，意为"必须""当"等，可用于所有人称。

(二) 科技合同常用句型

1. 多使用陈述句

科技合同是在多轮谈判达成一致协议后才签订的法律文件，其中明确规定签约双方的权利和义务，而不是提出问题或进行磋商，因而合同中的基本句式是陈述句，而没有疑问句和感叹句，也没有使用一般的修辞手法。例如：

The Seller is not liable under subparagraph (a) to (b) of the preceding paragraph for any lack of conformity of the goods if at the time of the conclusion of the contract the Buyer knew or could not have been unaware of such lack of conformity.

译文：如果在订立合同时买方知道或不可能不知道货物与合同不符，卖方就无须按前款（a）项至（b）项负有此种不符合同的责任。

2. 多使用定语从句

使用定语从句同样是由法律英语语言的精确性所决定的，使条款意义明确、清晰，以利于排除误解的可能性。例如：

The Seller must deliver goods which are free from any right of claim of a third party based on industrial property or other intellectual property.

译文：卖方所交之货物必须是任何第三方均不能根据工业产权或其他知识产权提出索赔权要求的货物。

3. 多使用状语从句

在合同中，为了使条款明确、清晰，排除一切可能产生的歧义与误解，严格界定缔约各方履行义务、承担责任及享有权利的时效、方式、条件等，常常使用大量时间、条件、目的和方式等状语从句。例如：

If the technical documentation is found lost, damaged or mutilated during air transportation, Party B shall supply Party A free of charge with a second set of the documentation within the shortest possible time but not later than thirty days after it has received from Party A the written notice.

译文：若技术资料在空运中丢失、损坏或残缺，乙方须于收到甲方的书面通知后30天内尽快免费向甲方重新邮寄该技术资料一份。

4. 常用现在时和被动语态

因为合同是具有法律效力的文书，合同条款中规定的是双方的权利和义务，这些权利和义务具有一定的通用性，大多成为相关方面的国际惯例。因此，合同语句经常采用现在时。当不强调动作执行者或出于体谅和礼貌时常采用被动语态，包括用在 shall 和 must 等情态动词之后的被动语态。例如：

Any developments improvements, modifications or inventions concerning the licensed products and equipment made by Foreign Side during the term of this contract shall become part of the technology and shall be disclosed and conveyed by Foreign Side to the AFF. CO. at no additional charge in accordance with the terms and conditions of this contract promptly after Foreign Side's use of same in its commercial manufacture of the licensed products and equipment.

译文：外方在合同期限内对特许产品和设备所做的任何发展、改良、修改或发明，应该成为这个技术的一部分，并应在把它们使用于特许产品和设备的商业性生产之后，立即按照本合同的条款向 AFF. CO. 透露和转让，而不另外加收费用。

5. 多用完整句、长句

合同规定缔约双方的权利和义务具有法律效力，为防止任何歧义和日后纠纷，合同意思要完整，结构要严谨，因此合同文书中一般采用的是主语、谓语都具备的完整句，而不用省略句或单部句。而完整句往往含有多重限制、修饰成分，或者是多种复合句或并列句，因此句子结构变得复杂而冗长。合同中大部分句子都是复合长句，一般超过英文句子的平均长度（17 个单词），有时一个句子就构成一个段落。例如：

The two parties agree that Party B shall, at request by Party A, provide personnel specified by Party A with technical instructions and trainings or technical services concerning usage of research and development results after these research and development results are delivered to Party A.

译文：双方确定，乙方应在向甲方交付研究开发成果后，根据甲方的请求，为甲方指定的人员提供技术指导和培训，或者提供与使用该研究开发成果相关的技术服务。

四、科技合同翻译技巧

（一）掌握合同结构

科技合同是具有一定法律效力的正式文件。从内容、条款上看，合同较为具体、详尽，着眼于微观，签订合同后，要全面明确各项具体的细节。从涉及范围上看，合同的标的往往比较单一集中，也很明确，通常都是一事一议，就事论事。从书写格式上看，合同已基本格式化了，大家可以看到许多的合同范本。所以，这里对合同结构进行简要的介绍，使读者对合同有个基本的了解，从而在翻译中能够灵活运用。

一份完整的英文合同通常可以分为标题（title）、前言（preamble）、正文（habendum）、附录（schedule）及证明部分（结尾辞，attestation）五大部分。前言部分，目的在于很简略地介绍合同内容之人、事、时、地、物等背景，让阅读合同的人在接触冗长复杂的正文前，先有一个基础的认识与心理准备。正文部分也称为 body，具体约定当事人的权利和义务。各式各样的正文条款是合同中最核心的部分，也是篇幅最大的部分，与当事人的权利义务关系发生最直接、最密切的牵连。这里我们将英文科技合同中的正文条款分为两类：特殊条款与一般条款。特殊条款指的是只有在某些特定性质的合同中才会出现的条款，如某些合同中约定的保密条款、竞业禁止条款等。反过来说，不管合同性质如何，通常都会出现一般性的约定，比如买卖、合资、技术转让等合同，尽管缔约目的不同，却都少不了违约与免责、解除与终止等一般性的约定，记载这些一般性约定的条款就叫作一般条款。附录部分作为对正文条款的补充，不是所有英文合同中都有的一项。英文合同中的最后一部分就是结尾辞，指的是在当事人签名之前经常会出现的一段文字，相当于中文合同中的"双方签字盖章，特此为证"。至于签名档的部分，如果当事人是公司的话，除代表人的签名外，还要加盖公司印鉴，并且通常会注明代表人的职务（title）。

（二）熟练使用专业术语

科技合同中有大量的正式用语（formal words）、科技词汇及合同的专业用语，正因为如此，才能保证其科技性、专业性和法律性。在翻译选词时，要依照目的语合同的通用语进行翻译，保证合同能够被目的语读者看懂。

中文合同中常见瑕疵（defect）、不可抗力（force majeure）、管辖（jurisdiction）、转让（transfer、assignment、conveyance）、提交（submission）、同意（consent）等专业用词。例如：

Any such assignment, transfer or conveyance shall be without other consideration than the mutual covenants and considerations of this agreement.

译文：任何转让除根据本协议相互契约和对价外不得考虑其他。［transfer、assignment 和 conveyance 都有"转让"的含义，其中 assignment 和 conveyance（多用于不动产）尤为专业。］

The submission to and consent by the engineer of such programs or the provision of such general descriptions or cash flow estimates shall not relieve the Contractor of any of his duties or responsibilities under the contract.

译文：向工程师提交并经同意的上述进度计划或提供上述一般说明或现金流量估算，并不解除合同规定的承包人的任何义务或责任。

（三）意义对等优先，点面结合

英文科技合同的翻译也是关于科技产品买卖、科技转让、科技施工、科技咨询、劳务承包等的翻译。英文科技合同属于信息型文本，特点是专业性强、逻辑严密、表达比较严谨，译文应反映这些特点。科技合同翻译时，应注重点面结合。点就是科技合同中词汇、句法的特点，其中包括术语的使用、正式语的近义词并用、情态动词使用及句子结构翻译等；面则是结合句式篇章的特点在翻译时考虑到意义的传达与文本的特点。

为使源语和目的语之间的转换有一个标准，减少差异，奈达从语言学的角度出发，根据翻译的本质，提出了著名的"动态对等"翻译理论，即"功能对等"。"动态对等"中的对等包括四个方面：词汇对等、句法对等、篇章对等、文体对等。奈达认为，"意义是最重要的，形式其次"（郭建中，2000）。在英文科技合同翻译时，译者除要考虑到意义的准确传递外，还要考虑到英文科技合同本身的法律性质及合同本身的要求和特点，即要遵循一定的合同模式。这里英文科技合同翻译在遵循奈达的"功能对等"的同时，还应兼顾形式对等。

在英文科技合同翻译过程中，先理解合同的意思，根据意思进行词汇对等、句法对等、篇章对等及文体对等。

1. 要实现词汇对等

一般来说，专有名词的翻译是英汉一一对应的，例如，credit guarantee 译为"信用担保"，unfair competition 译为"不公平竞争"，floating interest rate 译为"浮动汇率"，shipping details 译为"装运细节"，chartering agent 译为"租船代理人"，comparative advantage 译为"相对优势"，management training 译为"管理培训"。

2. 要实现句法对等

句法对等主要关注的是对句子语法结构及句子情态的翻译处理及转换。英文更强调客观性，更多地运用被动语态，而在中文中，习惯使用主动语态，给句子赋予一个主语，所以被动、主动转换是句法对等的一个关注点。例如：

If a written agreement for the extension of the period of existence of the JVC is not signed by three years prior to the expiration of the then current period of existence, the JVC shall terminate at the end of such current period of existence and the provisions of clauses 4. 5 and 4. 6 shall then apply.

译文：如不在本合资公司（JVC）现有合同期到期前3年签订书面续约协议，自本合同到期之日起本合资公司即告终止，并开始实施（执行）本协议第四条第五款、第六款的规定。

The efficiency of machines has been more than doubled or trebled.

译文：这些机器的效率已增加了一倍或两倍多。

另外，句子语法结构还包括句子结构搭配，比如英文原文是动宾结构，在翻译成中文时，由于意义的需要，可能会调整成名词性结构或主谓结构等，这些都要在实际翻译中进行操作。英文句子注重"形合"，由逻辑关系统辖，句子有可能特别长（特别是在科技合同文本中）；而中文句子注重"意合"，长短句相间，结构松散，逻辑暗含在意思当中。这就要求译者在适当的位置对英文句子进行拆分调整，有时甚至需要对小句的顺序进行调整。

句法对等的另一个关注点是句子情态的翻译。英文的情态主要由句式和情态词体现，句式在前面已经提到，情态词（如表示许可的 may、表示命令的 shall 及表示禁止的 must not 等）在科技合同中常常出现。在翻译时，需要译者体会文本要表达的情态是命令还是许可，是主观情态还是客观要求，语气是轻还是重。

（四）科技合同翻译方法与策略

1. 熟悉全文，掌握大意

译者在拿到一份科技合同时，必须先通读几遍全文，了解原文的结构及每个词语在具体上下文中的含义，着重领会合同中法律语言的确切性，必要时还需要查阅有关资料，切不可一拿到文件就提笔翻译，这种草率的工作态度是绝对要不得的。熟悉全文，掌握大意有两大益处。一方面有利于选择词义。例如，合同文件中经常出现的一句话"具有同等效力"译为"to have the same effect"就欠妥，"effect"是指"效果；效应"，而合同文件中的"具有同等效力"是指具有同等的法律效力，应译为"to be equally authentic"；又如"合同双方中的任何一方"应译为"either of the parties to the contract"，这里的介词"to"不能用"of"代替，因为"to"是指"作为一方参加合同"。另一方面有利于行文造句，不同种类文章，其文体不同，译文的表达方式、行文造句也就随之不同，如果译者在翻译之前没有领会原文的内容，那么其译文就不会是"行话"。

请看下面条款："本合同于一九八七年五月十三日在××签订，一式两份，每份用英文和中文写成，两种文本具有同等效力。"原译为"This contract was signed in ×× on May 13, 1987, each copy was written in English and Chinese with two copies in each, the two texts have the same effect."。原文是在"合同的生效、终止"一条中出现的，由于译者对这一条款的行文不了解，加之对主要意思也不太懂，所以把这一结构严谨的条文翻译成了"大白话"。原文应译为"Signed in ×× on the 13th day of May, 1987 in duplicate in English and Chinese languages, both texts being equally authentic."。原文中的"于一九八七年五月十三日在 ×× 签

订"译为"signed in ××on the 13th day of May, 1987";"一式两份,每份用英文和中文写成",讲的是用两种文字写成的合同双方各执一份,内容完全相同,因此直接译为"in duplicate in English and Chinese languages",这里的"in duplicate"指文件一式两份。

2. 推敲词义,分析结构

合同文件中的词语是构成合同文书最基本的单位,对合同文书中的一些重要词语的正确理解,是英译合同文件的基础。要透彻理解原文,必须认真地推敲词义,那种孤立、片面地去理解词义的态度,必定会出错。以"遵守"一词为例,英文中有"observe""obey""abide by""comply with"等可供选择,但这几个单词或短语并不都是同义词,在不同上下文和不同的条文中就有不同的处理方法。例如:

(1) "全体人员应遵守项目所在国的法律和法令,尊重当地风俗习惯",在这一条款中执行者是人,表示人对法律的遵守,英译时谓语动词应选择"abide by",译为"All the personnel shall abide by the laws and decrees in the project – host country and respect the local customs and traditions"。

(2) "双方的一切活动都应遵守项目所在国的法律、法令和有关条例规定",这一条款中的主语是"活动",英译时谓语动词应选择"comply with",不能用"abide by"、"obey"或"observe",故译为"All the activities of both parties shall comply with the provision of laws, decrees and pertinent regulations in the project – host country",如果把以上几个词套用就会使译文词不达意,也不像法律语言。

又如合同中经常使用的一个词"承担",英文中有"bear""accept""undertake""take""respond in""shoulder"等可供选择,但"承担"在不同句子中有不同"承担"效果。例如:

(1) "由乙方承担法律上和经济上的全部责任"译为"Party B shall bear full of the legal and financial responsibility which may arise","bear"主要表示"承担"法律上和经济上的责任(注:"法律上和经济上的责任"不能译为"law and economic responsibility")。

(2) "双方承担风险"译为"Both parties shall accept the risk"。

(3) "乙方承担后果"译为"Party B shall take the consequence"。

(4) "出让方承担赔偿费用"译为"licensor shall respond in damages"。

(5) "承担受让方的赔偿责任"译为"to honor licensee's liability for compensation"。

以上例句中如果"承担"一成不变地译为"bear",那么就会令人费解。当然其他词也具有类似的情况,译者一定要注意结合上下文仔细推敲词义,勤查专业工具书,做到词不离句、句中求词。

除仔细推敲词义外,还应认真地分析原文的结构及各种成分之间的制约关系,英译时,译者必须吃透原文含义,对句子本身的内部结构、句间的逻辑关系要反复加以分析,如有必要,可打散原文结构,按英文习惯重新调整,使译文具有可读性。请看下面条款:"本合同自签字之日起六个月仍不能生效,双方有权取消合同。"原译为"If the contract cannot come into effect within six months after the date of signing the contract, both parties shall have the right to cancel the contract."。这句译文不太像"行话",力量也不足,确切的行文应该是把主句和从句的主语调整成一个,从句的谓语动词使用"come into force",将连词"if"改为"in case"表示双方希望合同按期履行。主句谓语动词改用"be binding neither to Party A, nor to

Party B", 说明一旦合同在规定时间内不能生效的话, 合同对双方就不具有约束力, 并不是取消不取消的问题。所以, 原文应译为 "In case the contract cannot come into force within six months after the date of signing the contract, the contract shall be binding neither to Party A, nor to Party B."。

合同中的各条款间一般都有着普遍的制约关系, 当某一条款受到其他条款制约时, 英译时一定要注意突出主要内容, 保持主要内容的独立性与完整性。再看下面两个条款: ①"在合同有效期内, 双方对合同产品设计的技术如有改进和发展, 应相互免费将改进和发展的技术资料提供给对方使用。" ②"改进和开发的技术, 其所有权属于改进和开发一方。"第一条主要规定了相互许可使用的问题, 英译时一定要突出"双方相互提供使用"这个关键结构 (both parties shall provide each other with...), 其他成分都是围绕这一结构进行的, 故译为 "Within the validity period of the contract, both parties shall provide each other with the improvement and development of the technology related to the contract products free of charge."。原文中的谓语动词"提供"一词, 有人译为"supply", 就不如"provide"确切, 尽管这两个词都表示"供给", 但牵涉金钱时, "provide"表示"免费供给", "supply"则不太明确, 一般来说需要给钱。第二条的重点在于规定改进和开发的技术所有权归属问题, 所以其中的"属于"一词是很关键的, 有人译为"belong to", 也有人选择"possess", 还有人用"own"。"belong to"的含义通常是指财产"属于……"; "possess"和"own"虽然都着重于所属关系, 但"possess"只是指目前属于某人, 并没有讲清是如何得到的, 而"own"含有"对……合法占有"的意思, 与原文意思一致。故把这一条译为 "The improved and developed technology shall be owned by the party who has improved and developed the technology."。如果将"属于"改用"belong to", 其主语就应该用"ownership"表示"所有权属于", 因为知识产权是与物质财产直接相联系的, 译为 "The ownership of any improved and developed technology shall belong to the party who..."。

3. 厘清层次, 逐条翻译

科技合同的条款主要有四大类, 即定义条款、基本条款、一般条款和结尾条款, 英译时一定要首先厘清层次, 突出重点, 对合同条款本身的内部结构、各条款间的制约关系应仔细琢磨, 吃透其含义, 然后按英文合同问题的语气逐条翻译。请看下面三个仲裁条款: ①"在执行本合同中所发生的与本合同有关的一切争议, 双方应通过友好协商解决。如通过协商不能达成协议时, 则提交仲裁解决。" ②"仲裁裁决是终局裁决, 对双方均有约束力。" ③"除在仲裁过程中进行仲裁的部分外, 合同应继续执行。"以上三个条款是互相联系、不可分割的。

现就以上三个条款进行逐个英译。

(1) 以上①这一条款重点讲明了, 只要双方当事人订立仲裁协议, 发生争议首先通过友好方式解决, 一旦解决不成也只能提交仲裁。英译时应分成两句。第一句译为 "All the dispute arising from the execution of, or in the connection with the contract shall be settled between both parties through friendly consultations."。翻译"在执行本合同中所发生的与本合同有关的"时, 两个定语之间的连词应选择"or", 不能用"and", 第一个定语"在执行本合同中所发生的"译为 "arising from the execution of..." 显得更明确、具体。第二句译为 "In case no settlement to the disputes can be reached between both parties through such consultations, the

disputes shall be submitted for arbitration."。对于"不能达成协议"的翻译，中文原文否定动词，译文否定名词；加译"between both parties through such consolations"使全句完整、明晰。

（2）以上②这一条款中的"仲裁裁决"不能译为"the arbitration ruling"或"the arbitration adjudication"，因为这里的"裁决"是指由仲裁员做出的决定，应译为"the arbitration award"；"对……有约束力"使用"be binding on/upon"结构。全句译为"The arbitration award shall be final and binding on both parties."。

（3）以上③这一条款主要说明了合同双方当事人应当履行那些没有提交仲裁的条款，所以译成主动语态显得关系更明确。"在仲裁过程中"用"be under arbitration"。全句译为"In the course of arbitration, both parties shall continue to execute the contract except the part of the contract which is under arbitration."。

4. 校改译文，润色词语

在校改合同英译文时，要对译文的词语、行文进行更进一步的推敲，要着眼于译文的严谨性和准确性。所以，译者在搁笔之前一定要逐段、逐句、逐词仔细修改、润色译文，做到词义精确、结构严谨。请看下面例句和译文。

例1：按照本合同第二条规定的合同内容和范围，甲方向乙方支付的合同总价为××美元（大写：××××）。其分项价格如下。

译文：The total contract price to be paid by Party A to Party B according to the content and scope stipulated in Article 2 of the contract is ×× US dollars (say：××××). Their classified prices are as follow.

在该译文中有几处值得商讨。第一，"按照……"译为"according to..."欠妥，"according to"的主要意思是"on the authority of"，而原文所表达的是"按双方同意在平等基础上签订的合同条款为根据"，应选用"in accordance with"或"pursuant to"较为贴切。第二，在"本合同第二条"这个短语中，"合同"和"第二条"的关系是一种归属关系，介词应用"to"，不能用"of"，该短语应译为"Article 2 to the contract"。第三，"合同总价为……"译为"The total contract price is..."，不像合同语气，正确的译文应是"The total contract price shall be..."。第四，"其分项价格如下"应译为"Their breakdown prices are as follow"，这里"breakdown"是指把成本价格、总数分成细目，而"classified"主要表示把货物分类或分等级。另外，"甲方向乙方支付的"应改用定语从句"... price which shall be paid by Party A to Party B"才能体现出甲方所负有的义务。经过校改，原文应改译为"The total contract price which shall be paid by Party A to Party B in accordance with (or pursuant to) the content and scope stipulated in Article 2 to the contract shall be ×× US dollars (say：××××). Their breakdown prices are as follow."。

例2：本合同有效期从合同生效之日起××年，有效期满后，本合同将自动失效。

译文：The contract shall be valid for ×× years from the date of signature, after the expiry of the validity period of the contract, the contract shall become null and void automatically.

原文似乎没有什么难懂的地方。但是，这里的"从合同生效之日起"不能译为"from the date of signature"，因为合同双方授权的代表在合同上的签字日期，并不是合同生效日期，合同生效日期往往是最后一方政府当局的批准日期，所以应该译为"from the effective date of the contract"。"有效期满后"中的介词译文用"after"显得范围有些大，用"on"来代替可以弥补这一不足；"有效期"在合同中的"行话"应是"the term of validity"。经过校改，原文应改译为"The contract shall be valid for ×× years from the effective date of the contract, the contract shall become null and void automatically on the expiry of the contract's term of validity."。

例3：乙方保证是本合同规定提供的一切专有技术和技术资料的合法所有者并有权向甲方转让。如果发生第三方指控侵权，由乙方负责与第三方交涉并承担法律上和经济上的全部责任。

译文：Party B guarantees that he is the legitimate owner of the know – how and technical documentation supplied to Party A in accordance with the contract, and that he has the right to transfer them to Party A. If the third party accuses Party B of infringement, Party B shall take up the matter with the third party and bear all the legal and economic responsibility arising therefrom.

原文是一个较长的条款，而且其中关系也比较复杂，要处理好译文并不容易。在校改时，译者应首先厘清原文中的各种关系，尽力使译文在用词上准确，在结构上严谨。下面来分析一下以上的英语译文。第一，"Party B guarantees that he..."中宾语从句的主语用"he"显得有些含糊，应重复"Party B"。第二，"supplied to Party A"，应在分词"supplied"后加译"by Party B"，使全句完整、明晰。第三，"in accordance with the contract"对应于原文的含义是指按合同中规定的条款，所以应改译为"in accordance with the stipulation of the contract"，这样才能体现出与原文的一致性。第四，"he has the right to transfer them to Party A"这句译文不太像"行话"，况且句中的两个代词也应换成名词，因为在英文合同文件中，出现过的名词尽量不用代词代替。"乙方有权向甲方转让"应改译为"Party B is lawfully in a position to transfer..."，这更强调乙方对以上技术和技术资料占有的绝对合法性。此句应改译为"Party B is lawfully in a position to transfer the know – how and technical documentation to Party A"。第五，"If the third party accuses Party B of infringement"这句译文明显有两处错译："the third party"仿佛给人一种甲乙双方都已知道的"第三方"，应改为"any/a third party"；"accuse sb of sth"一般是指控诉某人触犯刑律，而这句中的"指控"仅指一般的民事侵权，故应用"bring a charge of infringement"。第六，"Party B shall take up the matter with the third party"这句译文中的"take up sth with sb"，是指口头或书面向某人提出某事，没能确切地传达出原文的"由乙方负责与第三方交涉"，应改译为"Party B shall be responsible for dealing with the third party"。第七，"bear all the legal and economic responsibility arising therefrom"。这句译文看上去是没有问题的，但译者在校改时，不能只停留在语言的表层上，一定要进入语言的深层。其实，该译文至少有两处不妥：第一，"全部责任"指的是由于上述原因而发生的乙方应承担的责任，所以"全部"应选用"full of"，不用"all"；第二，合同中的"经济责任"主要指合同当事人在违约时应承担的财产责任，因此这句译文中的"economic responsibility"应改为"financial responsibility"。第八，"responsibility arising therefrom"中的分词短语最好改用含有情态意义的从句，因为改用从句才能传达出发生承担上述责任是一种"或然性"而不是"必然性"，故译为"responsibility which may arise"。经过校改，原文应改译为"Party B guarantees that Party B is the legitimate owner of the know – how and technical documentation supplied by Party B to Party A in accordance with the stipulation of the contract, and Party B is lawfully in a position to transfer the know – how and technical documentation to Party A. If a third party brings a charge of infringement, Party B shall be responsible for dealing with the third party and bear full of the legal and financial responsibility which may arise."。

根据以上探讨不难看出，科技合同的翻译涉及许多方面的问题。在英译过程中，译者必须深入领会原文中的每个词语在具体上下文中的含义，着重理解词语的"准确性"，对每个经过精心挑选的词句进行反复推敲，严格按翻译程序行事，而不可一拿到文件就动笔翻译，在定稿前，一定要不厌其烦地反复校改，润色译文，有条件的要请行家再校，直到译文能准确而完整地把原文内容表达出来为止。在此过程中，译者通常采用的翻译技巧如下。

（1）顺译法。不同民族在思维方式和语言表达习惯上既有差异，又有相同之处。例如，有些英语长句的内容是按逻辑顺序排列的，也有些英语长句所叙述的一系列事件是按时间顺序排列的。这类英语句子的表达方式恰好与汉语的表达习惯相吻合。例如：

In the event that the company's operations are reduced substantially from the scale of operation originally anticipated by the parties, or the company experience substantial and continuing losses resulting in negative retained earnings not anticipated by the parties in the agreed business plan, or in any other circumstance permitted under applicable laws or agreed by the parties, the parties may agree to reduce the registered capital of the company on a pro rata basis.

译文：如果公司的经营规模比双方原来预期的规模有大幅度缩减，或者公司持续遭受严重亏损，导致在双方商定的业务计划中出现未预期的负留存收益，或者在相关法律允许或双方一致同意的其他情况下，双方可以协商按原有出资比例减资。

该英语长句由三个条件从句和一个主句构成，按先条件后结果的逻辑顺序排列，这恰好与汉语的表达习惯相吻合，因而可采用顺译法翻译。

（2）逆译法。汉语语序通常按照时间的先后、先因后果、由假设到推论、由事实到结论的逻辑顺序逐层推进；而英语在表达多层逻辑关系时，往往将重要信息放在句首，次要信息放在句末。这就造成了英汉两种语言在语序上的差异。正是由于存在这种差异，翻译时就要采用逆译法。例如：

The Buyer shall have the right to claim against the Seller for compensation of losses within 60 days after arrival of the goods at the port of destination, should the quality of the goods be found not in conformity with the specifications stipulated in the contract after re – inspection by the State Administration for Market Regulation.

译文：若货物经国家市场监督管理总局复检后发现质量与本合同之规定不符，买方有权于货物抵达目的港后的60天内向卖方提出索赔。

此复合句主句在前，条件状语从句在后，并且主句含有时间状语"within 60 days after arrival of the goods at the port of destination"。将此复合句译成汉语时，依照汉语的表达习惯，应采用逆译法，将条件状语从句置于主句之前，而时间状语放在动词前面作修饰和限定之用。

（3）分译法。当英语长句中主句与从句或主句与修饰语之间的关系不十分紧密时，可把长句中的从句或短语转换成句子，根据汉语句子结构简洁的特点进行翻译。为使语义连贯，有时还可以适当增加词语。例如：

Payment of penalty made by the Licensor to the Licensee in accordance with the stipulation in Clause 8.4 to the contract shall not release the Licensor from his obligations to continue to deliver the technical documentation and software which is subjected to penalties for late delivery.

译文：出让方按合同第 8.4 款向受让方支付迟交罚款后，并不能免除出让方继续交付（迟交的）技术资料和软件的义务。

（4）综合法。有些英语长句比较复杂，仅仅采用顺译法、逆译法或分译法中的某一种译法很难得到合乎汉语表达习惯的译文。对这种情况应进行细致的分析，先译出句子的主干部分，再补充各细节部分，有顺有逆、有主有次地对全句进行综合处理。例如：

The Contractor shall indemnify the Owner in respect of all damage and injury occurring before the issue of the final acceptance certificate to any property and person and against all actions, suits, claims, demands, charges and expenses arising in connection therewith which shall be occasioned by the negligence of the Contractor or any of his subcontractors or by defective design, materials or workmanship but not otherwise.

译文：在最后验收合格证书签发之前，凡因承包商或其分包商的疏忽或设计、材料、工艺上的缺陷而造成的任何财产损失或人身伤害，承包商应无条件向业主赔偿，并保护业主免受由此而产生的各种诉讼、索赔、要求或费用支出。

从上面这个翻译示例可以看出，为了使汉语译文忠实、通顺，在翻译过程中译者综合了逆译法和分译法，从而译出了合乎汉语表达习惯的译文。

（五）科技英语中长句翻译的方法

英语长句一般指的是各种复杂句，即语法结构复杂、从句和修饰语多、包含的内容层次在一个以上的英语句子。在英语中，一般习惯用长的句子表达较为复杂的概念；而在汉语中，却常用若干个短句对同一复杂概念进行层次分明的表述。这种差别正体现了英语和汉语两种语言句子结构的差异，那就是英语注重"形合"，汉语注重"意合"。所谓"形合"，是指词语或分句之间用语言形式手段（如关联词）连接起来，表达语法意义和逻辑关系。英语句子主干结构突出，除句子的中心谓语动词外，句子的其他语法成分是依靠连词、分词、代词、副词等进行连接而成的。英语句子的结构比较清晰，能较为明了地表达句子的意思和逻辑关系。所谓"意合"，是指词语或分句之间不用语言形式手段连接，其中的语法意义和逻辑关系，通过词语或分句的含义表达。与英语不同的是，汉语是用隐含的方式来表达语法意义和逻辑关系的。汉语句子倾向于采用多个动词的连用形式，句子之间的关系多数由读者根据语境推测出来。

在复合句中，英语的主句一般放在句首，主句是整个句子的重心，是句子的主要部分。汉语习惯于按照时间的先后顺序和推理方式安排内容，因此一般将主要部分置于句尾，句子的重心在后。正是由于英语和汉语这两种语言在句子结构上的差异，长句的英译汉在转换方面的困难通常采用分切法（cutting）和拆句法（splitting）来解决。

1. 分切法

所谓分切法，就是汉译时将英语长句化整为零，在原句中的关系代词、关系副词、主谓连接、并列或转折连接、后续成分与其主体的连接等处按意群切割，将切割后的意群译成汉语的简单句，然后考虑意义、形式及逻辑关系对译成的简单句重新排列组合。分切法是最常用、最便利、最有效的长句汉译法。例如：

Plastics is made from water which is a natural resource inexhaustible and available everywhere, coal which can be mined through automatic and mechanical process at low cost and lime which can be obtained from calcinations of limestone widely present in nature.

译文：塑料由水、煤和石灰制造而成。水是无处不在、取之不尽的自然资源；煤可以通过自动化和机械化方式开采，成本较低；石灰可以通过燃烧自然界广泛存在的石灰石获得。

原文是一个主从复合句，主句是"Plastics is made from..."，其中介词"from"带出三个宾语"water""coal""lime"，这三个名词后各自都有由"which"引导的定语从句。这种句子在英语中很常见，结构并不复杂，但对在思维模式上早已习惯了言简意赅、铺排流散结构的汉语读者而言，往往不知从何译起。在这种情况下，可以先总括主题，再进行分述。对于三个定语从句，可采用重复主题词，引出定语从句的方法。这样原文的长句在汉语译文中就分解成若干短句，符合汉语的行文习惯。

有时英语长句非常复杂，为了较好地去表达原文所蕴含的内容，单独采用一种翻译方法是不够的，必须结合多种译法，对原句的结构和内容进行重组，以求译文能忠实地表达原句的含义。在弄清楚英语长难句在句子结构和语序上的特点后，翻译时就可采取相应的策略，以求做到忠实、通顺。这一策略便是拆句法。拆句法主要是符合了汉语链式结构的特点，将长句中的从句或短语译成句子，分开来叙述，使意义连贯。拆句法的目的就是将长句分解成短句，使译文连贯，有整体感。

2. 拆句法

具体来说，拆句法就是根据英汉双语之间的差异，在理解原文大概意思的基础之上，将妨碍译者整体地、一气呵成地翻译句子的部分分离出来，即将那些作为修饰成分的从句、短语或单词从中提炼出来，译成单独的一个小句，形成独立的主谓宾。可以将这种小句放置于主语之前，作为背景进行交代；又可以将其放置于后面作为补充说明；或是可以将其顺序完全颠倒，打散结构，重新组合。例如：

An outsider's success could even curiously help two parties to get the agreement they want.

译文：说来奇怪，一个局外人取得的成功竟然能够促使双方达成一项他们希望得到的协议。

"curiously"一词与其所修饰的动词之间关系不是很密切，又妨碍句子的整体译出，因此可以将其从主句中拆解出来，单独处理，以保证后面所译句子的完整性。

This land, which once barred the way of weary travelers, now has become a land for winter and summer vacations, a land of magic and wonder.

译文：这个地方现在已经成了冬夏两季的休假胜地，风光景物，蔚为奇观；而从前精疲力竭的旅游者只能到此止步。

原文中有"which"一个非限制性定语从句，修饰"This land"；"a land for winter and summer vacations, a land of magic and wonder"这一宾语中，"a land of magic and wonder"作为同位语，做进一步解释说明。翻译时，采用拆句法，将非限制性定语从句从主句中拆解出来，将主句和定语从句译为两个独立句，两个独立的句子采用对比的手法进行关联。这样使译文表达清晰，有层次感。

拆句法是最常见的译法。这些可拆解的成分可以是单词、词组或从句。这也明显地体现了英汉的区别，英语强调结构的完整性，而汉语则强调语义的连贯性。因此，拆句法不失为最佳的处理方式。

五、课后练习

（一）思考题

简述英文科技合同的翻译技巧与方法。

（二）翻译练习

1. 中译英

系统技术开发（委托）合同

委托方（甲方）：_____
受托方（乙方）：_____

甲方委托乙方设计智能收发验证系统技术方案，乙方将设计开发方案的验证电路及负责整体思路的建立，为此订立以下协议，并由双方共同恪守。

第一条　定义

智能收发验证系统技术方案（以下简称"方案"）是指设计满足附件要求的总体技术方案所需要的解决方案。该解决方案包括全部设计方案资料及关键电路验证技术资料。

技术资料指研发解决方案所必需的资料，包含乙方在设计方案的过程中所使用的全部有关验证技术资料。

第二条　合同内容和范围

2.1　合同技术方案要求。

2.1.1　技术内容。

（1）设计技术方案。

（2）验证关键电路。

（3）详细技术要求见技术协议附件。

2.1.2　技术方法和路线。

（1）采用 SoC 和 ASIC 技术。

（2）采用智能收发组件系统对关键电路进行实验验证。

2.2　双方义务。

2.2.1　乙方应在本合同生效后两个月内向甲方提交研究开发计划。

2.2.2　乙方应按下列进度完成方案设计工作。

1）第一阶段。

（1）启动阶段：晶圆厂的选定、设计文件的获得、工艺文件的分析、计算机系统的建立、EDA 软件的租用及购买、验证电路与整体方案设计思路的初步沟通。

（2）设计阶段：芯片的模块划分、原理设计、计算机仿真、版图设计。

(3) 测试阶段：初测及继续测试。

2) 第二阶段：根据第一阶段测试结果对方案进行修改，同时配合总体设计进行修改。

2.2.3　双方确定，乙方应在合同方案的关键电路验证合格后，根据甲方的请求，为甲方指定的人员提供技术指导和培训，或者提供与完成方案相关的技术服务。

2.2.4　双方确定，在本合同有效期内，甲方指定_____为甲方项目联系人，乙方指定_____为乙方项目联系人。一方变更项目联系人的，应当及时以书面形式通知另一方。未及时通知并影响本合同履行或造成损失的，应承担相应的责任。

2.3　交付。

乙方应按本合同条款2.2.2规定的内容，将合同方案技术资料交付甲方。

2.4　合同方案的验收。

双方确定，按所签订的验收标准对乙方完成的合同方案技术进行验收。

为保证乙方提供合同方案的正确性、可靠性和先进性，由甲乙双方技术人员一起，按本合同条款2.1、2.2、2.3及技术协议附件规定，共同对技术方案设计和核心电路进行考核和验收。考核验收合格后，双方代表要签署验收合格证书一式两份，双方各执一份为凭。

2.5　合同技术方案研发成果及相关知识产权的归属。

双方确定，因履行本合同所产生的研究开发成果及其相关知识产权权利归属，按以下方式处理。

2.5.1　甲方享有申请专利的权利。专利权取得后的使用和有关利益分配方式如下：专利权为甲方所有，利益归甲方所有。

2.5.2　有关使用和转让的权利归属及由此产生的利益按以下约定处理。

（1）技术秘密的使用权：归甲方所有。

（2）技术秘密的转让权：归甲方所有。

（3）相关利益的分配办法：归甲方所有。

2.5.3　乙方利用研究开发经费所购置与研究开发工作有关的设备、仪器等实物固定财产，归乙方所有。

2.5.4　双方确定，甲方有权利用乙方按照本合同约定提供的研究开发成果，进行后续改进。由此产生的具有实质性或创造性技术进步特征的新的技术成果及其权属，由甲方享有。

第三条　合同价格

3.1　按第二条所规定的合同内容和范围，乙方所提供的合同方案包括设计方案、设计图纸、技术服务和技术培训等的全部资料总价格为_____美元。

3.2　上述合同的价格为固定价格，包括本合同第二条所规定的全部技术资料。该价格包括乙方在本合同中所承担的其他义务的全部费用在内。

3.3　本合同内的一切费用均以美元计算和结算。

第四条　支付与支付条件

4.1　合同签订后启动费用：甲方支付乙方8万美元。

4.2　乙方整个初步方案通过后，甲方支付乙方12万美元。

4.3　选定晶圆厂后，甲方支付乙方15万美元。

4.4　提供设计方案及模拟结果后，甲方支付乙方30万美元。

4.5 提供测试电路并验证了电路后，甲方支付乙方剩余的 87 万美元。

第五条 侵权和保密

5.1 乙方保证所提供的总体方案不受任何第三者干涉和指控。如果发生第三者干涉和指控，则由乙方负责同第三者进行交涉，并由其承担法律上和经济上的全部责任和损失。

5.2 在本合同终止后，甲方仍有权继续使用乙方提供的技术方案和全部技术文件进行相应产品的生产。

第六条 保证和索赔

6.1 双方确定：任何一方违反本合同约定，造成研究开发工作停滞、延误或失败的，按以下约定承担违约责任。

（1）甲方违反本合同第四条约定，应当按合同总额的 10% 支付违约金。

（2）乙方违反本合同第二、四或五条约定，应当按合同总额的 10% 支付违约金。

6.2 在本合同履行中，因出现在现有技术水平和条件下难以克服的技术困难，导致研究开发失败或部分失败，并造成一方或双方损失的，双方按约定承担风险损失。

本合同项目的技术风险按双方认可的专家权威机构确认的方式认定。认定技术风险的基本内容应当包括技术风险的存在、范围、程度及损失大小等。认定技术风险的基本条件如下。

（1）本合同项目在现有技术水平和条件下具有足够的难度。

（2）乙方在主观上无过错且经认定研究开发失败为合理的失败。

6.3 一方发现技术风险存在并有可能致使研究开发失败或部分失败的情形时，应当在发现技术风险后 5 日内通知另一方并采取适当措施减少损失。逾期未通知并未采取适当措施而致使损失扩大的，应当就扩大的损失承担赔偿责任。

第七条 税费

甲乙双方将各自承担所在国（地区）所应支付的税费。

第八条 不可抗力

对于战争、严重水灾、火灾、台风、地震等人力不可抗力，以及双方同意的其他人力不可抗力因素，发生人力不可抗力的责任方应尽快将发生人力不可抗力事故的情况，用电传或电报通知另一方，并于事后 14 天内，以航空挂号信将有关政府当局出具的证明文件给另一方，予以认证。因发生不可抗力而影响了合同的执行，如果事故延续 20 天以上，则双方应尽快通过友好协商方式协商合同的进一步执行问题。

第九条 仲裁

所有与此合同有关的争议应通过友好协商解决。若通过协商无法达成一致，此争议应提交中国国际经济贸易仲裁委员会进行仲裁。仲裁应在北京进行，其结果对双方均有约束力，任何一方均不应向法院或其他政府部门申请以改变仲裁结果。仲裁费由败诉方负担。

第十条 合同的生效及其他

10.1 本合同由双方代表签订后，双方分别向各自的政府或主管部门申请批准，以最后批准一方的日期为合同生效日期。双方均应尽最大努力在 60 天内获得批准，并用邮件通知另一方。

10.2 本合同条款的任何改变、修改或增减，均需要经双方协商同意后双方授权各自的代表签署书面文件，作为本合同不可分割的一部分，与合同其他条款一样具有同等的效力。

10.3 双方确定,出现下列情形,致使本合同的履行成为不必要或不可能的,一方可以通知另一方解除本合同。

(1) 因发生不可抗力。

(2) 技术风险出现。

10.4 合同有效期为两年。

10.5 本合同的附件为本合同不可分割的组成部分并具有与正文同样的法律效力。如果附件与合同正文不一致,合同正文效力优先。

本合同一式两份由中英文写成,双方签字并各持有一份。执行中如有异议,以中文为准。

甲方(签名盖章):　　　　　　　　乙方(签名盖章):

法定代表人(签字):　　　　　　　法定代表人(签字):

签署日期:　　　　　　　　　　　签署日期:

2. 英译中

Technology Development (Commission) Contract

Entrusting Party (Party A): _____

Entrusted Party (Party B): _____

Clause 1 Requirements for research and development in the contract are as follows.

1. Technical objectives: on foundation of the existing envelop printing system facilities of Party A, by using high – speed H – resolution video head, to develop the postal envelop characters automatic identification system on the equipment transport tape.

2. Technical contents: details can be seen in the attached sheet of Party B.

3. Technical method and route: details can be seen in the attached sheet of Party .

Clause 2 Party B shall provide a detailed research and development plan to Party A within 30 days after the contract takes effect, including details can be seen in the attached sheet of Party B.

Clause 3 Party B shall conduct research and development in accordance with the progress.

Clause 4 Party A shall provide Party B with the following technical materials and cooperation affairs.

1. Technical materials: (1) detailed descriptions on system requirements; (2) all standards of envelop samples used by final user of the system at present, with each kind of envelop having at least 2 samples.

2. Other cooperation affairs: to assist manufacturing the transport tape and envelop speed detection and automatic positioning equipment, with the accuracy error of envelop label automatic positioning less than 5mm; also, to provide a triggering signal to the high – speed camera for photo taking and scanning.

3. Term and method of supply.

After the contract expires, the aforesaid technical materials shall be disposed as follows: transport tape and envelop speed detection and automatic positioning equipment shall be returned to Party A, and the rest technical materials shall be disposed by Party B freely.

Clause 5 Party A shall pay expenses and rewards for research and development in the following ways.

1. The total expenses and rewards for research and development are 340,000 RMB Yuan. Including:

(1) Development costs: 100,000 RMB Yuan.

(2) Equipment costs: 150,000 RMB Yuan.

(3) Software costs: 40,000 RMB Yuan.

(4) Management costs and others: 50,000 RMB Yuan.

High – speed digital cameras for image catching shall be imported, with resolution above 2 million pixels and speed more than 30 frames/second.

If domestic high – speed digital cameras are adopted for image catching, the total expenses and rewards for research and development shall be 300,000 RMB Yuan.

2. Party A shall pay expenses for research and development to Party B in installments, with concrete payment methods and terms as follows.

(1) After the contract takes effect, Party A shall pay Party B 40% of the total contract expenses for the latter to initiate development on the item.

(2) After accepted qualified, Party A shall pay 60% of the total contract expenses.

Name, address and account number of Party B's bank of deposit are as follows. (omitted)

Notes: please write "002" in the "Purpose" column of your cash remittance.

Clause 6 Party B shall independently manage the expenses for research and development in the contract and assume sole responsibilities for his profits or losses. Party A shall give no interferences.

1. If any significant changes to the system requirements, Party A may provide request to modify or terminate this contract, and Party B shall not return his received expenses for research and development.

2. If any technical obstacles which cannot be solved by existing technologies occur in the research and development process of the system, Party B may provide request to terminate the contract and return 85% of his received expenses for research and development to Party A.

3. If the system cannot pass the acceptance test on August 15, Party A may provide request to terminate the contract, and Party B shall return 85% of his received expenses for research and development to Party A.

Clause 7 If any technical problems which cannot be solved under current technical level and conditions occur during fulfillment of the contract, which makes research and development fail or partially fail and causes losses to either party or both parties, the two parties agree to bear risks and losses as follows.

(1) If the system fails to meet technical indices stated in the contract owing to Party B's reasons, the latter shall bear relevant expenses for labors and development, and return 85% of the initial payment to Party A.

(2) If the system fails to meet technical indices stated in the contract owing to inaccurate posi-

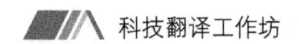

tioning of Party A's envelop speed detection and automatic positioning equipment, Party B shall not bear any responsibilities.

The two parties agree that the primary contents of the contract project's confirmed technical risks here of should include existence, scope, degrees and losses of such technical risks. The essential conditions for confirming technical risks are as follows.

1. The contract project has adequate degree of difficulties under current technical level and conditions.

2. Party B has no subjective faults, and failures for research and development are confirmed to be rational.

If any technical risks which may make research and development fail or partially fail occur, either party shall notify the other and take appropriate measures to reduce losses to the minimal within 7 days after the technical risks are found. Should either party fail to give notices or take adequate measures within the stated term, which has made losses deteriorated, the defaulting party shall bear corresponding liability for compensations.

Clause 8 Other items and conditions agreed by the two parties in the contract are as follows.

Party B shall provide a year maintenance free of charge for the developed system and charge fees for maintenance from the second year, with fee charging standard negotiated additionally.

Clause 9 The contract is signed in quadruplicate, with Party A and Party B processing two copies each. All copies shall be equally binding upon both parties.

Clause 10 The contract takes effect on the date signed by both parties with their seals.

Consignor: Consignee:

Legal representatives: Legal representatives:

Date: Date:

学思践悟

第四章 科技新闻

一、科技新闻文本介绍

(一) 科技新闻概述

科技新闻 (science and technology news) 是新闻的一个分支,主要内容包括科学技术最新成就和研究动态、科技界的重大活动、做出重大贡献的科技界人物、科技发展相关的新政策、自然界的新发现和趣闻等科技领域的事实报道。科技新闻以新闻报道的形式反映人类认识自然、改造自然的实践活动,迅速及时地报道国内外科技领域的新事件、新气象与新成就。网络技术的飞速发展,丰富了新闻传播的形式,加快了新闻传播的速度,拓宽了新闻信息的覆盖范围,从某种程度上来说打破了传统媒体在时间和地域方面的局限性。国内外科技新闻的及时报道可帮助广大受众了解科学知识,同时促进国内外科技交流,了解世界科技前沿动态。

科技新闻一般由标题、导语、正文三个部分组成。标题字体字号通常较为醒目,表述形象生动,用简短的句子概括新闻的主要内容。常见的科技新闻种类有科技消息、科技通讯、科技评论、科技人物专写、科技特写等。

(二) 科技新闻的特点

1. 语篇特点

(1) 时效性。科技新闻的选材在内容上一定是科技界最新的重大事件和研究动态。例如,有以下报道:2021年2月科学家们首次利用卫星图像统计非洲大象的数量,2021年10月英国科学家研发医疗机器人。

The scientists used a series of these images—pictures of one national park in South Africa, taken by a satellite—to test a new way of monitoring the African elephant population.

译文:科学家们利用了一系列由卫星拍摄的南非一个国家公园的图像,来测试一种用于监测非洲大象数量的新方法。

Here at Southmead Hospital in Bristol, they're experimenting with a robot programmed to interact with patients for simple forms of physio. Another possible use might be basic bedside checks.

译文:布里斯托尔的索斯米德医院正在试验一种机器人,它可以按照预设的程序与病人进行简单的理疗互动。它的另一个潜在用途是进行基本的床边检查。

(2) 科学性。科学性是指科技新闻所报道的内容要客观真实,符合事物的发展规律,有科学依据。同时,科技新闻选材要具有科普性,向大众普及基本的科学知识,为帮助大众理解,可能会插入一些背景知识。例如,《中国日报》在2021年7月22日对全国碳排放权交易市场的报道中插入了一段全国碳市场的知识介绍,美联社在报道世界上最大的运输机在平流层发射时回顾了机载发射的历史。

The national carbon emissions trading market launched online trading on July 16. The power generation industry was the first industry to be included, with 2,225 power companies taking the lead.

译文:7月16日,全国碳排放权交易市场启动上线交易。发电行业成为首个纳入全国碳市场的行业,纳入重点排放单位2225家。

[知识点]全国碳市场是利用市场机制来达到控制和减少温室气体排放、推动绿色发展的制度创新。碳交易市场,就是通过碳排放权的交易达到控制碳排放总量的目的。通俗来讲,就是把二氧化碳的排放权当作商品来进行买卖,需要减排的企业会获得一定的碳排放配额,成功减排可以出售多余的配额,超额排放则要在碳市场上购买配额。

Airborne launches date back decades, most famously to the X-15 program of the 1950s and '60s, when manned rocket planes were carried aloft under the wing of a B-52 bomber and released on hypersonic research flights.

译文:机载发射可以追溯至几十年前,最著名的是20世纪五六十年代的X-15计划,当时一架B-52轰炸机的机翼下搭载了载人火箭飞机,并在超音速飞行的测试研究飞行器上发射。

(3)规范性。规范性是指每个科学领域都有相应的科学术语,科技新闻应该使用正确的术语来描述所报道的科技内容,体现所报道的这一领域的专业性。这也是科技新闻与其他类型新闻区别比较大的一个特点。例如,科技新闻中常有 semiconductor(半导体)、digital wallet(数字钱包)、hypersonic weapon(高超音速武器)、gene silencing medicine(基因沉寂药物)、crippling pain(剧痛)等专业科技词汇。

对译者而言,要准确地再现原文内容,除具备专业知识和科学文化知识外,还应当秉持科学严谨的态度,善于利用网络资源查证术语,多关注国内外科技新闻网站。科技新闻网站一般是由一些科技协会或科研机构开办的,其中的新闻内容全部跟科技相关。国内科技新闻网站如中国科技新闻网、中国科技网等,国外科技新闻网站如英国皇家学会网、英国自然环境研究委员会网、美国《科学》杂志网等。同时,作为一名科技新闻翻译工作者,也应当关注翻译界和科技界、新闻界的最新研究动态和发展方向。具有良好的知识储备是保证科技文本翻译质量的首要因素。以下这些科技新闻标题翻译可供学习参考。

Japan plans long-term strategy to build semiconductor resilience
日本计划制定半导体长期战略

Facebook says ready to launch digital wallet
Facebook 准备推出数字钱包

Gene silencing medicine transforms crippling pain
基因沉寂药物有望改善卟啉病患者生活

(4)简洁性。科技新闻读者范围广,层次不同,文化程度不同,这要求文章必须通俗

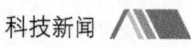

易懂，节约读者的宝贵时间；另外，报纸或杂志自身版面有限，要在有限的版面上传播尽可能多的信息，这就要求文章言简意赅，语言生动形象，增强文章的可读性。科技新闻为符合读者的阅读习惯，要尽量采用简练的表达。对于一些新的术语和词语，通过添加备注或使用隐喻来帮助读者理解，使读者有顺畅的阅读体验。试着体会以下例句。

SpaceX launches world's first "amateur astronaut" crew to orbit Earth
SpaceX 将首批业余宇航员送上太空，绕地球飞行

Korean designer creates "Third Eye" for smartphone zombies
韩设计师为智能机僵尸打造"第三只眼"

2. 文本特点

（1）科技新闻一般采用陈述语气，考虑到大众读者的阅读体验，多采用简单句式，描述多形象生动，善用比喻手法。相比于汉语语篇多动词的特点，英文科技新闻语篇多名词化结构，多使用缩略词。为体现文章的专业性，多采用直接引语和间接引语增强说服力。例如：

Sicilian Rouge High GABA is a special type of tomato designed to contain high levels of gamma–aminobutyric acid (GABA), an amino acid believed to aid relaxation and help lower blood pressure.

译文：高 γ–氨基丁酸西西里胭脂是一种特殊的番茄，它含有大量的 γ–氨基丁酸，据称，这种氨基酸可以让人放松，并有助于降低血压。

Alexander Zhanovich Medvedev is currently in charge of the project and leads a lab at the Siberian Branch of the Russian Academy of Sciences (SBRAS). Researchers tested the mask against the influenza A virus as well as the staphylococcus and E. coli bacteria, according to Nikolai Zakharovich Lyakhov, chief researcher at the Institute of Chemistry and Technology at the SBRAS.

译文：梅德韦杰夫目前负责该项目，并管理俄罗斯科学院西伯利亚分院一个实验室的工作。该院化学和技术研究所首席研究员尼古拉·扎哈罗维奇·利亚霍夫说，研究人员测试了这种口罩抵御甲型流感病毒及葡萄球菌和大肠杆菌的性能。

Tara Hall, a spokesperson for Hillarys, said: "Getting a good night's sleep is about more than simply going to bed early—it's about waking up at the right time, too. Using a formula based on the body's natural rhythms, the Sleep Calculator will work out the best time for you to rise or go to sleep.

译文：Hillarys 公司发言人塔拉·霍尔表示："睡个好觉不仅仅是早睡，还需要在正确的时间醒来。睡眠计算器根据人体生物钟，计算出你起床或入睡的最佳时间。"

（2）科技新闻因其独特的专业性，经常出现大量的图片、公式、表格作为辅助说明，避免使用大段的抽象文字，而是使用具体的图表加以代替。例如，在介绍美国疾病控制与预防中心（CDC）在其官网公布的 8 个与病态肥胖有关联的基因（变体）时列出表格（如表 1 所示），并配以文字说明；在报道嫦娥五号月球样品时辅以样品照片（如图 1 所示）。使专业术语和具象的图形能够"一一对应"，提高了信息传达的精准度，同时也对读者的阅读进行了有意识的引导，增强了图表的易读性，使读者能够以最小的代价获得最多的信息，可以

说，在信息的专业性和图表的易读性之间找到了平衡。

Genes contribute to the causes of obesity in many ways, by affecting appetite, satiety (the feeling of fullness), metabolism, body-fat distribution, and the tendency to use eating as a way to cope with stress.

译文：基因可以多方面影响肥胖的成因，它会影响食欲、饱腹感、新陈代谢、体脂分布，以及将饮食作为减压方式的倾向。

表1 与病态肥胖有关联的基因（变体）

Gene symbol	Gene name	Gene product's role in energy balance
ADIPOQ	Adipocyte-, C1q-, and collagen domain-containing	Produced by fat cells, adiponectin promotes energy expenditure
FTO	Fat mass- and obesity-associated gene	Promotes food intake
LEP	Leptin	Produced by fat cells
LEPR	Leptin receptor	When bound by leptin, inhibits appetite
INSIG2	Insulin-induced gene 2	Regulation of cholesterol and fatty acid synthesis
MC4R	Melanocortin 4 receptor	When bound by alpha-melanocyte stimulating hormone, stimulates appetite
PCSK1	Proprotein convertase subtilisin/kexin type 1	Regulates insulin biosynthesis
PPARG	Peroxisome proliferator-activated receptor gamma	Stimulates lipid uptake and development of fat tissue

An analysis of moon rocks brought back to Earth by China's Chang'e 5 mission suggests the samples are a new type of lunar basalt, different from those collected during previous Apollo and Luna missions.

译文：研究发现，嫦娥五号月球样品为一类新的月球玄武岩，不同于美国阿波罗和苏联月球系列采样任务返回的月球样品。

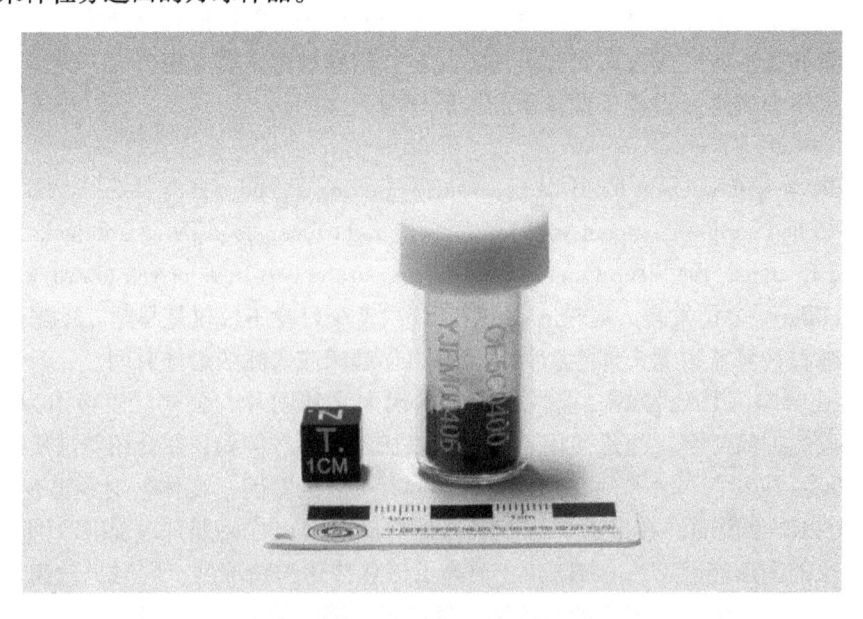

图1 嫦娥五号带回的月球土壤样品

二、科技新闻范文展示

范文一

"Small Data" Are Also Crucial for Machine Learning
The most promising AI approach you've never heard of doesn't need to go big

When people hear "artificial intelligence", many envision "big data". There's a reason for that: some of the most prominent AI breakthroughs in the past decade have relied on enormous data sets. Image classification made enormous strides in the 2010s thanks to the development of ImageNet, a data set containing millions of images hand sorted into thousands of categories. More recently, GPT-3, a language model that uses deep learning to produce humanlike text, benefited from training on hundreds of billions of words of online text. So it is not surprising to see AI being tightly connected with "big data" in the popular imagination. But AI is not only about large data sets, and research in "small data" approaches has grown extensively over the past decade—with so-called transfer learning as an especially promising example.

Also known as "fine-tuning", transfer learning is helpful in settings where you have little data on the task of interest but abundant data on a related problem. The way it works is that you first train a model using a big data set and then retrain slightly using a smaller data set related to your specific problem. For example, by starting with an ImageNet classifier, researchers in Bangalore, India, used transfer learning to train a model to locate kidneys in ultrasound images using only 45 training examples. Likewise, a research team working on German-language speech recognition showed that they could improve their results by starting with an English-language speech model trained on a larger data set before using transfer learning to adjust that model for a smaller data set of German-language audio.

Research in transfer learning approaches has grown impressively over the past 10 years. In a new report for Georgetown University's Center for Security and Emerging Technology (CSET), we examined current and projected progress in scientific research across "small data" approaches, broken down in terms of five rough categories: transfer learning, data labeling, artificial data generation, Bayesian methods and reinforcement learning. Our analysis found that transfer learning stands out as a category that has experienced the most consistent and highest research growth on average since 2010. This growth has even outpaced the larger and more established field of reinforcement learning, which in recent years has attracted widespread attention.

Furthermore, transfer learning research is only expected to continue to grow in the near future. Using a three-year growth forecast model, our analysis estimates that research on transfer learning methods will grow the fastest through 2023 among the small data categories we considered. In fact, the growth rate of transfer learning is forecast to be much higher than the growth rate of AI research as a whole. This implies that transfer learning is likely to become more usable—and therefore more widely used—from here on out.

Small data approaches such as transfer learning offer numerous advantages over more data-intensive methods. By enabling the use of AI with less data, they can bolster progress in areas where little or no data exist, such as in forecasting natural hazards that occur relatively rarely or in predicting the risk of disease for a population set that does not have digital health records. Some analysts believe that, so far, we have applied AI more successfully to problems where data were most available. In this context, approaches like transfer learning will become increasingly important as more organizations look to diversify AI application areas and venture into previously underexplored domains.

Another way of thinking about the value of transfer learning is in terms of generalization. A recurring challenge in the use of AI is that models need to "generalize" beyond their training data—that is, to give good "answers" (outputs) to a more general set of "questions" (inputs) than what they were specifically trained on. Because transfer learning models work by transferring knowledge from one task to another, they are very helpful in improving generalization in the new task, even if only limited data were available.

Moreover, by using pretrained models, transfer learning can speed up training time and could also reduce the amount of computational resources needed to train algorithms. This efficiency is significant, considering that the process of training one large neural network requires considerable energy and can emit five times the lifetime carbon emissions of an average American car.

Of course, using pretrained models for new tasks works better in some cases than others. If the initial and target problems in a model are not similar enough, it will be difficult to use transfer learning effectively. This is problematic for some fields, such as medical imaging, where certain medical tasks have fundamental differences in data size, features and task specifications from natural image data sets such as ImageNet. Researchers are still learning about how useful information is transferred between models and how different model design choices hinder or facilitate successful transfer and fine-tuning. Hopefully, continued progress on these questions through academic research and practical experience will facilitate wider use of transfer learning over time.

AI experts have emphasized the significance of transfer learning and have even stated that the approach will be the next driver of machine learning success in industry. There are some early signs of successful adoption. Transfer learning has been applied for cancer subtype discovery, video game playing, spam filtering, and much more.

Despite the surge in research, transfer learning has received relatively little visibility. While many machine learning experts and data scientists are likely familiar with it at this point, the existence of techniques such as transfer learning does not seem to have reached the awareness of the broader space of policy makers and business leaders in positions of making important decisions about AI funding and adoption.

By acknowledging the success of small data techniques like transfer learning—and allocating resources to support their widespread use—we can help overcome some of the pervasive misconceptions regarding the role of data in AI and foster innovation in new directions.

"小数据"对机器学习的重要作用
你也许闻所未闻,这是最有发展前景的人工智能,它甚至不需要做大
（长期被忽略的"小数据"人工智能潜力不可估量）

当人们听到"人工智能"时,很多人都会联想到"大数据"。这并非空穴来风:过去十年中在人工智能方面取得的重要突破都依赖于庞大的数据集。由于 ImageNet（ImageNet 是一个计算机视觉系统识别项目名称,是目前世界上图像识别最大的数据库,是由美国斯坦福大学的计算机科学家模拟人类的识别系统建立的）的发展,图像分类在 21 世纪 10 年代取得了巨大的进步。ImageNet 是一个包含数百万张图片的数据集,这些图片被手动分类成数千个类别。最新出现的 GPT-3 是一种使用深度学习生成类人文本的语言模型,它的发展得益于对数千亿单词的在线文本训练。因此,在大众的印象中,人工智能与"大数据"紧密相连也就不足为奇了。但人工智能不仅仅是关于大数据集的,过去十年中,"小数据"方法的研究得到了广泛的发展,所谓的迁移学习就是一个很有发展前景的例子。

迁移学习也被称为"微调",在对感兴趣的任务几乎没有数据,但对相关问题有大量数据的情况下,迁移学习是很有用的。其工作方式是,首先使用大数据集训练模型,然后使用与特定问题相关的较小数据集进行小幅度的重新训练。例如,从 ImageNet 分类器开始,印度班加罗尔的研究人员使用迁移学习训练一个模型,仅使用 45 个训练示例就在超声图像中实现了肾脏定位。与此相似,一个研究德语语音识别的研究团队表明,他们可以先从一个更大的数据集上训练英语语音模型开始,然后使用迁移学习来调整该模型,以适应更小的德语音频数据集,从而改善最终结果。

在过去十年中,迁移学习方法的研究取得了令人瞩目的成就。在乔治城大学安全与新兴技术中心（CSET）新发布的一份报告中,我们通过"小数据"方法研究了当前和未来的科学发展,大致分为五个类别:迁移学习、数据标注、人工数据生成、贝叶斯方法和强化学习。我们分析发现,对迁移学习的研究是自 2010 年以来年均增长最稳定也是最高的一类,甚至超过了近年来引起广泛关注、更大且更成熟的强化学习。

而且,可以预见,对迁移学习的研究在未来会继续增长。通过一个三年增长预测模型,我们分析预计,到 2023 年,在我们提到的小数据分类中,对迁移学习方法的研究将实现最快增长。实际上,预计迁移学习的增长率远高于人工智能研究的整体增长率。这意味着从现在起,迁移学习可能会变得更有用,因此也会得到更广泛地运用。

与数据密集型方法相比,迁移学习等小数据方法具有更多优势。在数据较少的情况下使用人工智能,在一些数据很少或没有数据的领域,研究人员也可以取得进步,比如预测发生频率较低的自然灾害,或者预测没有数字健康记录的人群疾病风险。一些分析人士认为,到目前为止,我们已经将人工智能成功地运用于解决一些数据较容易获取的问题。在这种背景下,迁移学习等方法将变得越来越重要,因为越来越多的组织希望将人工智能应用领域多元化,并尝试开发一些以前未被开发的领域。

迁移学习的另一种价值是普遍化。人工智能使用过程中反复出现的挑战是模型需要"概括"训练数据之外的内容,即对更一般的（相比于它们所受的专业训练）"问题"（输入）给出好的"答案"（输出）。因为迁移学习模型的工作原理是将知识从一项任务转移到另一项任务,所以即使只有有限的数据,迁移学习模型也可以使新任务变得更加普遍化。

此外,通过使用预训练模型,迁移学习可以缩短训练时间,也可以减少训练算法所需的

计算资源。提高效率是很必要的,因为训练一个大型神经网络的过程需要相当大的能量,碳排放量可达一辆美国普通汽车寿命碳排放量的5倍。

当然,有时,在新任务中使用预训练模型比其他情况效果更好。如果模型中一开始的问题和目标问题不甚相似,就很难有效地使用迁移学习。这对某些领域来说是有问题的,比如医疗成像,有些医疗任务在数据大小、特征和任务规格方面与自然图像数据集(如ImageNet)存在根本性差异。研究人员仍在学习如何在模型之间传递有用的信息,并研究不同的模型设计选择如何阻碍或促成成功的传递和微调。希望随着时间的推移,通过在学术研究和实践经验上的努力促进迁移学习得到更广泛的应用。

人工智能专家也强调了迁移学习的重要性,甚至表示这种学习方法将是机器学习在工业领域取得成功的下一个驱动因素。早期已有一些成功案例,比如迁移学习已经应用于发现癌症亚型、玩电子游戏、拦截垃圾邮件等。

尽管对迁移学习的研究激增,但对迁移学习的关注度仍然相对较低。虽然如今的许多机器学习专家和数据科学家可能已经对此比较熟悉,但在人工智能的资助和选择方面,决策者和商业领袖并没有在众多的决策选择中意识到迁移学习等技术的存在。

通过迁移学习等小数据技术的成功,以及分配资源以支持它们的推广,我们可以帮助克服一些关于数据在人工智能中作用的普遍误解,并在新的方向进行创新。

范文二

NASA's Perseverance Rover Lands on Mars

NASA has unveiled the first pictures from its fifth Mars rover, Perseverance, after a successful landing on the red planet's Jezero crater at approximately 3:55 pm Thursday (Feb. 18).

"This landing is one of those pivotal moments for NASA, the United States, and space exploration globally—when we know we are on the cusp of discovery and sharpening our pencils, so to speak, to rewrite the textbooks," acting NASA Administrator Steve Jurczyk said in a press release. "The Mars 2020 Perseverance mission embodies our nation's spirit of persevering even in the most challenging of situations, inspiring, and advancing science and exploration."

Perseverance, the most technologically advanced robot NASA has sent to date, traveled 293 million miles to reach Mars over the course of more than six months after launching on a United Launch Alliance Atlas V rocket from Cape Canaveral Space Station on July 30. It will remain on Mars for nearly two years, searching for signs of ancient life and exploring the planet's surface.

The mission will help prepare the agency for future human exploration on Mars in the 2030s.

The $2.7 billion rover, built in NASA's Jet Propulsion Laboratory in Pasadena, Calif., is about 10 feet long, 9 feet wide, 7 feet tall and about 2,260 pounds, roughly 278 pounds heavier than its predecessor, Curiosity.

Perseverance is designed to drive an average of 650 feet per Martian day and features seven scientific instruments, a robotic arm that reaches about 7 feet long, a rock drill. It is nuclear powered, using a plutonium generator provided by the US Department of Energy.

The scientific instruments on the rover include a camera designed to take high-definition video, panoramic color and 3D images of the Martian surface and features in the atmosphere with a zoom lens to magnify distant targets, a group of sensors to measure weather and monitor dust on the planet's surface, a system that will be used to produce oxygen from the Martian carbon-dioxide atmosphere, an x-ray and camera system that can measure the chemical makeup of rocks and analyze features as small as a grain of salt, a ground-penetrating radar system to analyze geologic features under Mars' surface, a group of cameras, spectrometers, and a laser to search for organics and minerals as well as take close up images of rock grains and surface textures, and a camera to identify the chemical composition of rocks and soils, including their atomic and molecular makeup.

In addition, the rover carries a commemorative plate to honor COVID-19 healthcare workers and has the names of 10.9 million people stenciled into three of its silicon chips with the words "Explore as one" written in Morse code.

Scientists hope to find biosignatures embedded in samples of ancient sediments that Perseverance is designed to extract from Martian rock for future analysis back on Earth—the first such specimens ever collected by humankind from another planet.

Two subsequent Mars missions are planned to retrieve the samples and return them to NASA in the next decade.

Apart from NASA, missions from the UAE and China to Mars also kicked off last year. In 2023 the European Space Agency is expected to land on Mars its Rosalind Franklin rover, which will carry a drill capable of reaching metres below the surface, where biomolecules may survive protected from the harsh conditions above.

美国"毅力"号火星车成功着陆火星

"毅力"号火星车于本周四（2月18日）下午3点55分左右在火星的耶泽罗陨石坑成功着陆后，美国宇航局发布了第一批由其传回的照片。"毅力"号是美国宇航局的第五辆火星巡游车。

美国宇航局代理局长史蒂夫·尤尔奇克在一份新闻发布稿中称："此次着陆是美国宇航局、美利坚合众国和全球太空探索的关键时刻之一。我们正处于发现和磨炼的时期，而这次着陆改写了教科书。火星2020'毅力'号任务代表了美国在最具挑战性的形势下坚忍不拔的精神，启发和推动了科学与探索。"

"毅力"号是美国宇航局迄今为止发射的技术最先进的遥控设备。去年7月30日从卡纳维拉尔角空间站搭载联合发射联盟的阿特拉斯5号火箭发射升空后，"毅力"号在六个多月的时间里飞行了2.93亿英里（4.7亿千米）才到达了火星。"毅力"号将在火星上停留近两年，寻找远古生命迹象，探索火星表面。

该任务将有助于美国宇航局为2030年后人类探索火星做好准备。

这个耗资27亿美元（约合人民币175亿元）的火星车是在美国宇航局位于加州帕萨迪纳市的喷气推进实验室制造的，车体长10英尺（约3米），宽9英尺，高7英尺，重2260磅（约1025千克），比它的前任"好奇"号火星车大约重278磅。

按照性能要求，"毅力"号平均每个火星日可以行驶650英尺，包含七个科学设备、一个能伸到7英尺长的机器手臂，还有一个凿岩机。这个核动力火星车使用的是美国能源部提供的钚发电机。

火星车上的科研设备包括一台可拍摄火星表面和大气特征的高清视频、全景全色三维影像并配备放大远处目标物体的变焦镜头摄像机,一组可以观测天气和监控火星表面尘土的传感器,一个能用火星大气中的二氧化碳制造氧气的系统,一个可以测定岩石化学成分和分析细微特征的 X 光摄像系统,一个分析火星表面下的地质特征的地面穿透雷达系统,一组搜寻有机物、矿物质并给岩石颗粒和表面纹理近距离拍照的摄像机、光谱分析仪和激光器,以及一台识别包括原子和分子组成在内的岩石和泥土化学成分的摄像机。

此外,这辆火星车还携带了一块致敬新冠肺炎医护人员的纪念牌,把 1090 万人的名字刻在了纪念牌的三块硅晶片上,并用摩尔斯电报编码写上了"齐心探索"。

科学家希望能在"毅力"号从火星岩石提取以供未来地球分析的古老沉积物样本中找到生命信号,这是人类首次从其他星球采集到这种样本。

下一个十年计划开展两次火星任务,从火星取回样本送到美国宇航局。

除美国宇航局外,去年阿联酋和中国也执行了太空任务。2023 年欧洲航天局的罗莎琳·富兰克林漫游车将在火星着陆,这辆火星车上将搭载一个可钻入地下数米的钻孔机,那里可能存活着远离地表恶劣环境的生物分子。

范文三

Apple introduced three new iPhones—the Xs, Xs Max, and XR. The Xs is the successor to last year's iPhone X, and the awkwardly named Xs Max, is a larger version of the Xs. The company also introduced the colorful iPhone XR, which will be a slightly more affordable model. The Xs looks nearly identical to last year's X, but includes a range of internal upgrades. It has a 5.8–inch "Super Retina" HD screen. Apple says its battery life is also about a half–hour better than the iPhone X.

苹果公司此次发布了三款新 iPhone 手机,分别是 iPhone Xs、iPhone Xs Max 和 iPhone XR。iPhone Xs 在去年推出的 iPhone X 基础上进行了升级改造。iPhone Xs Max 名字有些拗口,是增大版的 iPhone Xs。苹果公司还推出了配色绚丽的 iPhone XR,价格也更亲民。iPhone Xs 与去年推出的 iPhone X 外观几乎相同,但多种功能都有提升,搭载 5.8 英寸超清视网膜屏。苹果公司称其待机时间也比 iPhone X 长出大约半小时。

范文四

Cars of the Future Could Use 5G to "Talk" to Each Other

Cars could soon be communicating with each other using 5G to make drivers aware of upcoming hazards, scientists claim. The ultra–fast mobile internet would allow for rapid information transmission and could make drivers aware of black ice, pot holes or other dangers up ahead.

Several car manufacturers are already integrating 5G into their vehicles, including as a tool to help usher in the generation of self–driving vehicles. Experts at Glasgow Caledonian University (GCU) believe the high–speed connection will also improve the reliability and capability of automated vehicles to the point where they will be safer than the manual cars being driven today. They predict the number of road traffic accidents—which according to the World Health Organisation account for more than 1.3 million deaths and up to 50 million people injured worldwide every year—will drop drastically as a result. Dr. Dimitrios Liarokapis, a member of the research group, said: "To have a better idea of what the future will look like, think of having Tesla–like cars that not only use sensors to scan what's around them, they can also talk to each other and exchange safety – related information about their surroundings over an area that covers several square miles." "I'm sure anyone who has had a bad experience on frozen roads would have benefited from knowing about the dangerous conditions in advance, so they could have adjusted their speed or, if possible, even avoided that route altogether. The same could be said of potholes." "With the help of 5G, a vehicle – generated early warning system that alerts drivers is feasible within the next few years. Cars that are close enough to the danger area will transmit warning messages to other cars around them using short – range communication technologies, but also to cars further away using 5G, fast and reliably. Then those cars will send the same information to cars near them and so on, forming a joined – up, multi – vehicle communication chain that stretches far and wide." "5G is an exciting mobile technology, which will give a massive boost to smart cities and autonomous vehicles among many other things."

Automotive giant Ford is already working on connected cars. Earlier this year it revealed its intention to fit 80 percent of its vehicles made this year with technology that warns drivers about upcoming road accidents, bad weather and traffic jams. The system pools data from other connected road users, emergency services and the authorities and beams it from the cloud directly to the car. Alerts pop up on the car's dashboard display warning the driver about what lies around the corner.

5G 网络将助力实现车与车"对话"

科学家称，汽车很快就可以通过 5G 网络互相"对话"，让司机意识到前方的危险。超高速的移动互联网将使信息可以快速传输，并让司机意识到前方的黑冰、坑洞或其他危险。

几家汽车制造商已经将 5G 网络技术融入汽车生产，包括以此来帮助引领自动驾驶汽车时代的到来。格拉斯哥喀里多尼亚大学的专家认为，高速互联网还将提高自动化车辆的可靠性和性能，使其比目前使用的手动驾驶汽车更安全。他们预测，道路交通事故的数量将因此大幅下降。据世界卫生组织统计，全球每年因道路交通事故死亡的人数超过 130 万，受伤人数高达 5000 万。研究小组的迪米特里奥斯·利奥卡皮斯博士说："为了更好地了解未来会是什么样子，想想你有特斯拉那样的车，它不仅使用传感器来扫描周围，车辆之间还可以相互交流，并且交换附近几平方英里内的安全信息。""我相信任何在冰冻道路上有过糟糕经历的人都会从提前了解危险情况中受益，这样他们就可以调整车速，如果可能的话，甚至完全避开这条路线。路遇坑洞也是如此。""在 5G 网络的帮助下，未来几年内，可以向司机发出警报的车辆生成预警系统将会出现。距离危险区域足够近的汽车将使用短程通信技术向周围的其他汽车发送警告信息，同时使用 5G 网络向更远的汽车发送警告信息，快速且可靠。然后这些车将同样的信息发送给附近的车，以此类推，形成一个连成一体的多车通信链，延伸到很远很远的地方。""5G 网络是一项激动人心的移动技术，它将极大地推动智能城市和自动驾驶汽车等领域的发展。"

汽车巨头福特公司已经开始研发联网汽车。今年早些时候，该公司透露，计划在今年生产的 80% 的汽车中配备一种技术，用于提醒司机注意前方的交通事故、恶劣天气和交通堵塞。该系统将来自其他联网道路用户、应急服务机构和管理部门的数据汇聚在一起，并将数据从云端直接传输到汽车上。汽车仪表盘上弹出的警报将提醒司机前方的危险。

三、科技新闻常用词汇和句型

（一）新闻报道和科技新闻常用词汇

1. 新闻报道常用词汇

topical news　时政新闻
regular press conference　例行记者会
press briefing　新闻发布会
foreign correspondent　驻外记者
contributing editor　特约编辑
morning edition　晨报
public notice　公告
bureau/copy chief　总编辑
special correspondent　特派员
source of information　消息来源
news/press conference　记者招待会

2. 科技新闻常用词汇分类介绍

（1）生态能源相关科技新闻常用词汇如下。

carbon–reduction　碳减排
carbon–free　无碳
carbon footprint　碳足迹
global energy governance　全球能源治理
carbon peaking and carbon neutrality goals　碳达峰和碳中和目标（双碳目标）
green transition　绿色转型
carbon neutrality　碳中和
water conservation　节水
optimized allocation of water resources　水资源优化配置
forest chief scheme　林长制
flood control and drought relief systems　防洪抗旱体系
marine economy　海洋经济
nuclear fuel　核能燃料
charging point　充电桩
silicon solar cell　硅太阳电池
manned submersible　载人潜水器
fossil fuels　化石燃料

（2）互联网/通信相关科技新闻常用词汇如下。

mobile internet　移动互联网

quantum communication　量子通信
mobile data traffic consumption　移动互联网接入流量
the Beidou Navigation Satellite System　北斗卫星导航系统
industrial chain of big data　大数据产业链
digital industrialization　数字产业化
Metavers　元宇宙
Blockchain　区块链
Digital Twin　数字孪生
cyberspace governance　网络治理
online service　在线服务
searching engine　搜索引擎
digital divide　数字鸿沟
spam message　垃圾电子邮件
e‑commerce　电子商务
Internet‑based　以网络为基础的
cyber phobia　计算机恐惧症
online transaction　网上交易
electronic platform　电子化平台
community portal　社区门户
online consultation　在线咨询
electronic banking　电子银行
e‑government　电子政务

（3）航空航天相关科技新闻常用词汇如下。
manned space station　载人空间站
space lab　空间实验室
rendezvous and docking　交会对接
carrier rocket　运载火箭
booster rocket　助推火箭
radar echoes　雷达回波
space docking　空间对接
lunar exploration program　探月工程
low‑Earth orbit　近地轨道
core module of the space station　空间站核心舱
reconnaissance drone　无人侦察机
extravehicular activities（EVAs）　出舱活动
space‑based lecture　太空授课
unmanned probe　无人探测
the launch tower　发射塔
definitive orbit　既定轨道

external tank　外壳
lunar module　登月舱
the re-entry capsule　返回舱
emergency landing　紧急降落
command module　指令舱
multistage rocket　多级火箭

（4）生命科学相关科技新闻常用词汇如下。
Alzheimer's disease　阿尔茨海默病
allosteric modification　别构修饰
general biology　普通生物学
cell growth factor　细胞生长因子
multidrug resistant tuberculosis　多重抗药性结核病，耐多药性结核病
cortical functional column　皮层功能柱
neuronal recognition　神经元识别
anticoagulant/decoagulant　抗凝剂
hemopoietic stem cell　造血干细胞
platelet aggregation　血小板聚集
ABO blood group system　ABO血型系统
community immunity　群体免疫
adoptive immunity　过继免疫
immunogenetics　免疫遗传学
gene map　基因图谱
gene position effect　基因位置效应
gene library　基因文库
sister chromatid　姐妹染色单体
community-acquired infection　社区获得性感染
respiratory system　呼吸系统
nucleic acid　核酸
immunosuppressive therapy　免疫抑制治疗
Computed Tomographic Pulmonary Angiography（CTPA）　CT肺动脉造影
glucocorticoid/glucocorticosteroid　糖皮质激素
Extracorporeal Membrane Oxygenator（ECMO）　体外膜氧合器
viral mutation　病毒变异
pulmonary consolidation　肺实变
terminal disinfection　终末消毒
clinical trial　临床试验

（5）其他学科科技新闻常用词汇如下。
deep-sea mining　深海采矿
hybrid rice　杂交水稻

non – road mobile machinery　非道路移动机械
self – service parcel pickup machine　自助快递箱
thermal imaging　热成像技术
heat signatures　热信号，热标记
gas boiler　燃气锅炉
blanket bog　覆被沼泽

3. 科技新闻中常见机构

（1）常见新闻机构如下。

AP：Associated Press　美国联合通讯社，美联社
CBC：Canadian Broadcasting Corporation　加拿大广播公司
ABC：Australian Broadcasting Corporation　澳大利亚广播公司
CNN：Cable News Network　美国有线电视新闻网
Reuters：　英国路透社
AFP：Agence France – Presse　法国新闻社，法新社

（2）通信电子方面常见机构如下。

ACM：Association for Computing Machinery　美国计算机协会
IEC：International Electrotechnical Commission　国际电工委员会
ECMA：European Computer Manufacturers Association　欧洲计算机制造商协会
CERT：Computer Emergency Response Team　计算机应急响应小组
ITU：International Telecommunication Union　国际电信联盟
UPU：Universal Postal Union　万国邮政联盟
INTELSAT：International Telecommunications Satellitie Organization　国际通信卫星组织

（3）航空航天方面常见机构如下。

NASA：National Aeronautics and Space Administration　美国国家航空航天局，美国宇航局，美国太空总署
ESA：European Space Agency　欧洲航天局
ROSCOSMOS：　俄罗斯联邦航天局
DLR：Deutsches Zentrum für Luft – und Raumfahrt　德国航空太空中心
JAXA：Japan Aerospace Exploration Agency　日本宇宙航空研究开发机构
CNES：Centre National d'Études Spatiales　法国国家空间研究中心
ISRO：Indian Space Research Organisation　印度空间研究组织
ASI：Agenzia Spaziale Italiana　意大利航天局
ICAO：International Civil Aviation Organization　国际民用航空组织，国际民航组织

（4）生态能源方面常见机构如下。

WEC：World Energy Council　世界能源理事会
UNEP：United Nations Environment Programme　联合国环境规划署，联合国环境署
IAEA：International Atomic Energy Agency　国际原子能机构
IEA：International Energy Agency　国际能源机构
OPEC：Organization of Petroleum Exporting Countries　石油输出国组织

NSG：Nuclear Suppliers Group　核供应国集团
CAS：Commission of Atmospheric Sciences　大气科学委员会
IAH：International Association of Hydrogeologists　国际水文地质学家协会
SSA：Seismological Society of America　美国地震学会
RO：Royal Observatory　英国皇家天文台

（二）科技新闻常用句型

be due to　是由于
bring about/on　引起
be able to/be capable of　能够，有能力
be unable to/be incapable of　不能，没有能力
have the ability/capacity/power to　有……能力，能够
be defined as　被定义为
define … as …　把……定义为……
be determined/decided/governed/fixed by　取决于
according to/in accordance with　按照，根据……的说法
as an example/illustration　作为（一个）例子
compensate for　补偿
with an accuracy of/to an accuracy of　精度达
carry out　实现
in spite of the fact that　虽然
irrespective of/regardless of　不论，不管
to sum up/on the whole　总而言之，总的看来
on condition of/in the event of　在……条件下，条件是……
in the event of　如果发生……，万一……
A new report submitted to… has found…　一份提交给……的新报告显示……
The report states　报告指出
Analysts say　分析人士表示
advocated by experts　被专家倡导
It is assumed that　人们假设，可以假设
A comparison between A and B shows that　对 A 和 B 进行了比较表明
There is evident that/There is evidence to show（indicate/suggest）that　有证据表明
These data lead us to a conclusion that/These data lead us to conclude that/These data enable us to conclude that/On the basis of these data/From these data　这些数据使我们得出结论……
be inversely proportional to/depend inversely on　与……成反比
… is (are) widely used/is (are) in wide use/find (s) wide application (use)　……得到广泛应用
We have performed/done/made/conducted a number of experiments to prove/test/verify/check　我们做了许多实验来检验

We have succeed in doing/We have been successful in doing　我们成功实现了

This experiment failed to show (demonstrate) /This experiment has not shown (demonstrated)　该实验未能证明

This is due to/is caused by/results from/is the result of/arises from　这是由于

This gives rise to/leads to/results in/brings about　这引起了

This should arouse/attract/gain/have/receive our attention　这一点应受到/引起重视/关注

This is a problem concerned with/concerning/related/relating to/bearing on/dealing with　这是与……有关的问题

It can be said with certainty that　可以断言

fall into this category/belong in this categorg　属于这一类的有

be used as/serve as/be used for ＋doing　（被）用作，（被）用来

四、科技新闻翻译技巧

（一）与原文风格保持一致

1. 保持严谨

科技新闻英语语言正式程度适中，有时还带有一些会话语体色彩，所以译文语言不可太雅，也不可过俗，译文需要符合科技新闻的特点，既客观严谨又不失生动，保持原文的语言风格。例如：

The network is probably the most important (and is definitely the most expensive) element of the internet of things infrastructure, but another ongoing debate is about where the information collected by the thousands (millions?) of sensors we'll connect will be turned into action or aggregated to form meaningful insight.

译文：网络很可能是物联网最重要的（当然，肯定是最贵的）基础设施，但当前另一个争议是，我们连接起来的成千上万（也或许是以百万计？）的传感器收集的信息在哪里可以操作，或者可以整合成有建设性的观点。

从原文两处括号注释中可以看出科技新闻的严谨性，在翻译时译者也应加注。但如果仅对括号中内容直译，比如分别译成"而且肯定是最贵的"与"百万？"，意思则不够明确，而且有翻译腔的嫌疑。所以，译者对第一处采取断句，对第二处采取增译，以消除读者在语感上的不适。

2. 运用增译法

英汉两种语言具有不同的表达方式，翻译时需要酌情在词量上进行增减。增译法即为在原文基础上添加必要的词、词组、分句或完整句。这当然不是随意进行增词，而是增加原文中虽无其词但确有其意的一些词，从而使得译文在语法、语言形式上符合目的语的习惯，并在文化背景、词语联想方面与原文一致起来，这样译文与原文就在内容和形式等方面对等起来。翻译时，有时为了明确原文的含义，需要通过增译"们""一些""许多"等把英语中表示名词复数的概念译出。例如：

Consumers might be a bit disappointed to find that the smartwatch a partner device reliant on being paired with a Samsung Android smartphone and tablet, rather than being the completely autonomous media and communication device many consumers were expecting and hoping for.

译文：许多顾客可能会有点儿失望，因为他们发现这款智能手表只不过是一个与三星安卓系统智能手机和平板配对的配套设备，而不是他们正期待的那种完全独立的媒体通信设备。

此处"Consumers"采用复数形式，那么在翻译时就应增加一些形容词来表示其隐含的数目，所以采用"许多"来进行增译。

3. 处理好科技新闻中的新词和生造词

科技新闻用语与时俱进，新词不断出现，而且和其他学科有所交叉，译者也应当与时俱进，了解科技时事，给出最新译文。科技新闻英语有自己特定的一套惯用词汇，因此译者应准确理解这些词汇在科技新闻英语中的特定含义，在翻译前仔细查阅相关科技词典，否则极易造成误译。例如：

The pregnancy is the result of an embryo screening technique.

译文：这次怀孕是一项胚胎甄别技术的结果。

这里，译文中出现"结果"一词显得非常唐突，读起来显然不合逻辑。"result"一词除有"结果"的意思外，还有"成果""效果"的意思。显然，这里应选择后者。译者在翻译时可能习惯性地用"结果"来套用"result"的翻译。这里应改译为："这次怀孕是胚胎甄别技术的成果。"

又如：Alipay Shocks Online Spenders (Jan. 28, 2013, 21*st* Century)。其中，"Alipay"的意思是"支付宝"，是国内的第三方支付平台。该词是随着网络购物的盛行而出现的新词汇。再如：Is Twitter Really That Big? (Jun. 9, 2011, 21*st* Century) "Twitter"是一家美国社交网络及微博客服务的网站，近年来随着国内微博的广泛应用，该词也常被提到。

（二）灵活简洁，重点突出

1. 运用隐喻的译法

科技新闻通常简明扼要，为节约版面、吸引读者，科技新闻标题通常醒目生动，吸引眼球。科技新闻标题通常为一句话，很少采用复杂的复合句式。运用隐喻的译法能使科技新闻标题更形象生动。以下是一些科技新闻标题翻译的例子。

In Iceland, CO_2 sucked from the air is turned to rock
冰岛工厂"魔法"，将二氧化碳变成石头

Electric air taxis to make their debut in Brazil's most congested city
"空中出租车"将现身巴西交通最拥堵的城市

Giant hyper-realistic 3D cat billboard appears in Tokyo
超逼真！日本街头现巨型3D大花猫

以上三个标题在翻译时，在译出原意的基础上，采用了形象比喻，使得标题更具吸引力。例如，如果将"air taxis"直译为"空气出租车"会使读者费解，译者在翻译时译为

"空中出租车",并用双引号引起来提醒读者,此处的"空中出租车"并非字面意义,使得语言风格更加活泼有趣。又如第三个标题中的"Giant hyper-realistic 3D cat billboard",本意是"巨型超现实3D猫广告牌",但是译者将其译为"巨型3D大花猫",处理手法非常巧妙。

隐喻的译法不仅多见于标题的翻译,在正文的翻译中也很常见。例如:

Once the most-used web browser, Internet Explorer had been on a steady downward trajectory for nearly two decades. Its share of the browser market fell below the 50% threshold in 2010 and now sits at about 5%, according to browser usage tracker NetMarketShare. Google's Chrome is the browser leader, commanding a 69% share of the market.

译文:IE浏览器曾经是应用最广泛的网络浏览器,但近二十年来使用量一直不断下降。浏览器使用量追踪机构NetMarketShare的数据显示,2010年IE浏览器在浏览器市场的份额跌破50%大关,如今只剩下5%的份额。谷歌浏览器是浏览器中的领头羊,占据了69%的市场份额。

上述例子中的"leader"被译为"领头羊",比直译为"领导者"更生动。

2. 运用拆句法

举例如下。

Currently, IPv6 also provides a lot less geolocation data than IPv4 does. This is really just a temporary state of affairs, however, there is so such more IPv4 penetration that it is easy for geolocation database builders to identify the geographical location associated with an IPv4 address. (From IPv6 will allow them to track you down. Not!)

译文:目前,IPv6提供的地理定位信息远少于IPv4。不过这只是暂时的,因为IPv4的渗透率很高,地理定位数据库创建者很容易识别与IPv4地址相关联的地理位置。(通过IPv6将允许他们跟踪你。不!)

如果直译画线部分,则应译为"IPv4的渗透率如此之高以至于地理定位数据库创建者很容易识别与IPv4地址相关联的地理位置",译文过长,读起来吃力,明显不符合汉语科技新闻简明易读的特点。所以,译者将"so...that"转译为"因为……"加在最前面。

3. 巧用四字格

采用四字格,这完全是从四字格的优点出发的。四字格的运用可以使文章增添不少生花之笔,符合中国人的阅读习惯。在忠实于原文的基础上,发挥译文语言优势,运用四字格不但应在其他普通翻译中提倡,而且也应在科技新闻翻译中提倡。例如:

The contrast with fuel cells could not be greater.

译文:与燃料电池相比,差距再大不过了。

"could not be greater"意为"十分大""再大不过了",改译后为:"与燃料电池相比,简直有天壤之别。"用四字格"天壤之别"来进行替换,改译后译文就更能突出差别之大,将原文的意思淋漓尽致地表达清楚,译文也更具文采和可读性。

4. 引语的翻译

新闻报道中的直接引语是"未经任何介质改变的现实",新闻工作者转述或引述某人说话或写作的内容,会使得新闻内容更加客观,使新闻报道更具专业性和可信度。直接引语在形式上有引号标识,报道者必须准确无误地完整转述说话者的语言和思想。例如:

In an interview ahead of the BloombergNEF London Summit, Mr. Llewellyn said: "We already see massive increases in the amount of renewable energy being produced across the world. Wind energy production has multiplied by two over the last five years and solar energy production has multiplied by four."

译文：在彭博新能源财经伦敦峰会开幕前接受采访时，卢埃林称："我们已经看到世界各地产出的可再生能源在大量增加。过去五年风能产出量增加了一倍，太阳能产出量增加了三倍。"

"At 1.5℃, there's a good chance we can prevent most of the Greenland and west Antarctic ice sheet from collapsing," said climate scientist Michael Mann at Pennsylvania State University.

译文：宾夕法尼亚州立大学的气候科学家迈克尔·曼恩说："如果升温幅度在1.5℃以内，我们还有可能防止格陵兰岛和南极西部的大部分冰盖融化崩塌。"

间接引语是由新闻编译者转述别人的意思或陈述说话者的意图。英汉科技新闻编译过程中，或者将源语直接引语合并，合二为一或合多为一，或者将源语直接引语转换为间接引语，以达到融合相近信息、缩短篇幅的效果。这种处理可以使句子变得更为简洁，加快语流，实现他者话语和叙述者话语之间的顺畅过渡。使用无引号型间接引语代替标准型直接引语，可以使读者的阅读过程更为流畅。例如：

中国科学院深海科学与工程研究所称，中国深海载人潜水器"奋斗者"号11月10日在西太平洋马里亚纳海沟最深处成功下潜至10 909米，创下国内新纪录。被称为"挑战者深渊"的马里亚纳海沟最深处深度为11 000米。

译文：China's deep - sea manned submersible Fendouzhe, or Striver, set a national diving record of 10, 909 meters on November 10 in the Challenger Deep, a 11, 000 - meter chasm located at the bottom of the Mariana Trench in the western Pacific Ocean, according to the Institute of Deep - sea Science and Engineering under the Chinese Academy of Sciences.

Critics argue it's a less sociable shopping experience. But during the pandemic, that's a bonus.

译文：批评者认为，这种购物方式缺乏社交互动。但在疫情暴发期间，这却是一件意外的好事。

Experts say the bug may have arrived on imported goods or as part of a yet undiscovered local population.

译文：专家认为，这种虫子可能是通过进口商品传入的，或者是本地尚未发现的一批蟑虫的一部分。

Mark Zuckerberg said the existing brand could not "possibly represent everything that we're doing today, let alone in the future", and needed to change.

译文：扎克伯格表示，现有的品牌名称无法"代表我们现在所做的所有业务，更不用说将来的业务",所以需要更改。

五、课后练习

(一) 思考题

(1) 科技新闻有哪些文本特点?
(2) 科技新闻中的引语有几种处理方式?

(二) 翻译练习

1. 词组翻译

cancer blood test trial
cogeneration unit
extended reality
a clean and upright cyberspace
the quantum revolution
illegal pop–up ads
holographic projections
origin–tracing of the novel coronavirus
unmanned aerial vehicle express/drone delivery
virtual local area network
General Packet Radio Service (GPRS)
compressed natural gas
liquefied natural gas
polymer concrete
process flow diagram
raster image processor
magnetic resonance imaging
Convention on International Trade in Endangered Species of Wild Fauna and Flora
reinforced reaction injection molding
computational fluid dynamics
pressurized water reactor
partial differential equation
Fast Fourier Transformation (FFT)
Acquired Immune Deficiency Syndrome (AIDS)

2. 标题和句子翻译

(1) 标题翻译。

Facebook changes its name to Meta

Could a chewing gum reduce COVID – 19 spread? Researchers believe it can

Supermarkets using cardboard cutouts to hide gaps left by supply issues

（2）句子翻译。

①Instead of being on a computer, people in a metaverse might use a headset to enter a virtual world connecting all sorts of digital environments.

②Chinese scientists have created the world's first integrated space – to – ground quantum network that can provide reliable, ultrasecure communication between more than 150 users over a total distance of 4,600 kilometers across the country, according to a study published in the journal *Nature* on Jan. 7.

③Dogs, for their part, have more than eight different antigens that can attach to their red blood cells, most of them labeled Dog Erythrocyte Antigen (DEA 1.1, 1.2, 3, 4, 5, 6 and 7). Often individuals within a specific breed of dog will have the same blood type — for instance, 60 percent of greyhounds fall into the DEA 1.1 negative blood group. But new canine blood groups are still being detected — the recently discovered Dal blood group, for example, is only found in Dalmatians.

④A large asteroid striking our planet is an extremely rare event, but the consequences of a direct hit could be catastrophic. A rock measuring 150 meters across could release the energy of several nuclear bombs. Even larger objects could affect life across the world.

⑤The solution fills the rock's cavities and the solidification process begins—a chemical reaction turning it to calcified white crystals that occurs when the gas comes in contact with the calcium, magnesium and iron in the basalt.

⑥The CR400BF – G Fuxing bullet trains are designed to withstand the climate in extremely cold areas that may be hit by blizzards and temperatures as low as $-40℃$. Designers chose materials with better airtightness for the equipment compartment, which can keep out the snow and cold air.

⑦Using motion capture, force plates and electromyography, which record the electric activity of muscle tissue, the team showed the device reduced activity in the lower back muscles by an average of 15 to 45 percent for each task.

⑧Mandatory fortification will mean everyone eating foods like bread in the UK, will get a dose. There was a concern that for some people, particularly the elderly, boosting folic acid might have unintended negative consequences, such as covering up the symptoms of vitamin B12 deficiency.

⑨第三代杂交水稻最突出的一个特点是生长周期缩短。中国此前的高产量杂交水稻从种植到收割需要160天，甚至180天，而第三代的生长周期缩短到约125天。

3. 篇章翻译

（1）中译英。

①华为发布鸿蒙手机操作系统　有什么特点？

6月2日，华为正式发布HarmonyOS 2（鸿蒙）及多款搭载该操作系统的新产品。

鸿蒙是一个开源操作系统，可用于多个设备和场景。2019年8月，鸿蒙操作系统发布，主要用于可穿戴设备和平板等物联网设备。鸿蒙系统是新一代智能终端操作系统，为不同设备的智能化、互联与协同提供统一的语言。鸿蒙更便捷、更流畅、更安全。

　　武汉大学计算机学院助理教授赵小刚表示，鸿蒙创造出了"超级终端"，使多设备互联更加简化、高效。赵小刚说："鸿蒙大大提升了多设备的交互速度，提高了设备算力的利用率，从而提供了一个更加优化的多设备用户体验。"华为常务董事、消费者业务CEO余承东表示："随着每个人身边的智能设备越来越多，我们已步入万物互联时代。没有人是一个孤岛，每个人、每个设备都是万物互联大陆的一部分。我们希望与更多合作伙伴、开发者共同繁荣鸿蒙生态。"覆盖家电家居、运动健康、出行、娱乐、教育等领域的中国产业界对鸿蒙系统及其产业生态表现出积极态度。今年5月，中国家电业巨头美的集团宣布计划在年内推出近200款搭载鸿蒙系统的产品进入市场。华为计划在2021年实现将鸿蒙系统覆盖3亿台设备，其中2亿台为自有设备。

　　②中国建星地量子通信网

　　中国科研团队成功构建出全球首个星地量子通信网，可为用户提供可靠的、"原理上无条件安全"的通信。整个网络总距离4600公里，目前已接入150多家用户。该成果已于1月7日在英国《自然》杂志上刊发。在中国科学技术大学潘建伟的带领下，科学家们为此进行了数年研究。这项研究的审稿人评价称，这是地球上最大的量子密钥分发网络，这一成就"令人惊叹"且"具有前瞻性"，也代表着科学家朝着构建实用性广域量子通信网络迈出了重要一步。研究人员在论文中指出，在量子通信网络领域，视频通话、音频通话、传真、文本传输和文件传输等多项服务已实现技术验证和实用展示，预计不久将用于商业用途。

　　③我国的嫦娥五号探测器将择机实施月面软着陆，开展我国首个地外天体采样任务。来自国家航天局的消息，11月30日4时40分，嫦娥五号探测器着陆器和上升器组合体与轨道器和返回器组合体顺利分离，截至目前，探测器各系统状态良好，地面测控通信正常。着陆器和上升器组合体将择机实施月面软着陆，进行自动采样等后续工作。轨道器和返回器组合体将继续在高度约200公里的环月轨道上飞行并等待和上升器交会对接。嫦娥五号探测器于11月24日发射升空，是我国航天史上最为复杂、挑战最大的任务之一，也是40多年来世界首个月球采样返回任务。

　　④两会期间，虚拟现实和增强现实产品吸引了前来报道的记者们。这些产品在两会报道中心所在宾馆的一层展出。展出方是今日中国网络电视台。虚拟现实/增强现实产业在中国蓬勃发展。中国东部南昌市去年建立了虚拟现实产业基地，发起了10亿元人民币的天使投资基金，并落实100亿元人民币规模的虚拟现实产业投资基金。全国人大代表、南昌市市长郭安表示，虚拟现实技术能像互联网和智能手机那样改变人们的生活。他说："我们已经建立了完整的产业链，有超过50家企业和机构入驻产业基地。"为了培养人才，南昌市与高校合作，培训相关人员。"我们希望能在未来三年培训出一万名有虚拟现实技术基础知识的人员。"他说。虚拟现实技术可广泛应用于人工智能、教育培训、医疗、游戏、旅游和虚拟社区等领域。

（2）英译中。

①Online Calculator Developed That Can Predict When Older Adults Will Die

An online calculator that can help predict when older adults will die has been developed.

The algorithm is called "Risk Evaluation for Support: Predictions for Elder – Life in the Community Tool (RESPECT)".

The data is based on more than 491,000 older adults who used home care between 2007 and 2013 and is focused on people who are likely to die within the next five years.

The calculated life expectancy can be as low as four weeks for people who are very frail.

People are asked whether they have been diagnosed with diseases like stroke, dementia, or hypertension and whether abilities to carry out tasks over three months had decreased.

The ability to make decisions is also asked and whether they have suffered vomiting, swelling, shortness of breath, unplanned weight loss, dehydration or loss of appetite.

Researchers found declines in a person's ability to carry out activities of daily living were stronger predictors of six-month mortality than the diseases that a person has.

Dr. Amy Hsu, an investigator at the Bruyere Research Institute and at the University of Ottawa in Canada, said: "The RESPECT calculator allows families and their loved ones to plan." "For example, it can help an adult child plan when to take a leave of absence from work to be with a parent or decide when to take the last family vacation together."

Dr. Peter Tanuseputro, at the Ottawa Hospital, said: "Knowing how long a person has to live is essential in making informed decisions about what treatments they should get and where they should get them." "As a person gets closer to death, the balance shifts from having curative care as the primary goal, to care that maximizes a person's quality of remaining life."

②A Pressing Problem: The Pros and Cons of Placebo Buttons

Over the years, New York City authorities phased out most of the city's buttons that controlled crosswalk lights, but kept it quiet. They decided that computer-controlled timers worked better. By 2004, fewer than 750 of the 3,250 buttons still worked. But the city did not remove the buttons that no longer worked, leaving countless fingers in vain.

Initially, the buttons survived because of the cost of removing them. But it turned out that even inoperative buttons serve a purpose. Pedestrians who press a button are less likely to cross before the green man appears, says Tal Oron-Gilad of Ben-Gurion University of the Negev, in Israel. Having studied behaviour at crossings, she notes that people more readily obey a system which purports to heed their input.

Inoperative buttons produce placebo effects of this sort (the word placebo is Latin for "I shall be pleasing") because people like an impression of control over systems they are using, says Eytan Adar, an expert on human-computer interaction at the University of Michigan, Ann Arbor. Dr. Adar notes that his students commonly design software with a clickable "save" button that has no role other than to reassure those users who are unaware that their keystrokes are saved automatically anyway. Think of it, he says, as a touch of benevolent deception to counter the inherent coldness of the machine world.

That is one view. But, at road crossings at least, placebo buttons may also have a darker side. Ralf Risser, head of factum, a Viennese institute that studies psychological factors in traffic systems, reckons that pedestrians' awareness of their existence, and consequent resentment at the deception, now outweighs the benefits.

Something which happened in Lebanon supports that view. Crossing buttons introduced in Beirut between 2005 and 2009 proved a flop. Pedestrians wanted them to summon a "walk" signal immediately, rather than at the next appropriate phase in the traffic – light cycle, as is normal. The authorities therefore disabled them, putting walk signals on a preset schedule instead. Word spread that button – pressing had become pointless. The consequent frustration increased the amount of jaywalking, says Zaher Massaad, formerly a senior traffic engineer for the Lebanese government.

Beirut's disabled buttons are, says Mr. Massaad, now being removed. They should all be gone within three years. New York has similarly stripped crossings of non – functioning buttons, says Josh Benson, the city's deputy commissioner for traffic operations, though it does retain about 100 working ones. These are in places where pedestrians are sufficiently rare that stopping the traffic automatically is unjustified. However, internet chatter about placebo buttons has become so common that doubt, albeit misguided, seems to be growing about even these functioning buttons' functionality. This suspicion, says Mr. Benson, has spread beyond New York, to include places such as Los Angeles, where almost all the crossing buttons have always worked, at least during off – peak hours.

Truth be told, though, the end may be nigh for all road – crossing buttons, placebo or real. At an increasing number of junctions, those waiting to cross can be detected, and even counted, using cameras or infrared and microwave detectors. Dynniq, a Dutch firm, recently equipped an intersection in Tilburg with a system that recognizes special apps on the smartphones of the elderly or disabled, and provides those people with 5 to 12 extra seconds to cross. That really will be pleasing.

学思践悟

第五章 科技专利文献

科技专利文献是记载科技方面专利申请、审查、批准过程中所产生的各种有关文件的文件资料。随着全球经济一体化进程的加速，世界各国在科学技术、经济、文化、教育等方面的交流日益增多。无论是专利技术的引进，还是产品进出口，都需要查阅相关专利文献。因此，专利文献的翻译需要译者高度重视。

狭义的专利文献指包括专利请求书、说明书、权利要求书、摘要在内的专利申请说明书和已经批准的专利说明书的文件资料。广义的专利文献还包括专利公报、专利文摘，以及各种索引与供检索用的工具书等。专利文献是一种集技术、经济、法律三种情报为一体的文件资料。专利文献是发明、技术创新、新产品、新工艺和新方法的真实书面记录，是必要的技术文件。它汇集了世界上大多数发明和尖端技术，覆盖面广，内容详细，具有实用性、可靠性和及时性等特点。

一、科技专利文献文本介绍

（一）科技专利文献概述

作为非文学文本的一种，科技专利文献采用书面文本，按一定的格式撰写，有着其特定的表达方式，内容布局大同小异。科技专利文献涉及的通常是某一专业领域的发明创造、新技术、新材料、新方法、新产品或新设备，要求发明人必须将发明的内容全部公开，并达到该专业领域的技术人员凭借其内容即可基本上付诸实践的程度。科技专利文献是一种非常重要的文献资料。

（二）科技专利文献文本结构

科技专利文献行文格式严谨、规范化、国际化，可起法律文件和交流技术成果的作用。其特点是具有完整性、先进性和规范性。科技专利文献通常包括三个部分：标头、正文和权项。

1. 标头

标头具体地著录该专利的日期、分类、批准等事项，是申请专利、查阅专利必不可少的。标头用语简洁明了，具有规范性，即固定的格式。标头均位于首页。标头通常有专利号、国别标志、申请日期、申请号、国际专利分类号、专利标题和申请人七项。

2. 正文

正文是专利发明的核心部分，由前言（introduction）、发明背景（background of the invention）、发明概要（summary of the invention）、对发明的详细解释（detailed description of the invention)、实例（examples）几个部分组成。

3. 权项

权项通常为科技专利文献的最后部分，是专利申请人要求专利局对他的发明给予法律保护的项目，要求清楚、完整、准确。请求权项是申请人请求专利保护的技术范围，即申请人为其发明而要求享有独占权的实质性内容。在专利被批准后，权项具有直接的法律作用，故而权项文字庄重、正式、简明、准确。

(三) 科技专利文献的作用

专利制度的作用完全体现在科技专利文献中。同时，科技专利文献在专利审查和全球交流中发挥着关键作用。

1. 是体现专利制度根本目的的媒介

专利制度的基本功能：一是有效地保护发明创造，发明人将其发明申请专利，专利局依法将发明创造向社会公开，授予专利权，给予发明人在一定期限内对其发明创造享有独占权，把发明创造作为一种财产权予以法律保护；二是可以提高公民、法人搞发明创造的积极性，充分发挥全民族的聪明才智，促进国家科学技术的迅速发展；三是有利于发明创造的推广应用，促进先进的科学技术尽快地转化为生产力，促进国民经济的发展；四是促进发明技术向全社会的公开与传播，避免对相同技术的重复研究开发，有利于促进科学技术的不断发展。专利制度的根本目的是推动科学技术的进步，这一根本目的是通过在法律保障下以专利文献为媒介公开通报新的发明创造体现出来的。只有连续不断公开、出版新的专利文献，促进发明创造技术的传播，才能更好地推进专利制度的实施。

2. 为制定经济、科技发展规划提供依据

利用专利文献可以在制定科研计划及确定科研课题时，帮助我们去伪存真，明确研究方向，提高技术创新活动的起点，避免盲目性和重复性研究。通过查阅专利文献还可以开阔思路、激发灵感，借鉴前人的智慧，在现有技术的基础上做出新的发明创造，进而起到促进全社会科技进步的作用。充分利用专利文献可以缩短60%的科研周期，节约40%的科研经费。

通过对专利文献的分析研究，可以为政府机构科学地制定我国经济、科技发展规划及其重大战略提供决策依据，为技术创新、产业结构调整，以及生产、经营及科研发展提供科学参考。利用专利文献可以帮助我们在产品出口和日益频繁的贸易壁垒中规避侵权、正确应对；帮助我们在技术引进过程中正确选择、准确评估，避免造成不必要的经费损失。专利文献在知识产权战略与重大事项知识产权预警制度的制定与实施中发挥着重要的支撑作用。

3. 为发明创造提供借鉴和法律保护依据

企业是创新的主体，专利是创新的成果。在建设创新型国家过程中，企业不能盲目跟进，要借鉴前人的智慧，站在巨人的肩膀上，进行再创造。在这方面，专利文献可以起到一定借鉴作用。专利文献含有每一项申请专利的发明创造的具体技术解决方案（说明书）。在专利文献中记载了从航天、生物等高科技到生活日用品各方面的发明创造。研究自身领域的专利文献，对企业创新具有非常重要的作用：不仅可使企业避免重复研究，同时还可启迪企业研究人员的创新思路，提高创新的起点，实现创新目标。

专利制度以公开为条件，依法给发明创造以法律保护，并以专利文献作为印证发明创造受法律保护的文件。每件特定的专利文献确定了每项发明创造的权利保护范围及法律效力，通过实施其受保护的发明成果获得商业利益最大化。不论是在法院审理专利侵权案件时，还是在地方专利管理部门处理侵权纠纷和查处假冒专利行为时，都必须以专利文献作为实施专利保护的根据。

4. 提醒竞争对手，保护知识产权

人们申请专利的目的是寻求对其发明创造的保护。绝大多数专利申请人是基于以下认识申请专利的：专利制度赋予申请人专利权以保护其劳动成果，因此他们可以在专利制度的保护下，推广其专利发明以获得最大的商业利益。然而，专利权人最担心的是竞争对手侵犯其专利权。所以，专利权人寄希望于通过专利文献信息公布，提醒竞争对手。专利文献不仅向人们提供了发明创造的技术内容，同时也向竞争对手展示了专利保护范围。甚至许多人为其专利产品设计了独特的标记，帮助使用者很快找到产品说明书，了解其专利保护的内容，从而达到保护知识产权的目的。

5. 借鉴权利信息，避免侵权纠纷

任何竞争对手都要尊重他人的知识产权，杜绝恶意侵权行为，避免无意侵权过失，以形成良好的市场竞争氛围。专利文献可以起到这方面的借鉴作用。专利文献含有每一件专利的保护范围信息（权利要求书）、专利地域效力信息（申请的国家、地区）、专利时间效力信息（申请日期、公布日期）。专利文献信息恰似一面镜子，只要随时照一照（检索专利的法律信息），就可以实现自我约束，避免纠纷发生。

（四）科技专利文献的特点

1. 语篇特点

科技专利文献无论是在形式上还是在内容上都具有区别于其他文本类型的特殊之处。科技专利文献的语篇特点如下。

（1）专利文献集技术、法律、经济情报为一体。每一件专利文献都记载着解决一项技术课题的新方案，同时也包含发明所有权和权利要求范围的法律状况。通过查阅专利文献，可以找到购买许可证的对象和地址；也可根据专利的分布情况，分析技术销售的规模、潜在市场行情、发明的经济效益和国际间的竞争范围。

（2）专利文献的技术内容新颖、先进和实用。专利文献记载了从生活日用品到尖端科技的一系列应用技术和技术科学，反映了国内外首创的最新成果和技术。专利法规定专利说明书的内容必须具备新颖性、创造性和实用性。新颖性指该项发明创造从未公开发表过、使用过，也没有被公众所知；创造性指该项发明创造比同领域的技术都要先进，具有独创性，且具有实质性的特征和显著的进步；实用性则指该项发明创造可以转化为生产力并产生良好的效果。

（3）专利文献的内容广泛、完整和详尽。专利文献是人类关于科学技术和应用的发展史的记录，各个时期的新发明、新技术、新工艺和新设备大多反映在专利文献中。因此，专利文献涉及的学科范围十分广泛。专利文献的内容记录要全面、具体、详细，便于使用者掌握专利的某项技术及细节，并据以实施。

2. 文本特点

科技专利文献兼具法律与科技语言的文本特点，在结构与表达上具有一定程度的程式化，正式程度高。程式化指同类文章的基本表述方式大致相同。科技专利文献的构成要素大致上是相似的。例如，专利说明书包含技术领域、背景技术、发明或实用新型技术的内容、附图说明及具体实施方式等项目，各个项目都有其各自的表达特征与方式。科技专利文献语言严谨、简练、概括性强，并且具有很强的专业性，在表达上有其独特的句式，如"本申请基于×××年×月××日提交的日本专利申请（申请号××××××）和×××年×月××日提交的日本专利申请（申请号××××××），它们的内容引入本文作为参考。本说明书中引用的文献的内容同样引入参考""在下文中，将参照附图详细地描述本案例的公开实施过程"等。科技专利文献的文本特点如下。

（1）新名词。由于绝大多数新技术首次出现在专利申请文件（主要是外国专利申请文件）中，对这些技术名词的准确翻译就成为一个很大的问题。我国科技名词审定主要是依靠广大专家和学者进行的，目前有2000多位科学家参加科技名词审定工作。译者在翻译专利申请文件时，一般情况下，要采用经全国科学技术名词审定委员会公布的科技名词。所以，建议译者在翻译技术名词时，在尽量准确表达的情况下，应将委托稿件中的原名称在括号中标出，以便日后检索。

（2）自定义词。专利申请文件的起草者通常会有这样的目标：在有可能被侵权的情况下，这份专利申请文件要能用来阻止怀有这种企图的人制造、使用、销售或进口要求保护的技术主题，或者是通过许可等方式为专利权人带来报酬。因此，准确、清楚地描述技术方案就显得非常重要了。如果他们认为现有的名词可能与其本意不完全一致，就会在专利申请文件中对这些词的含义进行限定。在专利申请文件的翻译中，时常会遇到专利申请文件起草者对词的重新定义，译者要特别注意对这种情况的处理，首先要理解原意，然后要用恰当的手段和适当的汉语表达出来。

（3）长句子。在专利文献中，句子最长的就是权利要求了。目前，各国都会要求一个权利是一句话，即只能有一个句号。如果在一项权利要求中有两个以上的句号，就意味着有两个以上技术方案。中国专利文献的情形则有所不同：在中国的专利文献中，权利要求书是一个独立的文件，每项权利要求前都有数字标出。但在翻译说明书时，译者就不必完全拘泥于英语的句式，可以按照汉语语言习惯，适当变换句子的结构，以求表述易于阅读、更加清楚。

（4）被动语态。在英文专利文献中，为了显示客观公正而大量使用被动句。但是，"be + 过去分词"并不一定都是被动语态，有时是系表结构表示状态。当"be + 过去分词"表示动作时为被动语态，be是助动词，be后面的过去分词是主要动词，动作的对象是主语；当"be + 过去分词"表示主语所处的状态时为系表结构。

（5）时态变化。专利文献中最常用的时态是一般现在时，这主要是由于专利文献是对结构、过程等的客观描述，这些客观性的内容通常是没有时间性的。在这种情况下使用一般现在时，给人以"无时间性"的概念，以排除任何与时间关联的误解。可以说，一般现在时正是适应了这些"无时间性"的"一般叙述"的需要。在专利申请文件中，一般现在时和一般过去时是两种主要的时态，有时还可能会有少量的现在完成时和一般将来时，而其他时态较少。虽然时态变化不多，但翻译时还得小心，已经发生的事情和将要发生的事情，还是有很大区别的。

（五）科技专利文献等级分类及类型

1. 科技专利文献等级分类

《国际专利分类法》（International Patent Classification, IPC）是用于专利文献分类的等级列举式分类法，又译《国际专利分类表》。IPC 将与发明专利有关的全部技术内容按部、分部、大类、小类、主组、分组等逐级分类，组成完整的等级分类体系。全表共分 8 个部、20 个分部，以 9 个分册出版。1~8 册为分类详表，第 9 册为使用指南及分类简表（至主组一级）。IPC 的部（一级类）用 A~H 表示。分部仅是分类标题，未用标记。其中，A 部为生活必需品，B 部为各种操作、运输，C 部为化学和冶金，D 部为纺织和造纸，E 部为永久性构筑物，F 部为机械工程、照明、加热、武器、爆破，G 部为物理学，H 部为电学。从上述来看，专利文献几乎涵盖我们生活的方方面面，专业性很强。由于科技专利文献涉及发明创造及科学技术领域，因此有极高的技术含量，包含大量的专业术语。

世界上每年发明创造成果的 90%~95% 可以在专利文献中查到，而且约 80% 的发明成果仅通过专利文献公开，并不见诸其他文献。各国专利机构每年公布出版的专利文献已接近 300 万件，全世界可查阅的专利文献已达到 7000 万件。

2. 科技专利文献类型

根据专利文献的内容性质和加工程度的不同，专利文献可以归纳成以下几类。

（1）一次专利文献。一次专利文献是指首次出版的各种专利文献，也称原始专利文献。一次专利文献是指各工业产权局、专利局及国际（地区）性专利组织出版的各种专利单行本，包括授权发明专利、发明人证书、医药专利、植物专利、工业品外观设计专利、使用证书、实用新型证书、补充专利或补充发明人证书、补充保护证书、补充实用证书的授权单行本及其相应的申请单行本。一次专利文献统称专利单行本。

专利单行本是专利文献的主体，有出版发行和内部查阅两种形式。专利单行本的主要作用是完整、清晰地公开新的发明创造，请求或确定法律保护的范围。

（2）二次专利文献。二次专利文献是指各工业产权局、专利局及国际（地区）性专利组织出版的专利公报、专利文摘和专利索引等出版物。

①专利公报有广义和狭义之分。广义上是指专利公报、实用新型公报、外观设计公报，或者指其总和，即工业产权公报。狭义的专利公报仅指报道有关专利申请的审批情况及相关法律法规信息的定期出版物。专利公报通常以著录项目、著录项目与文摘或著录项目与权利要求的形式报道新的发明创造，因而分为题录型、文摘型、权利要求型三种类型。

②专利文摘通常为题录型专利公报的补充出版物，作为报道最新专利申请或授权专利的技术文摘，它与题录型专利公报同步出版。

③专利索引是指各工业产权局、专利局及国际（地区）性专利组织以专利文献的著录项目为条目编制的检索工具，按出版周期划分为专利年度索引、专利季度索引、专利月索引等，按索引编制条目划分为号码索引、人名索引、分类索引等。

二次专利文献的主要作用是帮助用户快速、有针对性地从一次专利文献中寻找、选择所需要的文献，了解发明创造的主要内容，避免可能的侵权行为，跟踪有关专利申请的审批情况等动态的法律信息。

（3）专利分类资料。专利分类资料是指按发明技术主题对专利申请进行分类和对专利文献进行检索的工具。专利分类资料包括专利分类表、分类表索引、工业品外观设计分类表等。

(六) 专利说明书组成部分

目前，各国出版的专利说明书在格式上和内容上趋于统一，各种专利说明书基本上包括以下几个部分。

1. 专利文献著录项目

专利文献著录项目是刊在专利说明书扉页上的表示专利信息的特征。

2. 说明书

说明书是清楚完整地描述发明创造技术内容的文件部分。

3. 权利要求书

权利要求书是确定申请人请求专利保护的发明创造技术特征范围的文件，也是判定他人是否专利侵权的法律依据。

4. 附图

附图是用于补充说明书文字部分的文件。

5. 摘要

摘要是对说明书所述发明创造的重点内容或主要技术特征的简明介绍。

6. 检索报告

检索报告是审查员在对专利申请文件中描述的发明创造进行有关技术水平的文献检索后，报告检索结果的文件。

(七) 专利文献著录项目及其代码

专利文献著录项目（INID）是专利局为揭示每一项专利或专利申请的技术情报特征、法律情报特征及可供人们进行综合分析的情报线索而编制的款目。简单地说，专利文献著录项目实质上就是用以表示专利情报的特征。在专利文献著录项目所表示的专利情报特征中，技术情报特征是主要组成部分，它包括用以表示有关申请专利的发明创造的内容的各种标志，如专利分类号、发明题目、摘要、相关文献、关键词等；法律情报特征则是重要组成部分，它包括用以表示有关专利权的各种标志，如申请号、申请日期、优先权、申请人、发明人、专利权人、专利代理人等。专利文献著录项目还包括一些其他专利情报特征，如出版国家、出版日期、文件号等。

在"发明专利申请公开说明书"中会用到专利文献著录项目和代码，常见的如下（一字线左侧是代码，右侧是专利文献著录项目）。

［11］—文献号（如专利号、公开号、公告号）

［12］—文献种类

［19］—国别

［21］—申请号

［22］—申请日期

［31］—优先权申请号

［32］—优先权申请日期

[33]—优先权申请国家或组织代码

[43]—未经审查或尚未授权的专利文献的公开日

[51]—国际专利分类号

[71]—申请人

[72]—发明人

[74]—专利代理机构和代理人

[85]—PCT 申请进入国家阶段日期

[86]—PCT 国际申请的申请数据

[87]—PCT 国际申请的公布数据

（八）专利文献公开号

专利文献公开时，会有一个公开号，公开号有四个部分。

（1）第一部分：国家代码。国家代码为国家英文名称的缩写，如表2所示。

（2）第二部分：第1位数字代表专利申请类型。

（3）第三部分：第2~7位数字代表流水号。

（4）第四部分：专利文献种类标识代码。公开号的末尾有一些字母，如A、B、S、U等，这些字母是专利文献种类标识代码，含义如下。

A—发明专利申请公布

B—发明专利授权公告

C—发明专利权部分无效宣告的公告

U—实用新型专利授权公告

Y—实用新型专利权部分无效宣告的公告

S—外观设计专利授权公告或专利权部分无效宣告的公告

表2 国家英文名称及代码

国别	中国	英国	日本	瑞典	俄罗斯	法国	美国	奥地利	澳大利亚
英文名称	China	The United Kingdom of Great Britain and Northern Ireland	Japan	Sweden	Russia	France	United States of America	Austria	Australia
代码	CN	GB	JP	SE	RU	FR	US	AT	AU

二、科技专利文献范文展示

范文一

技术领域：

本发明涉及一种试电笔，便于区分安危电压。

Technical field：

The invention relates to a test pen, which is convenient to distinguish the safe voltage from the dangerous voltage.

背景技术：

目前，公知的试电笔是由测试触头、限流电阻、氖管、金属弹簧和手触电极串联而成的。将测试触头与被测物接触，人手接触手触电极，当被测物相对大地具有较高电压时，氖管启辉，表示被测物带电。但是，很多电器的金属外壳不带有对人体有危险的触电电压，仅表示分布电容和（或）正常的电阻感应产生电势，使氖管启辉。一般试电笔不能区分有危险的触电电压和无危险的感应电势，给检测漏电造成困难，容易造成错误判断。

Background technology:

At present, the well-known structure of the test pen is composed of test contact, current limiting resistor, neon tube, metal spring and hand electrode in series. Contact the test contact with the object under test, and touch the hand electrode with your hand. When the object under test has a high voltage relative to the ground, the neon tube glows, indicating that the object under test is charged. However, the metal enclosures of many electrical appliances do not carry a voltage of shock that is dangerous to human body, only indicating that a distributed capacitance and/or normal resistance induce an electric potential to make neon tubes glow. General test pen cannot distinguish the dangerous electric shock voltage from the non-dangerous induced potential, causing difficulties in detecting leakage, easy to cause wrong judgment.

发明内容：

为了克服现有的试电笔不能区分有危险的触电电压和无危险的感应电势的不足，本发明提供一种试电笔，该试电笔不仅能测出被测物是否带电，而且能方便地区分是危险的触电电压还是无危险的感应电势。

Content of invention:

In order to overcome the deficiency that the existing test pen cannot distinguish the dangerous electric shock voltage from the non-dangerous induced potential, the present invention provides a test pen, which can not only measure whether the object under test is charged, but also can conveniently distinguish the dangerous electric shock voltage from the non-dangerous induced potential.

本发明解决其技术问题所采用的技术方案是：

在绝缘外壳中，测试触头、限流电阻、氖管和手触电极电连接，设置一条分流电阻支路，使测试触头与分流电阻一端电连接，分流电阻另一端与人体可接触的识别电极电连接。当人手同时接触识别电极和手触电极时，使分流电阻并联在测试触头、限流电阻、氖管、手触电极电路测试。当人手只和手触电极接触时，氖管启辉，表示被测物带电。当人手同时接触手触电极和识别电极时，若被测物带有无危险高电势，由于电势源内阻很大，从而大大降低了被测物的带电电位，则氖管不启辉；若被测物带有危险触电电压，因其内阻小，接入分流电阻几乎不降低被测物的带电电位，则氖管保持启辉，达到能够区分安危电压的目的。

The technical solution adopted by the invention to solve the technical problem is as follows:

In an insulated enclosure, the test contact, current limiting resistor, neon tube, and hand electrode are electrically connected. A shunt resistor branch is provided so that the test contact is electrically connected to one end of the shunt resistor and the other end of the shunt resistor is electrically connected to the identification electrode accessible to the human body. When the human hand contacts the identification electrode and the hand electrode at the same time, make the shunt resistor parallel to the test contact, current limiting resistor, neon tube, and hand electrode circuit for testing. When the human hand only contacts the hand electrode, and the neon tube glows, indicating that the object under test is charged. When the human hand contacts the hand electrode and the identification electrode at the same time, if the object under test has a non – dangerous high electric potential, because the internal resistance of the potential source is very large, which greatly reduces the charged potential of the object under test, the neon tube doesn't glow; if the object under test has a dangerous electric shock voltage, because of its small internal resistance, the connected shunt resistor hardly reduces the charged potential of the object under test, then the neon tube keeps on glowing, so as to distinguish the safe voltage from the dangerous voltage.

本发明的有益效果是：
可以在测试被测物是否带电的同时，方便地区分安危电压。分流支路中仅采用电阻元件，结构简单。

The invention has the following beneficial effects：
It is convenient to distinguish the safe voltage from the dangerous voltage while testing whether the object under test is charged. Only resistor element is used in the shunt branch, with simple structure.

范文二

Technical field：
The present invention relates generally to chewing gum compositions and methods for making same. More specifically, the present invention relates to chewing gum compositions that produce gum cuds having reduced adhesion as compared to typical chewing gum compositions.

技术领域：
本发明大体涉及口香糖组合物及其配制方法。更具体地讲，本发明所涉及的口香糖组合物所配制的口香糖与常规口香糖相比具有更低的黏附力。

Background technology：
Chewing gum – like substances have been enjoyed for hundreds of years. In the nineteenth century, the predecessor to today's chewing gum compositions were developed. Today, chewing gum is enjoyed daily by millions of people world wide.

背景技术：
口香糖类物质已经有数百年的历史了。19 世纪时，人们就已开发出了今天口香糖组合物的前身。今天，世界上每天有数百万计的人享用口香糖。

Summary of the invention:

The present invention provides chewing gums that are environmentally friendly. As used herein, the term "environmentally friendly" refers to chewing gum compositions that can be easily removed from indoor or outdoor surfaces.

本发明概述：

本发明提供环保型口香糖。此处所用的术语"环保型"是指可以从室内或室外物体表面上轻松除去的口香糖组合物。

Detailed description of the preferred embodiments:

The present invention provides improved gum bases. Moreover, the present invention provides improved chewing gum. To this end, the present invention provides gum bases that produce more environmentally friendly gum cuds.

By way of example, and not limitation, examples of the present invention are set forth below.

优选实施例详细描述：

本发明提供改进的胶基和改进的口香糖。为了达到这个目标，本发明提供的胶基让口香糖胶块更加环保。

仅通过以下实施例说明本发明。

Example1:

实施例 1：

…………

Claims:

1. A chewing gum comprising: a water insoluble portion not including a filler; a water soluble portion including a flavor; and approximately 3% to about 15% by weight lecithin.

2. The chewing gum of Claim 1 including an elastomer solvent.

权利要求书：

1. 一种口香糖，其中包括：水不溶性部分，其不包括填料；水溶性部分，其包括风味剂；以及按重量计约3%至约15%的卵磷脂。

2. 如权利要求1中的口香糖，其包括弹性体溶剂。

Gum bases, chewing gums, and methods of manufacturing same are provided. The gum bases do not include filler. It has been found that by using a gum base that does not include filler, a chewing gum is produced that results in gum cuds that have reduced adhesion.

本发明提供了胶基、口香糖及其配制方法。所述胶基不含填料。研究发现，采用不含填料的胶基配制的口香糖胶块具有更低的黏附力。

三、科技专利文献常用词汇和句型

（一）科技专利文献常用词汇

1. 科技专利文献常用复合词及词组

therebetween　介于两者之间
therefor　为此；因此
hereinabove　在上文
therefrom　由此；从那里
hereinto　在这里面，在此中
thereinto　在那里面；其中
hereon　关于此；在此
thereon　在其上
heretofore　直到此时；在此之前
thereout　从那里面；由此
hereupon　于是；此后
therethrough　借此；经由那个
thereto　向那里；随附
hitherto　迄今
thereunder　在其下；据此
thereupon　因此
thereunto　到那里
wherewith　用什么；用以
wherein　其中，在那里
whereat　在那里；在这里；对那个
whereupon　于是，因此
wherefore　为此
more particularly　特别，尤其

其他科技专利文献常用复合词可参考第三章。

2. 科技专利文献中涉及的文件类术语

annuity　年金
application fee　申请费
burden of proof　举证责任
case law　判例法
abstract　文摘，摘要
bibliographic data　著录资料
additional features　附加的特征

address for service　（文件）送达地址
caveat　警告
affidavit　宣誓书
cross license　交叉许可证
certificate of patent　专利证书
certificate of correction　更正证明书
certificate of addition　增补证书
appellation of origin　原产地名称
compulsory license　强制许可证
exclusive license　独占许可证
author's certificate　发明人证书
contractual license　契约性许可证；协议许可
claim to a method　方法权利要求
claim to a product　产品权利要求
gazette　公报
title of invention　发明名称
certified copy　经核证的副本
defensive publication　防卫性公告
confidential information　保密情报
application documents　申请案文件
complete application　完整的申请案
complete description　完整的叙述
application for patent　专利申请（案）
conflicting application　冲突申请（案）
confidential application　机密申请
divisional application　分案申请
continuation application　继续申请（案）
continuation-in-part application　部分继续申请（案）
content of the application as originally filed　原始申请的内容
application laying open for public inspection　公开供公众审查的申请
complete specification　完整的说明书
file copy　存档原件
final action　终局决定书
convention application　公约申请
data exchange agreement　数据交换协议
examiner's report　审查员报告
employee's invention　雇员发明
evidence　证据
counter pleadings　反诉状

art 技术
technical assistance 技术支持
technical data 技术资料；技术数据
know–how 诀窍；专门知识；实际经验
known technical solution 已知的技术方案
article of manufacture 制品
first–to–file principle 先申请原则
first–to–invention principle 先发明原则
force majeure 不可抗力

3. 科技专利文献中涉及的流程类术语

allowance 准许
amendment 修改
appeal 上诉
citation 引证
defense 辩护
expropriation 征用
experimental use 实验性使用
breach of confidence 泄密
claim 权项
arbitration 仲裁
development 发展
disclosure 公开
assignment 转让
examination 审查
counterclaim 反诉
dedication to the public 捐献于公众
fully comply with 完全遵守
file an application 提出申请
conflict award 冲突裁定
disclaimer 放弃权项；免责声明
dependent claim 从属权项
conflict procedure 冲突程序
contributory infringement 共同侵权
deferred examination 延迟审查
formal examination 形式审查
examination for novelty 新颖性审查

4. 科技专利文献中的"各种人"

assignor 转让人
assignee 代理人

defendant　被告人
examiner　审查员
applicant for patent　专利申请人
author of the invention　发明人
patent commissioner　专利局长
comptroller　审计官
co‒applicant　共同申请人
co‒inventor　共同发明人

5. 专利相关名称

design patent　设计专利
dependent patent　从属专利
invention patent　发明专利
expired patent　期满专利；过期专利
petty patent　小专利
basic patent　基本专利
precautionarl patent　预告专利
product patent　产品专利
patent family　同族专利
economic patent　经济专利
utility model patent　实用新型专利
patent of importation　输入专利/进口专利
interdependent patent　相互依存的专利
lapsed patent　已终止的专利，有效期已满的专利
patent act　专利法
patent application　专利申请
foreign patent application　外国专利申请
patent document　专利文件
patent documentation　专利文献
specification of patent　专利说明书
holder of a patent　专利持有人
patent procedure　专利程序
exploitation of a patent　专利实施
revocation of a patent　专利撤销
domination patent　支配专利
grant of a patent　授予专利权
misuse of a patent　滥用专利权
abandonment of a patent　放弃专利权
convention priority　公约优先权
exposition priority　展览优先权

exclusive right　专有权

copyright　版权

action for infringement of patent　专利侵权诉讼

community patent convention　共同体专利公约

convention establishing the world intellectual property organization　建立世界知识产权组织的公约

6. 科技专利文献中涉及的相关组织

examination countries　审查制国家

convention country　公约国

board of appeals　申诉委员会

BIRPI　国际保护知识产权联合局

FICPI　国际知识产权律师联合会

ESARIPO　非洲英语国家工业产权组织

European Patent Convention　欧洲专利公约

European Patent Office　欧洲专利局

7. 科技专利文献中涉及的日期类术语

application date　申请日期

date of issue　颁发日期；签发日期

date of patent　专利日期

date of publication　公布日期

grace period　宽限期

date of grant　授予日期

effective filing date　有效申请日

convention period　公约期限

convention date　公约日期

duration of patent　专利有效期

extension of term of a patent　延长专利期限

conception date　构思日期

8. 科技专利文献惯用词组

content of invention　发明内容

abstract of disclosure　披露摘要

the instant invention/the present invention　本发明

summary of the invention　本发明概述

detailed description of the invention　本发明详述

object of the invention　本发明的目的

background of the invention　本发明的背景

prior art　现有技术

state of the art　目前最高水平

background technology　背景技术

description of the prior art　现有技术介绍
field of the invention　本发明所属技术领域
method known in the art　此项技术中已知的方法
person skilled in the art　精通该技术的人
utility model　实用新型
description of the preferred embodiments　优选实施例描述
cross‐reference to related application　对相关申请的交叉引用
the drawing　图例
the accompanying drawing　附图
brief description of the drawing　附图简要说明
block diagram　方框图
plan view　平面图
perspective view　透视图
sectional view　剖视图/剖面图
elevational view　立面图
front view　正视图
Serial Number（Ser. No.）　流水号
application number　申请号
country code　国家代码
color coding　彩色编码
best mode　最佳方式
inventive step　创造性步骤
3 sheets‐sheet 2　第二页（图共三页）
sheet 3 of 5　第三页（图共五页）

9. 科技专利文献中的专有名词

vertical section　垂直切面
usage of stainless steel sheet and plate　不锈钢片、板的使用
surface treatment　表面处理
armature　衔铁；电枢
axial fan cooling tower　带轴流风机的冷却塔
air ejecting fan　抽气风扇；排气通风机
acoumeter　听力计
digital interface characteristic measurement　数字接口特性测量
optical time‐domain reflectometer　光时域反射仪
monitor‐capacitance control　监测器‐电容量控制
asbestos‐filled melamine plastics　三聚氰胺石棉塑料
curl flyer　毛刷式放线装置
travelling drum take‐up and pay‐off stand in port　移动式龙门收、放线架
shaftless constant tension pay‐off　无轴式恒定张力放线装置

dual pay-off 双盘放线装置
conduit-installation wire 管道安装线

（二）科技专利文献常用句型

（1）The invention will further be described with reference to the accompanying drawing, of which...

参照附图对本发明做进一步说明，图中……

（2）Preferred embodiments of the invention are as follows.

本发明的优选实施例如下。

（3）The claim of the patent is...

本专利的权利要求是……

（4）The object of one or more embodiments of the present invention is to...

本发明的一个或多个实施例的目标是……

（5）An advantage of the present invention is to provide...

本发明的一个优点是提供……

（6）Other disadvantages, objects, and advantages of the invention will be apparent from the description and drawings, and from the claims.

从描述和附图及权利要求中可更清楚地看出来本发明的其他缺点、目的和优点。

（7）Although specific embodiments have been illustrated and described, it will be obvious to those skilled in the art that various modification may be made without departing from the spirit which is intended to be limited solely by the appended claims.

尽管已经说明和描述了具体的实施例，但对本领域技术人员来说，显而易见的是，可以在不背离仅由附加权利要求限制的精神的情况下进行各种修改。

（8）According to another aspect of the invention, there is provided a complementary MOS type semiconductor device.

根据本发明的另一方面，提供了一种互补的 MOS 型半导体器件。

（9）Claim 1 is rejected on the basis that...

根据……驳回权利要求 1

（10）Figs. 8 to 10 are schematic diagrams illustrating different embodiments embodying the invention's concept.

图 8 至图 10 是说明体现本发明概念的不同实施例的示意图。

（11）Reference is made to our copending application No. 3379/58, filed 27th Aug. 1984.

请参阅我们在 1984 年 8 月 27 号提出的尚未批准的专利申请，其申请号为 3379/58。

（12）The application comprises a division of my copending application Ser. No. 584069, filed Mar. 17, 1994, now U. S. pat. No. 5380733.

本申请是我的尚未批准的专利申请的分案专利，该申请于 1994 年 3 月 17 日申请，申请号为 584069，美国专利号为 5380733。

（13）The invention described herein may be manufactured and used by or for the Government of the United States for all governmental purposes without payment of any royalty.

此处所述的发明因公务目的可以免费提供给美国政府制造并使用。

(14) The invention will be further explained in the following examples. All temperatures are in ℃ unless stated otherwise.

我们通过下列实例进一步解释本发明。除非另有说明，所有的温度都用℃作为单位。

(15) The invention relates to methods for audio signal amplification and to audio amplifier circuit and power supplies thereof.

本发明涉及音频信号放大的方法、音频放大器电路及其电源。

四、科技专利文献翻译技巧

由于科技专利文献跨越语言、法律和技术三个领域，很容易看出仅仅基于其中一个领域的翻译标准是片面和不完整的，科技专利文件翻译标准必须同时满足所有三个领域的要求。

(1) 语言方面。译文必须符合目的语的要求。清末启蒙思想家、翻译家严复提出"译事三难：信、达、雅"，成为后世公认的核心翻译理念或标准。"信"表示翻译准确，意思与原文不矛盾；"达"表示译文不拘泥于原文形式，通俗易懂；"雅"表示译文用词得体、简洁、雅致。此外，在翻译界，"直译"与"意译"的争论由来已久，一直持续到今天。英国著名翻译家彼得·纽马克（Peter Newmark）根据译者侧重译入语还是译出语，将翻译方法分为以下几类：侧重译出语（程度由高到低）分为逐字对译、字面翻译、忠实翻译、语意翻译，侧重译入语（程度由低到高）分为传意翻译、符合语言习惯的翻译、自由翻译、改写。要做好科技专利文献翻译工作，必须在"忠实原文"和"发挥译者的主观能动性"之间取得平衡。

(2) 法律方面。译文必须遵守专利法律法规的要求。由于科技专利文献的形式和内容受专利法律法规的约束，译文的形式和内容也必须符合国外申请所在国专利法的规定。为了准确翻译科技专利文献，一名专业的科技专利文献翻译人员必须首先掌握专利法律法规。

(3) 技术方面。科技专利文献的翻译必须使"本领域普通技术人员"能够理解。专利文献包括广泛的技术主题，分为八个IPC类别。然而，英语专业人员在专利文献翻译人员中占相当大的比例。许多专利文献翻译人员面临的重大问题之一就是他们缺乏学科知识。再者，即使是相关专业领域人员，也不可能掌握所有的专业信息，更不用说专利文献了，因为这些前沿专利文献内容瞬息万变。这迫使专利文献翻译人员必须对专利文献保持兴趣并定期学习新技能。事实上，这就是专利文献翻译行业的魅力所在。一方面，许多专利文献翻译企业根据翻译人员擅长的知识特征将翻译人员的工作按科技领域划分，这可帮助翻译人员更有效地工作；另一方面，帮助翻译人员减少工作量，有助于确保翻译质量和向客户负责。

下面介绍几种科技专利文献翻译技巧。

（一）词义选择

词义选择是科技英语文本翻译工作中一项重要的译前准备工作。英汉两种语言都有一词多义的现象，在实际语言应用中，词汇意义因语境的不同而变化。一个词的词义往往取决于与它搭配的词，有时还必须联系上下文，统观全句或全段才能确定这个词所具有的含义。在科技英语文本翻译的词义选择这一环节，必须首先弄清原句的基本结构，抓住句子的主干部分，确定关键词的词义；然后从应用语境入手，尽可能地从专业原理上理解相关术语，以期解决专业隔阂造成的词义选择的错误。

在选择和确定词义时通常从以下几个方面着手。

第一，依据词所处文本的专业语境选择词义。在科技英语中，词汇意义因专业语境的不同而变化。同形的词汇，用于专业语境下的意义往往不同于用于非专业语境下的意义。在某些专业技术领域中经常使用一些普通的词汇，但赋予这些词汇专门的意义。在这个过程中，译者既要处理好普通词汇的理解与表达，还需要从所应用的专业语境着手，判断该词汇是否应用在正确的专业语境下。译者必须通过具体的语境确定其意义。

第二，根据词在句中的词类选择词义。选择一个词在句中的具体词义，首先要判断这个词在原句中属于哪一种词类，根据词类再进一步确定具体的词义。翻译时应考虑句子涉及的专业术语。

第三，根据词在文中的搭配及其所处的上下文语境选择词义。英汉两种语言都有一词多义的现象，词义的灵活性是显而易见的，一个词的词义往往取决于与它搭配组合的词对它的制约。同一个词，在不同的搭配中，在不同的上下文中，在不同的专业场合中，其含义往往不同。因此，在翻译时，译者必须辨析专业与非专业语境下的词汇意义，利用上下文作为参照，来判断、确定、选择词义。

（二）增词法与省略法

增词法也就是增译法。省略法也就是省译法。作为翻译的一个普遍准则，译者不应该对原文内容随意增添或减少。但是，在翻译实践过程中，往往很难做到不增减词语。要准确地传达出原文的信息，译者难免要对译文做增添和删减。

奈达认为，好的译文一般略长于原文，由于语言文化上的障碍，译者常常要把原文中隐含的东西补充清楚，以便读者理解。增词的目的，或是为了补足语气，或是为了连接上下文，有的则是为了避免译文意义混乱。增词就是增加原文中虽然无其词但有其意的一些词。例如，英语中没有量词，汉译时应根据上下文的需要增加量词；汉语名词没有复数的概念，动词没有时态变化，汉译时如有必要应增加表达复数和表达时态的词；英语句中用一些不及物动词意义就很完整，这些词在汉语中为及物动词，必须增加宾语，否则意义就不完整。根据上述情况，英译汉时必须考虑增词。增词可遵循以下原则：原文有其意，但无其词，译文宜增补；增词不增意，译文增补无损于原意，有益于通顺，有助于传神。例如：

The leakage current of a capacitor is an important measure of its quality.

译文：电容器泄漏电流的大小，是衡量电容器质量好坏的重要尺度。

省略法与增词法相反。一般来说，汉语较英语简练，因此英译汉时，许多在原文中必不可少的词语要是原原本本地译成汉语，就会成为不必要的冗词。省略法是将原文中需要而译文中不需要的词语省去，其目的是使译文简洁明快。省略法不能随意更改、漏译原文内容。例如：

The jammer covers an operating frequency range from 20～500MHz.

译文：干扰机的工作频段为20～500MHz。

(三) 重复法

在翻译中，有时为了得到忠实于原文并且意义明确、文字通顺流畅、符合目的语习惯的译文，而将某一部分文字反复使用，这种翻译技巧就是重复法。例如：

The three most important effects of an electric current are heating, magnetic and chemical effects.

译文：电流的三种最重要的效应是热效应、磁效应和化学效应。

I had experienced oxygen and/or engine trouble.

译文：我曾碰到过，不是氧气设备出故障，就是引擎出故障，或者两者都出故障的情况。

(四) 拆译法

拆译法也称拆句法。在翻译科技专利文献时，通常有大量的长难句，为了使译文明了简洁，经常把一个句子译成两个或两个以上的句子。例如：

Computer simulation results show that with an antenna spacing as low as 1m, a DF error less than 1, can be obtained on a signal with a bandwidth of 100 MHz at a received power level lower than 100dB, using an integration time of a few milliseconds.

译文：计算机仿真结果表明，当天线间距短到1m时，可以在接收功率电平低于100dB时，仅用几个毫秒的积分时间，对带宽为100MHz的信号进行测向，其测向误差小于1。

(五) 被动语态的翻译

被动语态的广泛使用是英文科技专利文献的显著特征之一。这样做的原因是被动语态与主动语态刚好相反，可以更好地解释所论述的对象。此外，被动语态比主动语态客观，这正是科技专利文献所必需的。由于被动语态在英文科技专利文献中的广泛使用及英汉两种语言的差异，在英文科技专利文献中翻译被动语态更具有挑战性。英文科技专利文献中用被动语态表达的概念在汉语中经常用主动语态表达。因此，在翻译英文科技专利文献中的被动句时，不需要完全遵循原句的语态，而是需要使用灵活多样的翻译技巧。

1. 译为汉语的被动句

为了强调动作，或者不知道动作发出者，或者在特定的上下文中为了使前后分句的主语保持一致，或者为了使重点突出、语义连贯、语气流畅，可将英语的被动句译为汉语的被动句。在表达上除采用"被"字外，还可以根据汉语搭配的需要选用"遭""挨""给""受""加以""予以""为……所"等。例如：

Heat is regarded as a form of energy.

译文：热被看作能的一种形式。

2. 译为汉语的主动句

由于汉语习惯上多用主动句，因而许多英语的被动句都可以译为汉语的主动句。翻译时，既可以将其译为形式上主动而意义上被动的汉语句子，又可以改变原句的谓语动词将其译为汉语的完全主动句，还可以直接将原句的被动语态按主动语态译为汉语的完全主动句。

例如：

The objects have been identified as satellites.

译文：这些物体已查明是卫星。

3．译为汉语的判断句

那些着重描述事物的过程、性质和状态的英语被动句与系表结构十分相近，因而可译为汉语的判断句，由"是……的"结构加以表达。例如：

The volume is not measured in square millimeters. It is measured in cubic millimeters.

译文：体积不是以平方毫米计量的。它是以立方毫米计量的。

4．译为汉语的无主句

汉语的无主句是一种独特的句型。英语中的许多不需要或无法说出动作发出者的被动句往往可以译为汉语的无主句，而把原句中的主语译为汉语句子中的宾语。例如：

Methods are found to take these materials out of the rubbish and use them again.

译文：现在已经找到了从垃圾中提取这些材料并加以利用的方法。

5．采用习惯译法

对于"It + be + 及物动词过去分词 + 从句"中的被动语态，常采用习惯译法，将其译为汉语"据"字结构、无主句，或者增添泛指主语。例如：

It is known that left – ventricular end diastolic pressure and volume are reduced after treatment with nitroglycerin.

译文：据悉，经硝酸甘油治疗后，患者左心室舒张末期压力与血容量都会降低。

（六）定语从句的翻译

科技专利文献中带有定语从句的主从复合句比其他复合句更难理解。许多归因短语理解起来很简单，但表达起来却很复杂。这就要求译者掌握英汉两种语言的相似性和对比性，并运用适当的翻译方法将定语从句翻译成精确的短语。为了便于掌握，根据英语的语言结构与逻辑关系，分别论述以下几种不同的翻译方法。

1．后置前译

定语从句属于后置定语，即位于其所修饰的名词之后。而汉语的定语顺序则与之相反。因此，当定语从句较短时，可将英语的定语从句译成汉语的前置定语。例如：

The blood is a red, sticky fluid that circulates through the arteries, capillaries and veins.

译文：血液是一种通过动脉和毛细血管及静脉进行循环的、红色的黏性液体。

2．分句并译

当定语从句过长或内容与先行词关系不太紧密时，可以将其翻译成并列分句，放在所修饰的先行词后面。非限制性定语从句最好用这种翻译方法。例如：

There have been various advances in the design of reciprocating steam engines, which have greatly increased their efficiency and so enabled them to hold their own certain services, despite competition from turbine and diesel machinery.

译文：往复式蒸汽机的设计有了各种进步，这些进步大大提高了效率，所以说尽管有了涡轮机和内燃机这样的竞争对手，这种蒸汽机在某些部门仍然适用。

3．主从合译

有时可把定语从句和它所修饰的先行词结合在一起翻译，即将主句和从句合并翻译。例如，当主句是 there is（are）... 句型时，关系代词和先行词应合并成一个主语或宾语。例如：

There are so many skin eruptions which may resemble that of scarlet fever that it would be wrong to make the diagnosis on the basis of the skin eruptions alone.

译文：有很多皮肤疹可能类似于猩红热，因此仅根据皮肤疹做出诊断是错误的。

4. 偏正联译

在科技专利文献中，常以逻辑语义关系与主句密切的定语从句代替原因、结果、目的、条件等状语从句。这种语法定语从句也被称为状语化定语从句，因为它在本质上已经被状语化了。这类定语从句与主句的关系，恰似汉语的偏正复句，因此可将它们译成汉语的各种偏句。

（1）译成原因偏句。例如：

It is a fact of everyday experience that a conductor in which there is an electric current is thereby heated.

译文：某一导体因其中有电流通过而发热，这是平日常见的事实。

（2）译成结果偏句。例如：

This sets in motion the Bainbridge reflex by which vagal effect is diminished and the heart rate increases.

译文：这能引起班布里奇反射，因此迷走神经作用减弱，心率加快。

（3）译成目的偏句。例如：

We have to oil the moving parts of the machine, the friction of which may be greatly reduced.

译文：我们必须得给机器的可动部件加油，这样可以大大减少摩擦。

（4）译成条件偏句。例如：

An electric current begins to flow through a coil, which is connected across a charged condenser.

译文：如果线圈跨接于充过电的电容器上，电流就会流过线圈。

（七）状语从句的翻译

英汉两种语言分属不同的语系。汉语属于汉藏语系，结构松散，句与句之间的逻辑关系多以句序的先后来加以暗示，具有意合的特点，也有语言学家将汉语的结构比喻为竹节句法。而英语则属于印欧语系，句子突出主干结构，尤其是使用较多抽象名词，句子之间的逻辑关系用连接词连接起来，具有形合的特点，这种结构也被称为葡萄架句法。英语中的状语从句通常位于主句之前或之后，而在汉语中，状语一般位于谓语中心词之前。在翻译状语从句时，通常采用以下几种译法。

1. 顺序而下，顺序译法

当状语从句在主句之前时，按照原文的顺序翻译即可。例如：

Whenever we think hard, our brain produces special transport chemicals.

译文：当我们努力思考时，我们的大脑就会产生特殊的传输化学物质。

2. 逆流而上，逆序译法

当状语从句在主句之后时，需要按照汉语的表达习惯加以调整。例如：

The crops failed because the season was dry.

译文：因为气候干旱，作物欠收。

3. 移花接木，转换译法

在有的情况下，还可以将结果状语从句转译为并列句或一个汉语中的单句，唯有如此，译文才符合汉语的表达习惯。例如：

Some stars are so far away that their light rays must travel for thousands of years to reach us.

译文：有些恒星十分遥远，它们的光线要经过几千年才能到达地球。

4. 化整为零，分译译法

当采用顺序译法和逆序译法都无法解决问题时，可以考虑分译译法，把句子拆分为几个并列的分句或短语式的并列成分。例如：

He made notes as he prepared the experiment.

译文：他边准备实验，边做记录。

5. 递序而下，顺序译法

英语中的结果状语从句都放在主句之后，由连词 that 或 so... that、such... that 等结构引导。多数情况下，采取顺序译法翻译就可以了，但也不要拘泥于通通将其译为"因而""因此""如此……以致""致使……"等句式，需要译者灵活处理。例如：

The Sun is so much bigger than the Earth that it would take over more than a million Earth to fill a ball as big as the Sun.

译文：太阳比地球大多了，所以得有 100 多万个地球才能填满像太阳那么大的球体。

6. 纲举目张，变序译法

例如：

The lathe can turn holes even if it is not a driller.

译文：车床尽管不是钻床也能钻孔。

五、课后练习

（一）思考题

（1）科技专利文献的作用有哪些？

（2）翻译英文科技专利文献应遵循的标准是什么？

（二）翻译练习

1. 句子翻译

（1）The object of the present invention is to improve the power and applicability of the internal combustion engine.

（2）Another problem with using an ATM is that most of the keys look very similar to someone who is partially–sighted or cognitively–impaired, which may make it more difficult for these people to use ATMs effectively.

（3）Fig. 1 is a cross–section view of a prior art valve system connector incorporated into a control valve assembly.

(4) A variety of different chewing gums can be created pursuant to the present invention. Such chewing gums can include sugar gums, sugarless gums, bubble gums, coated gums, and novelty gums. Such chewing gums can be formed in the shape of pellets, sticks, tabs, or chunks.

2. 篇章翻译

(1) 中译英。

①本发明公开了一种安全温水瓶，它属于家庭生活用品，其解决了现有技术中温水瓶存在爆裂危险的技术缺陷。它主要包含内胆、外壳、立式提手、横式提手、瓶塞。内胆内测涂有保温材料，外壳用于保护内胆，立式提手用于提水，横式提手用于在使用开水时倒水，瓶塞用于保温。它主要用于家庭生活中开水的保存。

权利要求书

a. 本实用新型涉及一种安全温水瓶，它包含内胆、外壳、立式提手、横式提手、瓶塞。它的特征在于：内胆由外壳上部固定，外壳下部起到美观、平衡温水瓶的作用。

b. 根据权利要求 a 所述的安全温水瓶，其特征在于：在内外壳接口处，外壳下部设计成螺杆，外壳上部设计成螺母，通过外壳上部与外壳下部的相对旋转而固定。

c. 根据权利要求 b 所述的安全温水瓶，其特征在于：在内外壳接口处，外加四颗螺钉固定，外壳底部外侧有一个小孔。

d. 根据权利要求 a 所述的安全温水瓶，其特征在于：外壳上部设计有保护内胆的突出槽，内胆置于突出槽内，外壳上部与突出槽无缝连接，外壳上部与下部接口处于横式提手之间。

e. 本实用新型涉及一种安全温水瓶，为了换内胆方便，在外壳上部立式提手以上设计同样的开口。

说明书

安全温水瓶

技术领域

本发明涉及一种家庭生活用品，具体地说，涉及一种温水瓶。

背景技术

在日常生活中，很容易出现温水瓶破裂的情况。破裂事小，但伤人事大。一方面是热水伤人，另一方面是内胆碎屑伤人，内胆破裂后打扫也不方便。

发明内容

为了解决现有技术中温水瓶使用时易破裂，破裂后内胆碎屑容易伤人，不易打扫且所装开水也会产生一定伤害的技术问题，本发明提供了一种安全温水瓶。

现有技术中的温水瓶外壳也分上下部分，但上下部分的连接在底部，容易出现危险。本发明中的连接处在横式提手之间，且采用自身螺母、螺杆固定与外加螺钉固定两种方法，更安全。内胆由外壳上部的突出槽固定，组合的时候，突出槽就位于外壳下部内。若出现内胆破裂的情况，也可由外壳下部收集内胆碎屑，减小了危险。双开口，更方便。外壳下部还有装饰作用，它占主要外表面，因此可以在上面绘制一些图案，由合成革制作。

②真空烧结钛酸钡等铁电材料

发明背景

本发明介绍了一种提高铁电材料介电性能的方法，特别介绍了提高钛酸钡介电性能的方法。

通常，为了得到高介电常数的陶瓷，窑炉烧结和热压是外处理铁电材料的两种方法。就窑炉烧结而言，煅烧时使用了各种气氛，如氧气、二氧化碳等。在窑炉中煅烧铁电材料存在的问题是，所得到的陶瓷介电常数不够高，并且目标介电损耗大。现在采用热压铁电材料的方法，所生产的陶瓷已显示出优于窑炉烧结所生产的陶瓷，但热压价格昂贵，而目电子陶瓷的批量生产效率不高。

本发明概况

本发明总的目的是提供处理铁电材料的一种方法，以便获得电性能优良的陶瓷，同时提供一种经济的生产方法。现已发现，在铁电材料不发生还原的空气分压下煅烧已干压的铁电材料，就能获得价格便宜、介电性能良好的陶瓷。所使用的空气分压取决于具体的铁电材料，但必须在铁电材料不发生还原的空气分压的情况下进行。

最佳实施方案评述

首先把一种典型的铁电材料（即工业用的钛酸钡）的原料粉末进行干压，然后在可控气氛条件下燃烧。在本例中，干压的钛酸钡在 $1\sim1000\mu m$ 的中真空下，装在管式炉中煅烧。真空烧结的峰值温度为 $1200\sim1400℃$，达到峰值温度约 7h。然后冷却到环境温度。密炉开始工作时中真空便开始。

本发明也适于处理其他铁电材料，包括碱土金属钛酸盐，如钛酸锶。

（2）英译中。

①Get started

Plug EarPods into your iPhone, iPad, or iPod touch (iOS10 or later required), and put them in your ears.

Important safety information

Hearing loss

Listening to sound at high volumes may permanently damage your hearing. Background noise, as well as continued exposure to high volume levels, can make sounds seem quieter than they actually are. Check the volume before inserting EarPods in your ears. For more information about hearing loss and about how to set a maximum volume limit, see Apple's website.

WARNING: To prevent possible hearing damage, do not listen at high volume levels for long periods.

Driving hazard

Use of EarPods while operating a vehicle is not recommended and is illegal in some areas. Check and obey the applicable laws and regulations on the use of earphones while operating a vehicle. Be careful and attentive while driving. Stop listening to your audio device if you find it disruptive or distracting while operating any type of vehicle or performing another activity that requires your full attention.

Choking hazard

EarPods may present a choking hazard or cause other injury to small children. Keep them away from small children.

Skin irritation

Earphones can lead to ear infections if not properly cleaned. Clean EarPods regularly with a soft lint–free cloth. Don't get moisture in any openings, or use aerosol sprays, solvents, or abrasives. If a skin problem develops, discontinue use. If the problem persists, consult a physician.

Electrostatic shock

When using EarPods in areas where the air is very dry, it is easy to build up static electricity and possible for your ears to receive a small electrostatic discharge from EarPods. To minimize the risk of electrostatic discharge, avoid using EarPods in extremely dry environments, or touch a grounded unpainted metal object before inserting EarPods.

②**Impedance Transformer, Integrated Circuit Device, Amplifier, and Communicator Module**

Cross–reference to related application

[0001] The present application is based upon and claims the benefit of priority of the prion Japanese Patent Application No. 2010–34751, filed on February 19, 2010, the entire contents of which are incorporated herein by reference.

Field

[0002] The embodiment discussed herein relates to an impedance transformer, an integrated circuit device, an amplifier, and a communicator module.

Background

[0003] In integrated circuit devices used in communicator modules such as radar amplifiers and base station amplifiers, a plurality of integrated circuits, for example, are coupled in parallel and the widths of transistor gates of the integrated circuits are increased to establish a high output.

[0004] Furthermore, impedance transformers are coupled to both the input sides and the output sides of the plurality of integrated circuits coupled in parallel to match impedances through matching circuits in the impedance transformers. In particular, impedance transformers with quarter wave–lines coupled in multiple series are widely used in integrated circuit devices needed for wideband characteristics in order to obtain wideband characteristics by increasing the number of quarter wave–line stages.

[0005] As related art, for example, Japanese Laid–open Utility Model Registration Publication No. 5–65104, Japanese Laid–open Patent Application Publication No. 9–139639, Japanese Laid–open Patent Application Publication No. 10–209724S, S. B. Cohn, "Optimum Design of Stepped Transmission–Line Transformers", IRE Trans actions on Microwave Theory and Techniques, vol. 3, pp. 16–21, 1955., and E. J. Wilkinson, "An N–Way Hybrid Power Divider", IRE Transactions on Microwave Theory and Techniques, vol. 8, pp. 116–118, 1960 are disclosed.

第六章 科技论文摘要

科技论文摘要是对科技论文核心内容和基本观点的高度概括，是高质量科技论文的重要组成部分。优秀的科技论文摘要必须简明扼要地阐述论文的主旨大意，同时避免进行任何主观意义上的评述。因此，译者在翻译时必须遵守以上原则，谨慎选词，力求做到译文准确、忠实、简洁。

一、科技论文摘要文本介绍

（一）科技论文摘要概述

科技论文是介绍自然科学研究或技术开发工作成果的论说文章，通常基于概念、判断、推理、证明或反驳等逻辑思维体系，使用实验调研或理论计算等研究手段，按照特定格式撰写完成。科技论文可以分为期刊论文、学位论文等。科技论文摘要是位于科技论文正文前面的一段概括性文字，高度概括论文的内容，是正文实质性内容的精华，它突出了论文的核心，不加评论和补充解释。它通常位于标题、作者名之后，论文正文之前。

（二）科技论文摘要文本结构

科技论文摘要是科技论文关键的一部分，读者看论文必看两部分：标题和摘要。摘要存在的目的就是让读者在搜索文献的时候节省时间，读者通过摘要来评价这篇论文他是不是感兴趣。因此，摘要写得好不好，关系到论文能不能吸引住读者。

摘要的字数是有限制的，不能长篇大论，否则就失去它存在的意义了。一般来说，摘要的字数为300~400字。由于字数有限，在进行科技论文摘要写作时应字斟句酌，认真推敲。

科技论文摘要文本结构包含四个部分，分别是目的、方法、结果、结论。

（1）目的：是指出研究领域的研究重点，也就是这篇论文的写作目的。
（2）方法：是告诉读者解决这个重点的时候使用的是什么方法。
（3）结果：就是拿出实验或仿真数据来证实方法是有效的。注意，结果这一部分需要有数据来支撑。
（4）结论：就是一个简单的总结，说明这个方法有什么优缺点，等等。

（三）科技论文摘要的特点

科技论文摘要作为科技论文举足轻重的一部分，是科技论文的梗概，提供给读者实质性的内容，目的在于给予读者关于论文内容足够概括的信息，是读者判断论文价值和是否值得阅读的依据。科技论文摘要为科技情报文献检索数据库的建设和维护提供方便，以便索引和查找。论文摘要是读者检索文献的重要工具，论文摘要质量的高低直接影响论文的被检索率和被引频次。下面具体阐述科技论文摘要的语篇特点和文本特点。

1. 语篇特点

（1）正规。在科技论文摘要中，一般使用正式文体，面向专业读者，语法结构要求严谨规范。因此，摘要包含非常完整的句子，很少出现口语中的省略句或不完整的句子。词汇的使用也相对标准化。作者一般使用论文研究领域的规范术语。

（2）精练。在科技论文摘要中，一般不引用例子或与其他研究工作进行比较。语句也少有重复。为了使摘要更加紧凑连贯，主要使用更准确的专业术语。摘要的各个部分通过词汇的意义衔接并紧密地联系在一起，紧凑而富有共鸣。摘要更倾向于使用复合名词，因为它们使文章更加精简利落。

（3）具体。科技论文摘要中的每个论点、观点都要鲜明、具体、有根据。一般不会使用论文"与……有关"，而直接写论文"说明……"。用词要求明确具体，尽量避免使用一词多义、似是而非的词。

（4）完整。科技论文摘要本身要完整，一个规范的摘要应包含目的、方法、结果、结论。其中，目的要具体到研究、调查等的前提、目标和任务及所涉及的主题范围，方法要写明所用的原理、理论、条件、材料、对象、工艺、结构、手段、装备及程序等，结果要写明实验、研究的数据、确定的关系、观察结果、实际效果、性能等，结论要明确结果的分析、研究、比较、评价、应用等。

2. 文本特点

（1）人称特点。国家标准 GB 6447—86 中规定，文摘要用第三人称的写法。在很多中文期刊的投稿须知中，都会有"要用第三人称的写法""不要用 we...、I...、The author..."等明确要求。通常情况下，使用第三人称可以减少论文的主观性，强调论文的科学性和客观性，使读者更容易产生信服感和认同感。国内学者受此影响，并已习惯于此，无论是投稿给国内期刊还是投稿给国际期刊，他们都尽量避免在提交的材料中使用第一人称。

然而，近年来，很多国际期刊已开始倡导使用第一人称进行写作，尤其是英语本族语作者在提出议题、得出结果和结论时，常用第一人称"we"，表现出更主动地参与和互动的态度。很多学者认为，国际规范要求用第三人称实际上是要求作者客观公正地阐述观点，不能带有强烈的个人主观意识，并不拒绝使用第一人称。关于第三人称的使用实际上属于语用学范畴，是一种跨语言、跨文化的指示语问题，极易引起误解。

据研究显示，许多作者用"we + 谓语动词"和"our + 名词"来描述他们的个人工作进展情况。在部分摘要中使用这种形式来表达论文的观点，从作者研究的角度来看，这是作者工作的结果，表明这是一个突破性的、小规模的、阶段性的研究成果；在有些摘要中使用这种形式并不强调作者或作者的研究团队，只是为了缩小作者和读者之间的距离，邀请读者参与讨论，使他们更容易产生共鸣，并反映摘要的交际功能。

（2）语态特点。人称和语态是相互关联的，采取的主语人称决定了使用的语态。主动语态和被动语态在英语中有各自的应用规范和特点。主动语态表示主体是行为的执行者，被动语态表示主体是行为的接受者或客体。被动语态不仅可以省略施动者，避免出现"we..."表示式，而且还可以使需强调的事物做主语而突出它们的地位，有利于说明事实。同时，被动句在结构上有较大的调节余地，有利于使用恰当的修辞手法，扩展名词短语，扩大句子信息量。被动句的特点是集中大量信息于主语，并以直截了当的方式表达。目前来看，科技论文摘要中多用被动语态。但是，从结构上来说，主动语态更为简练、直接有力，可以通过简单的句型和自然的阅读来突出动词所表达的内容。一些权威的国际科技期刊越来越多地推荐使用主动语态。

（3）时态特点。一般来说，科技论文摘要要求作者使用尽可能少的行话，从目的、方法、结果和结论的角度客观、充分地表达文本的基本内容。由于摘要篇幅较短，而且学术文章的读者往往是具有类似研究兴趣的专家，因此作者不需要过多地介绍现有的研究工作。通常，作者只需要解释本文的工作事实和观点。因为这些内容大部分都是客观事实，所以英文科技论文摘要的时态主要是一般现在时，也会使用少量的现在完成时和一般过去时。

科技论文摘要中究竟应使用何种人称、语态和时态要依据客观情况而定。我们应该坚持具体问题具体分析，不应采取简单粗暴的统一形式进行硬性规定。随着国际学术交流的加深，国内作者现在可以看到大量的国际期刊出版物。建议在学习和借鉴先进学术成果的同时，更加关注这些出版物的写作风格，以便提高我们的语言表达能力，更加符合国际公认的语言特点。

（4）词汇特点。科技论文摘要的词汇特点首先体现在词义的单一性。科技工作者对客观事物现象和规律进行描述和论证时，必须做到概念明确、判断恰当、推理周密、论证严密，这就要求表义明确单一，排斥使用感情色彩浓郁、文化内涵丰富、词义模棱两可或隐晦不明的词汇。其次，使用术语及术语的缩写和代表符号是科技论文摘要词汇的另一个显著特征，这样语言表达简洁明快，信息容量大。再次，科技论文摘要中使用大量的专业词汇。这些词汇有些来源于拉丁语、希腊语，如由拉丁语词素构造出的词汇 insecticide（杀虫剂）、由希腊语词素构造出的词汇 graphite（石墨）；有些来源于法语、德语等，如 density（密度）等。正是由于这些来源和构词方式，科技论文摘要词汇有专业性强、合成词多、派生词多、缩略词多的特点。最后，在英文科技论文摘要中也存在普通词汇，但与它们在一般英语中的用词是不同的，这主要是由于科技论文摘要更加偏向于书面语，并且一些大词汇和长词汇出现的频率大大高于一般英语文体。科技论文摘要中的普通词汇常常借鉴外来语，或者广泛利用缀合法构词，广泛使用复合词、拼缀词等。

（5）句法特点。科技论文摘要是一种非常正式的书面文体，其句式比较完整，变化不多。科技论文摘要要求语言平实，很少或者几乎不使用一些华丽的修饰词汇，拟人、夸张等修辞手法也很少见到。英文科技论文摘要的句法特点主要体现在三个方面：一是形式主语和形式宾语的使用相对比较频繁，二是名词化结构的使用较多，三是常出现被动语态。

（6）语用特点。简而言之，英文科技论文摘要的语用特点，一是动名词的使用，二是模糊限制语的使用。就前者来说，在翻译英文科技论文摘要时，动名词的使用往往会给译者带来一定的困惑。这主要是由于这些动名词经过了动词的虚化手段，然后又用来表示动作或动作的过程、结果，或者是表示与某一事物相关的动作，所以有时候所表示的概念是比较抽象和含蓄的。就后者而言，令人困惑之处在于，既然科技论文摘要要求客观公正地描述人类科技发展的情况，就不应使用模糊限制语。其实，科技英语中适当地使用模糊限制语，不仅可以准确地陈述作者的观点，而且还能严谨地表达一些科技命题，这主要是由于科技工作者从实证性研究中得出的学术结论还没有得到学界的认可，具有不确定性、可探讨性等特点。所以，科技工作者要尽可能客观地表达命题，在措辞上可能就会相对选用一些模糊限制语。

（四）科技论文摘要的分类

科技论文摘要主要分为指示性摘要、报道性摘要、报道-指示性摘要。

1. 指示性摘要

指示性摘要也称作说明性摘要、描述性摘要或论点摘要，是指明论文的论题及取得成果的性质和水平的摘要，目的是使读者对该研究的主要内容有一个大致性的了解，一般只用两三句概括论文的主题，而不涉及论据和结论。创新内容较少的论文，其摘要可写成指示性摘要。篇幅以 100 字左右为宜。

2. 报道性摘要

报道性摘要常称作信息性摘要或资料性摘要，是指明论文的主题范围及内容梗概的简明摘要，相当于简介。这类摘要一般全面、简要地概括论文的研究目的、方法、主要结果和结论，或者简要提炼段旨句，达到扼要并逻辑性地揭示论文全貌的作用。通常，这类摘要可以部分地取代阅读全文。学术期刊多选用报道性摘要。篇幅以 300 字左右为宜。

3. 报道-指示性摘要

报道-指示性摘要以报道性摘要的形式表述论文中信息价值较高的部分，以指示性摘要的形式表述其余部分。篇幅以 100~200 字为宜。

二、科技论文摘要范文展示

范文一

本文探讨了在工作载荷作用下柔轮的畸变变形特点及规律，建立了柔轮原始曲线和工作载荷与柔轮畸变变形曲线的数学模型。针对柔轮的几何特征和运动特性，编制了求解柔轮畸变变形曲线的有限元程序，并利用计算机绘制出柔轮畸变变形曲线。该研究结果为谐波传动机构的设计提供了理论依据。

The characteristics and the regularity of the distortion of a flexspline under working loads are studied and the expressions of the original curve of the flexspline and the distortion curve of the flexspline against the working loads are derived in this paper. By using FEM (Finite Element Method), the programs for calculating the distortion curve of the flexspline are developed based on the geometric and kinetic characteristics of the flexspline. The distortion curve of the flexspline is plotted with the aid of a computer. The expressions, the programs and the curves provide a theoretical basis for the design of harmonic drives.

范文二

Color laser printers have fast printing speed and high resolution, and forgeries using color laser printers can cause significant harm to society. A source printer identification technique can be employed as a countermeasure to those forgeries. This paper presents a color laser printer identification method based on cascaded learning of deep neural networks. First, the refiner network is trained by adversarial training to refine the synthetic dataset for halftone color decomposition. Then, the halftone color decomposing ConvNet is trained with the refined dataset. The trained knowledge about halftone color decomposition is transferred to the printer identifying ConvNet to enhance the identification accuracy. Training of the printer identifying ConvNet is carried out with real halftone images printed from candidate source printers. The robustness about rotation and scaling is considered in training process, which is not considered in existing methods. Experiments are performed on eight color laser printers, and the performance is compared with several existing methods. The experimental results clearly show that the proposed method outperforms existing source color laser printer identification methods.

彩色激光打印机具有打印速度快、分辨率高的特点，使用彩色激光打印机进行伪造会对社会造成重大危害。源打印机识别技术可以消除这些弊端。本文提出了一种基于深度神经网络级联学习的彩色激光打印机识别方法。首先，通过对抗训练对细化器网络进行模拟训练，以细化合成数据集进行半色调颜色分解。然后，利用改进后的数据集对半色调颜色分解卷积神经网络（ConvNet）进行训练。将训练后的半色调颜色分解技术迁移到打印机识别ConvNet中，以提高识别精度。用从候选源打印机打印的真实半色调图像进行打印机识别ConvNet的训练。在训练过程中考虑了旋转和缩放的鲁棒性，这是现有方法没有考虑到的。在八种彩色激光打印机上进行了实验，并在性能方面与现有的几种方法进行了比较。实验结果表明，该方法优于现有的源彩色激光打印机识别方法。

三、科技论文摘要常用词汇和句型

（一）科技论文摘要常用词汇

abstract　摘要
acknowledgments　致谢
author(s)　作者
body　正文
conclusion　结论
discussion　讨论
result　结果
title　标题
proposal　提议
cover　包括
deal with　处理
relate to　关于
report　报道
describe　描述
discuss　讨论

examine　检查
study　研究
experiment　实验
apply　使用
assess　估算
improve　改进
indicate　指出
obtain　得出
prove　证明
embody　体现
provide　提供
record　标明
show　显示
resolve　解决
suggest　表明
reveal　揭示
give　给出
achieve　达到
construct　构造
derive　来自
plan　计划
establish　确立
measure　测量
be based on　基于
method　方法
by means of　用，依靠
observe　观察
calculate　计算
point out　指出
compare　比较
find　发现
approach　方法；路径
explain　解释
support　证实
analyze　分析
address　处理
test　检验
conjecture　推测
estimate　估计

use　使用
evaluate　评估
adopt　采用
exhibit　展现
illustrate　阐明
conclude　推断出
keywords　关键词
references　参考文献
seek　试图，设法
state　陈述，说明
demonstrate　证实；论证
introduction　引言；概述
elaborate　详细阐述，阐明
determine　决定
major object/primary goal　主要目标
chief aim/main purpose　主要目的
take... as reference　参考
with the assistance of　借助
aim at sth.　旨在
research objective　研究目的
research significance　研究意义
research method　研究方法
research background　研究背景
research conclusion　研究结论
overall objective　整体目标
agree with　同意，和……意思一致
accord with　与……一致
arrive at/come to the conclusion　得出结论
materials and methods　材料和方法
reduce to　归纳为
focus on　集中于
resume　简历；摘要
draw/reach definite conclusion from...　从……得出明确的结论

（二）科技论文摘要常用句型

（1）（Examples/ Experiments) demonstrate/confirm/show...
（实例/实验）显示/证实/表明……

（2）长主语（主语＋长修饰语）＋被动语态

（3）短主语＋被动语态＋分割修饰语

(4) The idea of our method of measurement is to analyze...
我们测量方法的思想是分析……

(5) Although a number of tests and comparison of the method have given satisfactory results, additional investigations to provide further justification and verification are required.
虽然对该方法进行了许多测试和对比，取得了令人满意的结果，但还需要进行额外的调查，以提供进一步的核实和验证。

(6) A principle of constructing... is considered in this paper.
本文研究了一种构建……的原则。

(7) A theoretical model has been developed to predict...
建立了一种理论模型来预测……

(8) A theoretical treatment of a new model of... is given.
给出了一种新的……模型的理论处理方法。

(9) Performance goals and various approaches to... are briefly described.
简述了性能目标和达到……的各种途径。

(10) The article contains some practical recommendations on...
本文包含了一些有关……的实际建议

(11) The performance characteristics of... are studied theoretically and experimentally.
对……的性能特点进行了理论和实验研究。

(12) This article discusses the reasons for... and offers an insight into...
本文讨论了……的原因并提出了对……的观点

(13) This paper/article is concerned with/aimed at/intended for the study/determination of...
本论文/文章涉及/旨在/意在研究/确定……

(14) This paper analyzes some important characteristics of...
本文分析了……的一些重要特征

(15) This paper determines the rational range of...
本文确定了……的合理范围

(16) This paper provides the quantitative background to...
本文为……提供了量化背景

(17) This paper reviews..., summarizes the theory of... from..., discusses..., and presents...
本文综述了……，从……总结了……理论，讨论了……，还提出了……

(18) This thesis explains/outlines/summarizes/evaluates/surveys/discusses/focuses on the results of...
本文解释/概述/总结/评估/调查/讨论/关注……结果

(19) The equation of... is formulated based on...
……的方程是以……公式为基础的

(20) The measurement of... is described.
描述了对……的测量。

四、科技论文摘要翻译技巧

在科技论文摘要翻译过程中,需要考虑到一些翻译技巧的使用。以下从时态、词汇、句法三个方面入手,简单阐述科技论文摘要翻译中可采取的翻译技巧。

(一)时态方面

通过时态的选择和在同一篇论文摘要中不同时态的搭配使用,译者可以很便捷地表达出各个研究行为间的时间先后次序及相互之间的影响与联系。然而,在许多科技论文摘要英译中存在着时态运用不当的问题,这会严重影响读者对论文的理解,也会降低科技论文的水平。科技论文摘要英译常选用的时态有一般现在时、一般过去时和现在完成时。

英文科技论文摘要中时态的使用规则和论文正文中时态的使用规则是相同的。下面介绍这些使用规则。

1. 使用一般现在时的情况

(1)在介绍背景资料时,需要用一般现在时叙述不受时间影响的普遍事实。例如:

This is especially true of natural gas which has excellent clean burning characteristic and low greenhouse gas production.

(2)在叙述研究目的或主要研究活动时,若采用"论文指引"型(如"This paper presents…"),则应该使用一般现在时。例如:

This paper investigates the issue of digging out differentially expressed groups of genes of a micro-array experiment.

(3)通过使用一般现在时来描述研究方法,如介绍数学模型、算法、分析方法或技术。例如:

We utilize FastCap to compute the values of the capacitances, and a closed form formula to obtain the inductance values.

(4)在叙述结论或推论时,应该使用一般现在时,可以使用 may、should、could 等情态动词。

2. 使用一般过去时的情况

(1)在叙述研究目的或主要研究活动时,若采用"研究指引"型(如"We discussed…"),则应该使用一般过去时。例如:

In this investigation, we explored vertical nutrient mixing in late summer in a test area 15 kilometers off the coast.

(2)在概述实验程序或方法时,通常使用一般过去时。例如:

Experiments on high-enthalpy water blowdown through a short converging nozzle were conducted out to measure the critical discharge flow rate under different stagnation conditions.

有的作者将研究目的和研究方法用一个句子表达,当研究目的用主句来表达时,可用短语描述研究方法;当采用主句来描述研究方法时,可用位于句首或句末的短语来表示研究目的。

(3) 在概述结果时，通常使用一般过去时。但是，若作者认为自己的研究结果普遍有效，而不只是在本研究的特定条件下才适用，则可以使用一般现在时。对结果的解释和说明，则用一般现在时。

3. 使用现在完成时的情况

在介绍背景资料时，需要使用现在完成时进行某种研究趋势的概述。例如：

The laboratory research has been conducted to determine actual decomposition rates of wheat and other small grain straws using l4C—labeled plant material.

（二）词汇方面

由于科技论文摘要在语言方面有着准确、简洁和明晰的特点，所以在科技论文摘要翻译时对词汇的选择也要遵循一定的原则：由于科技论文摘要书面化的特点，所以在选择词汇时要注意选择正式用语，避免出现口语中的非正式表达；在对特定领域中的某个研究过程或结果进行描述时，常常需要使用一些专业术语，对专业术语的翻译要确保准确。一般来说，许多领域都有自己的英汉和汉英翻译词典，通过查阅这些工具书，会使译者对翻译专业术语有更好的把握。

1. 名词的翻译

（1）不译的情况。汉语中有些较为抽象的名词，如"问题""情况""工作""条件""状况""现象"等，有时本身没有实质的含义，人们将这类词统称为"范畴词"（category words）。在将中文科技论文摘要翻译成英文时，这类范畴词通常不译。例如：

关于建筑产品质量问题的法律思考

译文：A Consideration from the Angle of Law on the Quality of Building Products

论劳动力市场需求约束条件下的经济模式

译文：The Economic Model Restrained by Market Demands of Labor Force

例子中，"产品质量问题"即"产品质量"的含义，"问题"不需要译出来；"……需求约束条件下的……模式"相当于"……需求约束下的……模式"，"条件"不需要翻译。总而言之，在遇到范畴词时，需要根据上下文语境仔细推敲词语，使译文能够准确地表达原文的含义。

（2）借助上下文逻辑关系明确多义词的词义。英语中很多词都是多义词，意义丰富。但是，不管一个词有多少个义项，它的每个义项都由于有了上下文的限制而成为明确而单一的语义。在特定的上下文中，确定一个词的准确词义，对英语学习者来说一直是一件令人困扰的事情。在科技论文摘要中，很多日常词汇更是蕴藏着全新的专业含义，无疑给译者设置了重重陷阱。例如，在"cell culture"中，"culture"不是我们常用的"文化；文明"的意思，而是"培养"之意，应当译为"细胞培养"。

（3）利用变译手段恰当表达词义。由于科技论文摘要用词庄重，多用比较抽象、具有概况性的词，因此译者在准确判断词在具体上下文中含义的基础上，利用增译、减译、转译等变译手段在汉译时恰当表达。

2. 连接词的翻译

由于科技论文摘要的高度概括性，许多复杂抽象的科学研究的目的、过程与成果都要在很短的篇幅中展示出来。所以，在科技论文摘要中，连接词的使用对读者更好地理解科学研究中的逻辑关系起着重要作用。无论是英语还是汉语，连接词都是衔接句子的重要成分。但是，根据逻辑关系，在翻译连接词时也经常会进行转换。例如：

JA delays GA-mediated DELLA protein degradation, and the DELLA mutant is less sensitive to JA for growth inhibition.

译文：茉莉酸可延缓由赤霉酸介导的 DELLA 蛋白降解过程，而 DELLA 突变体对茉莉酸的生长抑制作用不太敏感。

3. 缩略语的翻译

在科技论文摘要中，有些缩略语有多种含义，也有些缩略语不常用，应尽量避免使用这类缩略语。如果确实需要使用，应在正文中首次出现时，在括号内标出其全称以免造成误解。在翻译缩略语时，不仅要满足目标读者便于识别、熟悉并存储这些外来信息的需求，还应注意还原翻译音节等问题。因此，应根据科技论文摘要中缩略语的音节长短、使用频率、受众熟悉程度等不同情况采用相应的翻译方法。

（1）还原意译法：此法仅适用于汉语对应词的音节长短与英语原文的音节长短相差不多的缩略语翻译。例如，DME 的全称为 Distance Measuring Equipment（三个音节），可译为"测距仪"（三个音节）；RVR 的全称为 Runway Visual Range（三个音节），可译为"跑道视距"（四个音节）。

（2）音译法：对于汉语对应词很长但使用频率高的缩略语，通常采用音译法（又称借用法、直接引用法、形译法、合理零翻译法等），即在受众理解的条件下直接借用英语缩略语。例如，CDU 的全称为 Control Display Unit，汉语意思为"控制显示组件"，翻译时可直接用"CDU"。

（3）加字音译法：又称部分意译法，可作为上面直接引用法的延伸和补充，在借用的英语缩略语之后加上适当的说明词，更便于读者快速准确地识别其含义。例如，VNAV（Vertical Navigation，垂直导航）可根据情况译为"VNAV 键"。

（4）意译缩略法：即用汉语的缩略语来翻译英语的缩略语。目前各行业中，此类优秀翻译实例比比皆是。例如，EEC（European Economic Community，欧洲经济共同体）可译为"欧共体"。

（三）句法方面

1. 科技论文摘要英译时在句法方面的指导方向

在翻译科技论文摘要时，译者要意识到英汉不同的句子结构特点。相较而言，汉语句子是线性结构的，侧重语意的连贯性；而英语句子是空间结构的，侧重语法的完整性。

英汉句子的不同特点决定了科技论文摘要英译时在句法方面的指导方向：由于汉语句子侧重意合，而英语句子侧重形合，因此为了保持英译的科技论文摘要的完整性和准确性，常常把若干汉语句子合并成一个英语长句子，而且长句子里经常使用介词短语、分词结构和非限制性定语从句等；在科技论文摘要英译过程中，将一个汉语句子分译成多个英语句子的情况也经常发生。对于汉语原句的英译处理，无论是对其进行合并还是对其进行分拆，前提都是译者要对原文做透彻的分析和理解后再统筹安排句式，绝不能删减或省略原文的内容。英语句子中的省略现在虽然在口语中很常见，但由于科技论文摘要以专业研究人员为目标读者，因此要求句法结构严谨规范，英译科技论文摘要时，一般不出现口语中的省略句。

2. 双重被动句的翻译

英语双重被动句的译法类似于一般被动句的译法，但英语双重被动句在译成汉语时，不能生搬硬套，而要根据汉语的习惯，采取一些辅助、综合的办法，因为英语中有双重被动句，而汉语中没有相应的句式。

（1）英语双重被动句译成含有泛指主语的汉语主动句，译句前通常加上"人们""大家""有人""我们"等泛指词，原句中的主语与第二个不定式被动结构扩展为一个新的分句。例如：

The essay is known to have been plagiarized.

译文：人们知道，有人剽窃了这篇文章。

（2）英语双重被动句译成汉语无主句，原句中的主语与第二个不定式被动结构扩展为动宾结构。例如：

The guests were arranged to be met at the meeting room.

译文：安排在会议室见客人。

（3）英语双重被动句在汉译时，有时可以按顺序直接省略被动语言标志。例如：

That book is not allowed to be published.

译文：那本书不允许出版了。

（4）有的英语双重被动句在汉译时，可根据第一个被动结构的词汇意义译成"据……"，原句中的主语与第二个不定式被动结构扩展为一个新的分句。例如：

The research fruit is supposed to be announced soon.

译文：据估计，研究成果不久就公布。

一般来说，英语被动语态是由"be＋及物动词的过去分词"构成的，但是这并不是被动语态的唯一表示方式，除用助动词 be 外，有些动词也可以用来构成被动语态。弄清楚这一问题，有利于翻译英语被动句。例如，"动词 get、become 或 feel 的一定形式＋及物动词的过去分词"可表示被动语态，这一结构通常表示动作的结果而非动作本身，也常用来表示突然发生、未曾料到的态势。

3. 长难句的翻译

（1）适当增词、减词。翻译中适当地增词或减词是出于意义上、语法上、修辞上或逻辑上的需要，而非随意：增加的是原文中虽无其词却有其意的词，减少的是无须重复其意的冗余词，使译文无论从语气上还是从内容表述上都更显完整而又不烦琐。例如：

After finishing the last life, the non – destructive measurements was analyzed.

译文：最后一次疲劳寿命试验完成后，分析无损检测结果。

All simulation systems were designed to be modular and do not have to be permanently located in the chamber. The benefits of this are：easily reconfigured, expanded；minimize the number of penetrations through.

译文：所有的模拟系统实现模块化（积木式）设计，而且不必长期存放在环境室中。好处：系统易重组、扩充，减少打孔次数。

（2）综合顺序、变序、分译等方法。翻译含有多个从句修饰的长句时，可以综合使用顺序、变序、分译等方法。通过调整句子顺序、更改标点符号、采用意译等手法，使译句表达更符合汉语习惯和科技文本写作风格。例如：

The McKinley Climatic Laboratory is engaged in a major upgrade to develop an improved capability for simulating large scale icing conditions for both military and commercial full – scale aircraft.

译文：麦金利气候实验室正在进行重大升级，以提高军用和商用全尺寸飞机模拟大规模结冰条件的能力。

（3）语序转换法。语序转换法也称变序翻译法，指按照汉语的表达习惯对句子进行重新组织和排序。中西文化具有差异性，语言表达习惯也不相同。例如，汉语句子在时间上重先后顺序，在逻辑上重前因后果关系；而英语句子既可顺序排列句内语义成分，也可逆序排列句内语义成分。在翻译科技论文摘要中的长难句时，几乎不可能保留原文的语序，大多数情况下必须对语序做出调整，才符合汉语的表达习惯。例如：

（1）Aluminum remained unknown until the nineteenth century,（2）because nowhere in nature is it found free,（3）owing to its always being combined with other elements,（4）most commonly with oxygen,（5）for which it has a strong affinity.

译文：（3）铝总是跟其他元素结合在一起，（4）最普遍的是跟氧结合，（5）因为它跟氧有很强的亲和力，由于这个原因，（2）在自然界找不到游离状态的铝。所以，（1）铝直到19世纪才被发现。

五、课后练习

（一）思考题

（1）科技论文摘要有哪些特点？
（2）翻译科技论文摘要有哪些技巧？

（二）翻译练习

1. 句子翻译

（1）机器人技术的研究已从传统的工业领域扩展到医疗服务、教育娱乐、勘探勘测、生物工程、救灾救援等新领域。

（2）稀疏表示模型常利用训练样本学习过完备字典，旨在获得信号的冗余稀疏表示。设计简单、高效、通用性强的字典学习算法是目前的主要研究方向之一，也是信息领域的研究热点。

（3）The success of Alpha Go and corresponding thesis ensure the technical soundness of the parallel intelligence approach for intelligent control and management of complex systems and knowledge automation.

（4）Under the network – based control framework, the sampled measurements are transmitted through the communication networks, which may be attacked by energy limited Denial of Service (DoS) attacks with a characterization of the maximum count of continuous data losses (resilience index).

2. 段落翻译
(1) 英译中。

①Inter-satellite Link (ISL) scheduling is required by the BeiDou Navigation Satellite System (BDS) to guarantee the system ranging and communication performance. In the BDS, a great number of ISL scheduling instances must be addressed every day, which will certainly spend a lot of time via normal metaheuristics and hardly meet the quick-response requirements that often occur in real-world applications. To address the dual requirements of normal and quick-response ISL schedulings, a Data-driven Heuristic Assisted Memetic Algorithm (DHMA) is proposed in this paper, which includes a high-performance Memetic Algorithm (MA) and a data-driven heuristic. In normal situations, the high-performance MA that hybridizes parallelism, competition, and evolution strategies is performed for high-quality ISL scheduling solutions over time. When in quick-response situations, the data-driven heuristic is performed to quickly schedule high-probability ISLs according to a prediction model, which is trained from the high-quality MA solutions. The main idea of the DHMA is to address normal and quick-response schedulings separately, while high-quality normal scheduling data are trained for quick-response use. In addition, this paper also presents an easy-to-understand ISL scheduling model and its NP-completeness. A seven-day experimental study with 10,080 one-minute ISL scheduling instances shows the efficient performance of the DHMA in addressing the ISL scheduling in normal (in 84 hours) and quick-response (in 0.62 hour) situations, which can well meet the dual scheduling requirements in real-world BDS applications.

②With the increasing maritime activities and the rapidly developing maritime economy, the 5th-Generation (5G) mobile communication system is expected to be deployed at the ocean. New technologies need to be explored to meet the requirements of Ultra-reliable and Low Latency Communications (URLLC) in the Maritime Communication Network (MCN). Mobile Edge Computing (MEC) can achieve high energy efficiency in MCN at the cost of suffering from high control plane latency and low reliability. In terms of this issue, the Mobile Edge Communications, Computing, and Caching (MEC3) technology is proposed to sink mobile computing, network control, and storage to the edge of the network. New methods that enable resource-efficient configurations and reduce redundant data transmissions can enable the reliable implementation of computing-intension and latency-sensitive applications. The key technologies of MEC3 to enable URLLC are analyzed and optimized in MCN. The Best Response-based Offloading Algorithm (BROA) is adopted to optimize task offloading. The simulation results show that the task latency can be decreased by 26.5 ms, and the energy consumption in terminal users can be reduced to 66.6%.

③Recently, leading research communities have been investigating the use of blockchains for Artificial Intelligence (AI) applications, where multiple participants, or agents, collaborate to make consensus decisions. To achieve this, the data in the blockchain storage have to be transformed into blockchain knowledge. We refer to these types of blockchains as knowledge-based blockchains. Knowledge-based blockchains are potentially useful in building efficient risk assessment applications. An earlier work introduced probabilistic blockchain which facilitates knowledge-based blockchains. This paper proposes an extension for the probabilistic blockchain concept. The design of a reputation management framework, suitable for such blockchains, is proposed. The framework has been developed to suit the requirements of a wide range of applications. In particular, we apply it to the detection of malicious nodes and reduce their effect on the probabilistic blockchains' consensus process. We evaluate the framework by comparing it to a baseline using several adversarial strategies. Further, we analyze the collaborative decisions with and without the malicious node detection. Both results show a sustainable performance, where the proposed work outperforms others and achieves excellent results.

④AI acceleration is one of the most actively researched fields in IP and system design. The introduction of specialized AI accelerators in the cloud and at the edge has made it possible to deploy large scaled AI solutions that automate tasks that were previously not possible without AI. The growth of big data and the compute horsepower needed to process this data to provide business intelligence are keys for several companies in gaining a competitive edge. AI workloads are both data and compute intensive and improving the efficiency often requires an end-to-end solution. In this perspective paper, we identify key considerations for the design of AI accelerators. The focus of this paper is on Deep Neural Networks (DNN) and how to build efficient deep neural network accelerators through micro architectural exploration, energy efficient memory hierarchies, flexible dataflow distribution, domain-specific compute optimizations and finally hardware-software co-design techniques. The importance of interconnect topology and the impact of its scaling to the physical design of an AI accelerator is also a key consideration that is described in this paper. In the future, the energy efficiency of these accelerators may rely on approximation computing, compute-in-memory and runtime flexibility for significant improvement.

（2）中译英。

①电动汽车规模化应用后，其充电功率需求将对电网产生一定影响。分析了与电动汽车充电功率需求相关的各种因素。在一定假设条件下，根据燃油车的统计数据，考虑了部分随机因素的概率分布，建立了电动汽车功率需求的统计模型。用蒙特卡罗仿真方法求得单台电动汽车功率需求的期望和标准差，进而给出多台电动汽车总体功率需求的计算方法。以北京市和上海市夏季某日负荷曲线为例，计算得出不同规模电动汽车的功率需求对原负荷曲线的影响。计算结果表明，电动汽车的自然充电特性将使电网最大负荷发生一定增长。该统计模型为研究电动汽车对电网的影响提供了基础，也为电动汽车充电管理策略的设计提供了依据。

②视觉跟踪问题是当前计算机视觉领域中的热点问题，本文对这一问题进行了详细的介绍。首先，对视觉跟踪技术在视频监视、图像压缩和三维重构三个主要方面的应用进行了论述。其次，详细阐述了该技术的研究现状，介绍了其中的一些常用方法。为清楚说明这些方法，首先对视觉跟踪问题进行了分类，然后介绍了处理视觉跟踪问题的两种思路，即自底向上和自顶向下的思路，最后将具体的视觉跟踪方法分为四类进行了介绍，这四类分别是基于区域的跟踪、基于特征的跟踪、基于变形模板的跟踪和基于模型的跟踪。最后，从控制论角度给出视觉跟踪算法所面临的难点，即算法要满足鲁棒性、准确性和快速性要求时所遇到的困难，并对视觉跟踪问题的研究前景进行了展望。

③信息感知作为物联网的基本功能，是物联网信息"全面感知"的手段。信息交互是物联网应用与服务的基础，是物联网"物物互联"的目的。随着物联网研究热潮的兴起，以传统无线传感器网络为核心的感知网络研究迅速升温，并在信息感知与交互方面取得了大量研究成果。文章分析了物联网信息感知与交互方面的最新研究进展。在信息感知方面，从数据收集、清洗、压缩、聚集和融合几个方面，梳理归纳了数据获取和处理的主要方法。在信息交互方面，提出了物联网信息交互的基本模型，分析总结了信息交互涉及的主要技术。在此基础上，讨论了物联网信息感知与交互研究的热点问题，包括新的感知技术、能效平衡、信息安全和移动感知网络等。最后，指出了物联网信息感知与交互技术发展面临的问题和挑战，展望了未来的研究方向。

④本文综述了计算流体力学（Computational Fluid Dynamics，CFD），尤其是计算空气动力学的发展概况。从计算方法、网格技术、湍流模型、大涡模拟等方面分别总结了 CFD 所取得的成就，分析了当前存在的问题、困惑，展望了其发展趋势。在 CFD 计算方法中，主要介绍了中心格式、迎风格式、TVD 格式、WENO 格式、紧致格式及间断 Galerkin 有限元方法，对不同方法的原理和特性进行了系统阐述。网格技术包括结构网格、非结构网格、混合网格及重叠网格，重点讨论了重叠网格的若干关键技术。在湍流模型中，对目前的模型进行分类介绍，包括线性涡黏性模型、二阶矩模型、非线性模型等，还介绍了转捩模型、DES 方法及 SAS 方法等。在大涡模拟方法中，就其中若干相关的研究方向进行了探讨，包括滤波方法、亚格子模型、收敛标准、数值格式等。文中还包含了作者在相关领域的若干研究成果。

学思践悟

第七章 科技产品说明书

一、科技产品说明书文本介绍

(一) 科技产品说明书概述

科技产品说明书是以文字加图片、图表的方式对某科技产品进行相对详细的表述,使人认识、了解某科技产品。科技产品说明书是一种常见的说明文,是生产者向消费者全面、明确地介绍科技产品名称、用途、性质、性能、原理、构造、规格、使用方法、保养维护、注意事项等内容而写的准确、简明的文字材料。

(二) 科技产品说明书文本结构

科技产品说明书要全面地说明产品,不仅要介绍其优点,还要清楚地说明应注意的事项和可能产生的问题。科技产品说明书一般采用说明性文字,可根据情况需要,使用图片、图表等多样形式,以达到最好的说明效果。

科技产品说明书通常由标题、正文和落款三个部分构成。正文是科技产品说明书的主题、核心部分。

1. 标题

科技产品说明书的标题通常由产品名称或说明对象加上文种构成,一般放在第一行,要注重视觉效果,可以有不同的形式设计。

2. 正文

正文是科技产品说明书的主体部分,是介绍产品的特征、性能、使用方法、保养维护、注意事项等内容的核心所在。常见的主体项目有概述、指标、结构、特点、方法、配套、事项、保养、责任等。

由于科技产品说明书介绍的产品千差万别,因而不同科技产品说明书的正文内容侧重点也有所不同。

(1) 家用电器类说明书:此类说明书一般较为复杂,正文内容有产品的构成、规格型号、适用对象、使用方法、注意事项等。

(2) 日用生活品类说明书:正文内容有产品的构成、规格型号、适用对象、使用方法、注意事项等。

(3) 食品药物类说明书:正文内容有产品的构成成分、特点、性状、作用、适用范围、使用方法、保存方法、有效期限、注意事项等。

(4) 大型机器设备类说明书:正文内容有产品的结构特征、技术特性、安装方法、使用方法、功能作用、维修保养、运输、售后服务范围及方式、注意事项等。

3. 落款

落款是要写明生产者和经销单位的名称、地址、电话、邮政编码、电子邮箱等内容，为消费者进行必要的联系提供方便。

（三）科技产品说明书的作用

1. 宣传产品，指导消费

科技产品说明书的最基本作用是宣传产品以激发消费者的购买欲望，从而实现购买。促进科技产品流通，是科技产品说明书的基本属性。

2. 扩大信息量，促进消费

科技产品说明书在普及科学文化知识，扩大信息量，促进信息传播、复制、交流、利用、反馈等方面，作用十分重要。

3. 传播知识，创造品牌

科技产品说明书的内容常常涉及科技知识的普及、宣传和利用。消费者认识科技产品，往往是从认识科技产品说明书开始的。科技产品说明书是创造品牌的必需环节。品牌可以借助科技产品说明书推波助澜，使品牌有形象、直观的效果。

（四）科技产品说明书的特点

科技产品说明书兼具多种功能，具有如下六大特点。

1. 客观性

科技产品说明书主要面向普通消费者，介绍产品性能、特点等，以便他们做出正确的选择，并在购买后正确使用。这种语篇主要反映生产厂家与消费者之间的交流关系，所以说明书的内容必须通俗易懂、实事求是（即有一说一，有二说二），对产品进行客观的描述，采用恰当、准确的词汇如实地反映产品的真实情况，语气应客观冷静，不能使用过多的修辞手法或含有渲染、夸大成分。

2. 专业性

科技产品说明书是一种相当专业的应用文体，专业性是其一大突出特点。这种特殊文体有较固定的结构要素，如名称、型号、功能、特性、用途、注意事项、安装等，因此会频繁使用一些固定的名词词组或动词词组。由于描述对象覆盖面甚广，所以科技产品说明书会涉及各种专业和领域，常常会使用某个技术领域特有的行业术语。

3. 简明性

科技产品说明书的目标读者是普通大众而非专业人士，他们往往对产品所属的领域比较陌生，因此在语言上通常会尽量简洁。科技产品说明书的简洁性首先体现在名词化上，语篇中描述性的长句通常被浓缩为名词或词组；科技产品说明书的简洁性还体现在省略结构和缩略词的频繁使用上；科技产品说明书的简洁性又体现在结构形式上，它经常采用条款式，逐条列出。

4. 号召性

现代企业通常会在介绍客观信息的基础上，有效地利用产品说明书进行隐性的产品宣传，并拉近与消费者之间的距离，促使其产生购买欲望。

5. 科学性

科技产品是科学与生产实践结合的产物，在一定程度上体现了当代的科技水平，是科技的一种实用型代表。

6. 条理性

因文化、地理、生活、环境等的不同，人们对科技产品说明书的内容还存在着认识和理解上的差异，所以，科技产品说明书在陈述产品的各种要素时，要有一个由浅入深、循序渐进的顺序。

（五）科技产品说明书的分类

一张科技产品说明书除具有简明易懂的说明外，一般配有彩色插图，如机件位置和名称图、操作程序图等，电子产品还要附电路图。科技产品说明书是推销科技产品、传播新科技知识的一种重要媒介，一般会精心设计，印刷精美，引人注目。

按照不同的分类标准，科技产品说明书可分类如下。

（1）按对象、行业的不同分类，可分为工业产品说明书、农产品说明书、金融产品说明书、保险产品说明书等。

（2）按形式的不同分类，可分为条款（条文）式产品说明书、图表式产品说明书、条款（条文）和图表结合说明书、口述产品说明书等。

（3）按传播方式的不同分类，可分为包装式产品说明书和内装式产品说明书。包装式产品说明书就是直接写在产品外包装上的说明书。内装式产品说明书就是采用附件的形式，专门印制，有的甚至装订成册，装在产品的包装箱（盒）内的说明书。

（4）按功能的不同分类，可分为产品介绍说明书、产品使用说明书、产品安装说明书等。产品介绍说明书主要是针对某一产品基本情况进行介绍，如其组成材料、性能、存储方式、注意事项、主要用途等。产品介绍说明书的对象可以是生产消费品，如电视机；也可以是生活消费品，如食品、药品等。产品使用说明书是向人们介绍具体的关于某产品的使用方法和步骤的说明书。产品安装说明书主要介绍如何将一堆分散的产品部件安装成一个可以使用的完整的产品。为了方便运输，许多产品都是拆开分装的，用户在购买到产品之后，需要将散装部件合理地安装在一起，这时就需要有一张翔实的产品安装说明书。

二、科技产品说明书范文展示

范文一　以下为 NOHON 锂电池使用说明书

使用说明：

1. 我们建议您在首次使用此 NOHON 电池前，将电池完全充电。
2. 新电池只有在经过两次或三次完全的充放电周期之后才能达到最佳性能。
3. 请勿将充满电量的电池连接至充电器，因为过分充电可能会缩短电池寿命。
4. 如果您打算长时间不使用电池，请必须将电池拆下，并存放在低温、干燥的环境下，而且电池必须充满。电池即使未使用也会自动放电，因此应避免电池过分放电处于低电状态导致无法再度充电。
5. 请使用 NOHON 为此电池设计的充电器充电。

6. 请不要将电池放于充电器上充电超过24小时，不使用时将充电器从电源插座及装置中拔出。

7. 请不要把电池留在过热或过冷的地方（如夏天或冬天的密封车厢中），因为那样会减小电池容量，缩短电池寿命。

8. 请尽量将电池的温度保持在15℃至25℃之间。

9. 个别电池型号附有透明拉带，此拉带有助于将电池轻易拉出，请勿将电池拉带剪掉。

10. 如果装上过热或过冷的电池，即使电池完全充电，也可能暂时无法操作。

Operation instruction：

1. We recommend to fully chargethe NOHON battery before using it for the first time.

2. A new battery will achieve the best performance after 2 or 3 complete cycles of charge discharge.

3. Do not leave a fully charged battery connected to chargeras overcharging may shorten its lifetime.

4. If you're not going to use the battery for long period, you must remove the battery and keep it in dry and low – temperature environment with full charge. The battery will automatically discharge even if it is not in use. Therefore, it should be avoided excessive discharge to a low voltage condition that will lead to failure to charge again.

5. Use only NOHON chargers designed for its batteries.

6. Do not leave the battery on charger over 24 hours, and unplug the charger from the power socket and device while not using.

7. Leaving the battery in an overheated or cold place (such as a sealed cabin in summer or winter) will reduce the capacity and lifetime of the battery.

8. Always try to keep the battery between 15℃ and 25℃.

9. Some battery models have an additional transparency strap which helps to pull out the battery easily. Therefore, do not cut off the battery strap.

10. Installing an overheated or overcooled battery may not work temporarily even when the battery is fully charged.

范文二　以下为机械安装调试说明书

Cautions：

1. The entire installation and commissioning process must be carried out in accordance with the installation and commissioning manual, and personal safety must be paid attention to when powering on and debugging. Never adjust parts when the door is moving to prevent body injury.

2. After the end of commissioning, all fasteners on main components (such as mechanisms, door leaves, and emergency exit units) must be tightened to the rated torque, applied with thread lock glue according to the installation requirements, and marked with marking paint (to identify the tightening of screws).

3. Illustrations in the instruction manual are not detailed drawings, and are only used to provide required information for assembling. Unless specifically indicated, the installation work must be

done carefully by qualified mechanical and electrical personnel, especially for the technical information provided in the instruction manual.

4. The following power supply is required for the commissioning of the door system: DC110V (voltage fluctuation range DC77V – DC137.5V).

5. In the assembling of door system, ensure that the car body is in horizontal and braked state. If it is not horizontal, adjustment should be made. And the assembling of the passenger compartment door system can be stated only after completion of body horizontality adjustment.

6. It is not necessary to apply thread lock glue on the screws for connecting wire screw sleeve, but an appropriate amount of lubricating oil should be applied to prevent it from seizing (the lubricating oil mentioned in the instructions can be used).

7. The working environment must be clean and safe.

8. It is strictly forbidden to directly touch the long and shot guide pillar and screw surface with hand during the site installation or handling of mechanism, as such contact may lead to surface rusting! If contact is made, clean the surface promptly and apply anti – rust oil on it.

9. If the installation and commissioning method we provide has anything not suitable to the site conditions, treat it according to specific cases!

注意事项：

1. 整个安装和调试过程必须按照安装和调试手册进行，在通电调试时，必须注意人身安全。门在移动时，切勿调试零部件以防身体受伤。

2. 调试结束后，必须将主要部件（如机构、门扇和应急出口装置）上的所有紧固件按标定转矩旋紧，根据安装要求涂抹螺纹锁固胶，并用标记漆标记（用于识别螺钉的紧固性）。

3. 说明手册中的图示并非详细图纸，仅用于提供组装所需的信息。除非特别注明，安装工作必须由有资质的机电安装人员谨慎进行，特别是说明手册中提供的技术信息。

4. 调试门系统时需要以下电源：DC110V（电压波动范围为DC77V～DC137.5V）。

5. 在安装门系统时，要确保车体水平并处于制动状态。若车体不水平，则需要做相应调整。待车体水平调整完成后，才能进行乘客室门系统的安装。

6. 无须在用于连接钢丝螺套的螺钉上涂抹螺纹锁固胶，但应涂抹适量的润滑油以防止其咬死（可使用说明书中提及的润滑油）。

7. 工作环境必须清洁、安全。

8. 严禁在现场安装或搬运机械装置过程中用手直接触碰长导柱、短导柱和螺钉的表面，因为一旦触碰，则可能导致其表面生锈！如造成触碰，请立即清洁表面并涂抹防锈油。

9. 如果我们提供的安装和调试方法存在不适合现场条件之处，请根据具体情况进行处理！

范文三　以下为空调机组说明书

1　安装后检查

1.1　空调机组的安全操作

空调机组的操作和管理工作，必须由懂得制冷技术和电气技术的人员来担任。开机之前，必须认真检查电气系统的安全性，严格按照电工操作规则进行操作。在进行电气部件的检修时，必须切断电源，严禁带电作业。

空调机组必须确认下列项没有问题之后，方可开始运转。

1.2 密封检查

空调机组安装好后，需对机组出风口、回风口的密封进行检查。如无泄漏，则在风机运转下进行淋雨检查。雨水不得从出风口、回风口漏入车内。

1.3 配线检查

检查配线用的电气连接器是否确实接好。

检查电气回路是否正常。

检查主回路及控制回路的绝缘电阻是否均正常。

1.4 室内蒸发风机的确认

离心风机运转后，要确认机组出风口是否有风吹出。若出风口风量很小，则可能是离心风机反转。请将电源相序调整正确，即将三相中的任意两相对调。室外轴流风机、压缩机反转时，请直接将其电源三相中的任意两相对调即可。

注意：空调机组出厂时各电动机的相序已调好，请不要随意调换。

1.5 制冷运转确认

室外冷凝风机正常运转时，室外冷凝出风从冷凝器吸入机组，从机组顶部吹出。同时，从机组送风出口应有冷风吹出。请判断风机运转时，是否有异常振动或杂音。

制冷运转时，回风口、出风口温差约 8~10℃。

2 拆卸

参考文件：01.DP.01。

2.1 空调机组的拆卸顺序

切断电源—拔下空调机组连接器插头—拆除空调机组接地线—拆下空调机组紧固螺栓—将机组吊起—平移并缓缓放至地面—包装存放。

2.2 空调机组的拆卸

为了操作安全，空调机组拆卸时，必须先切断空调电源。然后，松开并拔下空调机组连接器插头。拔下连接器插头后，请将插座保护盖扣上以便保护连接器插座。接着，拆除空调机组接地线。松开并拆下空调机组与车体安装座之间的紧固螺栓，略外移机组，以便机组起吊。最后，用起吊装置将空调机组吊起，平移并缓缓放至地面。起吊时，注意不要碰伤机组。如果空调机组拆卸后短期内不对其进行有关检修或其他操作，请将其包装好并存放在干燥处。空调机组拆卸完毕，请将紧固螺栓保存好，以便下次使用。空调机组拆卸后，请检查送风出口密封胶条，如有破损或老化则需要更换。

1 Inspection after installation

1.1 Safety operation of air conditioning unit

Operation and management of the air conditioning unit must be undertaken by personnel knowing refrigeration technology and electrical technology. Before starting, the safety of the electrical system must be carefully checked, and the operation must be carried out strictly according to the electrician operating rules. The power must be cut off when the electrical components are inspected and maintained. And it is forbidden to operate when the power is on.

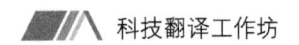

The air conditioning unit can only be operated after being confirmed that the following items are safe.

1.2 Sealing inspection

After the installation of air conditioning unit, it is necessary to check the seal of the air outlet and air return inlet. If no leakage is found, the rain test will be carried out with operation of fan. The rainwater shall not leak into the car from the air outlet and air return inlet.

1.3 Wiring inspection

Inspect whether the electrical connector for wiring is properly connected.

Inspect whether the electrical circuit is normal.

Inspect whether the insulation resistances of the main/control circuits are normal.

1.4 Confirmation of indoor evaporation fan

Please confirm whether the air is blown out from the air outlet of the unit after the centrifugal fan runs. It is possible that the centrifugal fan is reversely operating if the airflow at the air outlet is very low. Please correctly adjust the phase sequence of the power supply, that is, adjust any two of the three phases to each other. When the outdoor axial flow fan and compressor run reversely, please directly adjust any two of the three phases of the power supply to each other.

Note: please do not exchange the phase sequence of each motor at will, because it has been adjusted when the air conditioning unit leaves the factory.

1.5 Confirmation of refrigeration operation

Under the normal operation of outdoor condenser fan, the condensing air is taken into the unit from the condenser and blown out from the top of unit. Meanwhile, the cold air shall be blown out from the unit air supply outlet. Please take care whether abnormal vibration or noise is present during the operation of fan.

Temperature difference between the air return inlet and air outlet during the refrigeration is about $8 \sim 10\,^\circ\mathrm{C}$.

2 Disassembly

Referto 01.DP.01.

2.1 Disassembly sequence of air conditioning unit

Cut off the power supply – unplug the connector plug of air conditioning unit – remove the earth wire of air conditioning unit – remove the fastening bolts of air conditioning unit – lift the unit – translate and slowly put the unit on the ground – pack and store it.

2.2 Disassembly of air conditioning unit

In order to operate safely, the power supply of the air conditioner must be cut off first when the air conditioning unit is disassembled. Then, release and unplug the connector plug of air conditioning unit. After unplugging the connector plug, fasten the socket protection cover to protect the connector socket. Then, remove the earth wire of air conditioning unit. Loosen and remove the fastening bolts between the air conditioning unit and the vehicle body mounting base, and slightly move the unit outward to facilitate the lifting of the unit. Finally, use the lifting device to lift the air conditioning unit, translate and slowly put it on the ground. When lifting, be careful not to damage the unit.

If the air conditioning unit is not repaired or operated in a short time after disassembly, please pack it and store it in a dry place. After the air conditioning unit is disassembled, please keep the fastening bolts for next use. After the air conditioning unit is disassembled, please check the sealing strip at the air supply outlet, and replace it if it is damaged or aged.

三、科技产品说明书常用词汇和句型

(一) 科技产品说明书常用词汇

1. 使用普通词汇

科技产品说明书通常要面对不同阶层、不同文化程度的用户，其表达风格应当"就低不就高"，那些面向社会大众的民用产品说明书尤其如此。就科技产品说明书所用词汇而言，普通词汇应当占相当大的比例，充分体现了其用词简明、朴素的特点。例如：

Park the vehicle in a safe place and turn off the engine. Test the system as described on page 11. If the light does not go out after the test, or if it illuminates again, have your vehicle inspected at an authorized dealer. When this light comes on, the anti-lock braking system is not functioning and only the ordinary braking system is functioning.

译文：将汽车停于安全处，并关掉发动机。按照第11页上所介绍的方法对系统进行测试。如果在测试之后灯没有熄灭，或者又重新亮起，请将车辆交由经授权的经销商检查。灯亮说明此时防抱死制动系统未起作用，而只是普通刹车系统在发挥作用。

This camera has compact body featuring 28mm wide-angle to 200mm telephoto 7.1 x optical zoom lens that covers a wide range of shooting. This lens can be used to take pictures in various indoor and outdoor situations.

译文：这款相机机身小巧，配有一个28mm广角至200mm远摄7.1倍的光学变焦镜头，拍摄范围广。此镜头可用于在各种室内与室外场景下拍照。

2. 使用专业术语

科技产品说明书，特别是在不同领域的专用设备说明书，不可避免地要使用大量的专业术语，以体现这类产品说明书的严肃性、专业性和规范性。下面列出机械类、计算机类、通信工程类和医药类科技产品说明书常用专业术语。

(1) 机械类科技产品说明书常用专业术语如下。

automotive chassis 车架，汽车底盘
suspension 悬架
redirector 转向器
speed changer 变速器
mathematical model 数学模型
descriptive geometry 画法几何
mechanical drawing 机械制图

projection 投影
view 视图
profile chart 剖视图
standard component 标准件
part drawing 零件图
assembly drawing 装配图
size marking 尺寸标注
technical requirement 技术要求
rigidity 硬度，刚性
internal force 内力
displacement 位移
section 截面
fatigue limit 疲劳极限
fracture 断裂
plastic distortion 塑性变形
brittleness material 脆性材料
washer 垫圈
spacer 垫片
helical – spur gear 斜齿圆柱齿轮
straight bevel gear 直齿锥齿轮
kinematic sketch 运动简图
slide caliper 游标卡尺
micrometer caliper 千分尺
kinetic energy 动能
potential energy 势能
conservation of mechanical energy 机械能守恒
momentum 动量
axis 轴线
logic circuit 逻辑电路
pulse shape 脉冲波形
mechanical parts 机械零件
hardening and tempering 调质
abrasive grain 磨料颗粒
bonding agent 结合剂
grinding wheel 砂轮
turning 车削
grinder 磨床，研磨机
fluid dynamics 流体动力学
machining 机械加工

hydraulic pressure 液压
tangent 切线
stability 稳定性
medium 介质
hydraulic pump 液压泵
valve 阀门
invalidation 失效
intensity 强度
load 载荷
stress 应力
safety factor 安全系数
assembly line 流水线，装配线
barcode scanner 条码扫描器
fuse machine 热熔机
forklift 叉车
bolt 螺栓
stock age analysis sheet 库存货龄分析表
part number 零件代码
class 类别
difference quantity 差异量
spare molds location 模具备品仓

（2）计算机类科技产品说明书常用专业术语如下。

abstraction layer 抽象层
access 获取，存取
acoustic coupler 声音耦合器
Carbon Copy（CC） 复写本，副本；抄送
Cascading Style Sheets（CSS） 串联样式表
Central Processing Unit（CPU） 中央处理器
channel 信道，频道
composite formatting 复合格式化
concurrency conflict 并发冲突
concurrency model 并发模型
conditional compilation 条件编译
coupling number 耦合数
data access layer 数据访问层
Data Encryption Standard（DES） 数据加密标准
data integrity 数据完整性
data mining 数据挖掘
data pump 数据泵

Data Storage as a Service（DaaS） 数据存储即服务
Data Transfer Object（DTO） 数据传输对象
subdomain 子域
defensive programming 防御式编程
Document Object Model（DOM） 文档对象模型
Domain Name System（DNS） 域名系统
Domain – Driven Design（DDD） 领域驱动设计
failure domain 故障域
Hard Disk Drive（HDD） 硬盘驱动器
Human – Computer Interaction（HCI） 人机交互
hypertext – driven 超文本驱动
Intrusion Detection System（IDS） 入侵检测系统
Intrusion Prevention System（IPS） 入侵防御系统
load balancer 负载平衡器
load factor 负载因子
Optical CharacterRecognition（OCR） 光学字符识别
ServiceLevel Agreement（SLA） 服务等级协定
throughput 吞吐量
timestamp 时间戳
topology 拓扑
user analysis 用户分析
User Datagram Protocol（UDP） 用户数据报协议
Virtual Private Network（VPN） 虚拟专用网络
Virtual Reality（VR） 虚拟现实
weak generational hypothesis 弱分代假说
web scraping 网络抓取
zip disk 压缩磁盘

（3）通信工程类科技产品说明书常用专业术语如下。

Adjacent Channel Interference（ACI） 邻信道干扰
Analog – to – Digital Converter（ADC） 模数转换器
Asymmetric Digital Subscriber Line（ADSL） 非对称数字用户线
Authentication Header（AH） 鉴别头
Application Program Interface（API） 应用程序接口
bit – error rate 误比特率
Border Gateway Protocol（BGP） 边界网关协议
band pass filter 带通滤波器
channel impulse response 信道脉冲响应
constant modulus algorithm 恒模算法
Device – to – Device（D2D） 设备到设备

downlink 下行链路
digital predistortion 数字预失真
Digital – to – Analog Converter (DAC) 数模转换器
Fast Fourier Transform (FFT) 快速傅里叶变换
User Equipment (UE) 用户设备
frequency division duplexing 频分双工
Multiple – Input Multiple – Output (MIMO) 多输入多输出
heterogeneous – network 异构网络
in – phase and quadrature 同相和正交
inter – carrier interference 载波间干扰
Joint Transmission (JT) 联合传输
millimeter wave 毫米波
Non – Line – Of – Sight (NLOS) 非视距
Radio Frequency (RF) 射频
self – interference cancellation 自干扰消除
Spatial Channel Model (SCM) 空间信道模型
ultra – dense network 超密度网络
Ultra – Reliable and Low – Latency Communication (URLLC) 超高可靠低时延通信
phase noise 相位噪声
synchronization signal 同步信号

（4）医药类科技产品说明书常用专业术语如下。

urine analyzer 尿液分析仪
reagent 试剂
blood sugar (glucose) analyzer 血糖（葡萄糖）分析仪
test strip 测试条，试纸
semi – automatic biochemical analyzer 半自动生化分析仪
incubator 培养箱
automatic blood cell analyzer 全自动血细胞分析仪
microbiological incubator 微生物培养箱
disposable sterile injector 一次性无菌注射针
disposable venous infusion needle 一次性静脉输液针
disposable infusion set 一次性使用输液器
blood transfusion set 输血器
infusion bag 液袋
urine drainage bag 尿液引流袋
blood bag 血袋
medical catheter 医用导管
ICU monitor 重症监护仪，重症病人监护装置
stainless steel needle 不锈钢针

blood taking needle 采血针
needle destroyer 针头销毁器
medical injection pump 医用灌注泵，医用注射泵
multi-parameter monitor 多参数监护仪
respirator 呼吸机
maternal monitor/fetal monitor 母亲监护仪/胎儿监护仪
anesthetic equipment 麻醉机
electronic colposcope 电子阴道镜
smog absorber 烟雾吸收器
digital film room 数字胶片室
radiotherapeutic equipment 放射疗法设备
pharmaceutical equipments 制药设备
antibiotics 抗生素
sterilization and disinfection equipment 消毒灭菌设备
effective dose 有效剂量
emission spectra 发射光谱
gene mutation 基因突变
pacemaker 起搏器

（二）科技产品说明书常用句型

1. （情态动词）+ be + 形容词（或过去分词）+ 目的状语
这种句型常用于文章开头，说明产品是做什么用的。例如：
This mode is convenient for receiving both faxes and voice calls.
译文：此模式便于接收传真与接听语音呼叫。

This press is mainly suitable for cold working operations, such as punching, blanking, bending, shallow drawing, cutting and so on.
译文：本冲床主要适用于冲孔、落料、弯曲、浅拉深、剪切等冷加工操作。

This product can be used in hot water or steam line with the temperature limited to 225℃.
译文：该产品用于温度225℃以下的热水管或蒸汽管道上。

类似常见的句型还有 be used for...、be used to...、be used as...、be designed to...、be suitable to...、be available for (to)...、may be applicable to...、may be used to...、can be used as...、can be designed as...、be adapted for (to)...、be designed to... so as to...。

2. （情态动词）+ be + 介词短语
这种句型用于说明产品的特征、状态及计量单位等。例如：
CYJ15-18-18型抽油机结构简单紧凑。
译文：CYJ15-18-18 pumping unit is of simple and compact construction.

The motor and main shaft should be in correct alignment so as to avoid vibration and hot bearings.
译文：电动机与主轴应正确对齐，以免产生振动或引起轴承发热。

The new type of machine should be of simple and compact construction.
译文：新型机器必须是结构简单紧凑的。

3．be + 形容词 + 介词短语

这种句型举例如下。

It is reliable in usage, convenient in maintenance and able to work under very bad conditions.
译文：该机器操作时安全可靠，便于维修，能在恶劣条件下工作。

4．现在分词 + 名词

这种句型用于说明维修或操作程序及有关技术要求。例如：

When doing shallow drawings, care must be taken to ensure cleanness of the sheet and it is well lubricated.
译文：在进行浅拉深时，必须小心确保板材的清洁，并充分润滑。

在更换胶卷时，请使用 Sharp UX – 3CR 胶卷。一卷可以列印约 95 张 A4 大小的页面。
译文：When replacing the film, use Sharp UX – 3CR film. One roll can print about 95 A4 – size pages.

5．名词 + 过去分词（或形容词）

这种句型用在故障或原因说明。例如：

故　　障	原　　因
阀杆运动不灵活	1．阀杆弯曲 2．弹簧损坏 3．压盖填料压得太紧
压盖填料渗漏	1．压盖填料不足 2．压盖填料腐蚀、磨损
密封面渗漏	1．垫片压合不妥当 2．垫片损坏

译文：

Troubles	Reasons
The stem sticky	1. The stem bent 2. The spring broken 3. Gland packing pressed too tightly
Gland packing leaking	1. Gland packing not sufficient 2. Gland packing deteriorated
Sealing surface leaking	1. The gasket not pressed properly 2. The gasket worn

6. 名词性短语

英文科技产品说明书通常在其"产品技术指标和特点"这一部分大量使用名词性短语，这些名词性短语通常以醒目的条目形式出现，重点突出，简明扼要，用户可以轻而易举地了解到产品的性能和特点，从而大大提高文本的交际阅读效率。

下面这段产品说明书的文字可以直观地体现出名词性短语在展示产品技术指标和特点方面所具备的优势。

Long battery life and large memory capacity

Single cell AA size or C size battery operation

Pressure trigger

Sour service operation

Data separation for multiple jobs

Cable connection & Surface readout for real – time applications

Software for on – site configuration and data analysis

译文：电池寿命长，存储器存储容量大

正常工作仅需一节 AA 或 C 号电池

压力触发

可在酸性环境下工作

对多次工作的数据分类

电缆连接与表面读出实时处理

有用于现场参数配置和数据分析的软件

7. 祈使句

科技产品说明书中的祈使句属于指导用户进行某些操作或禁止用户进行某些操作的句子。祈使句的使用是否得当，对产品说明书的编写质量有较大影响。如果祈使句使用得当，那么用户在阅读产品说明书时就会有一种亲切感，从而会欣然按照说明书所指明的步骤进行操作。此外，科技产品说明书中的每个祈使句最好只提出一项要求，这样便于用户领会并遵照执行。例如：

Always switch device off and disconnect the charger before removing the battery.

译文：取出电池前，请务必关闭设备并断开与充电器的连接。

Connect the power cord for your monitor to the power port on the back of the monitor.

译文：请将显示器的电源线连接到显示器背面的电源端口。

Road safety comes first. Obey all local laws. Always keep your hands free to operate the vehicle while driving. Your first consideration while driving should be road safety.

译文：交通安全是第一位的。请遵守当地所有相关法律法规。请始终使用双手驾车。驾车时应首先考虑交通安全。

8. 省略句

在英文科技产品说明书中，省略句的使用十分普遍。这正是科技产品说明书简洁明了的文体特点的具体体现。省略句具有突出重点和提高交际效率的作用。例如：

Contraindications: none known.

译文：禁忌症：尚未发现。

The grinding machine you operate must be oiled, and that at once.

译文：您使用的那台磨床必须上油，而且要马上上油。

9. 被动句

在科技产品说明书中，被动句用得较多，这是由被动句的特点决定的。首先，被动句主观色彩少，更富于客观性，很适合描述客观事物；其次，被动句体现了一个"先入为主"的原则，从而能使被说明的对象引起人们的注意；最后，被动句的结构简洁，便于表达。例如：

The rate of condensation is governed by the rate at which latent heat is carried away from the surface to cooler vapor.

译文：冷凝速度由汽化潜热从该表面转入较冷蒸汽的速度决定。

The serial number and model number of both the keyboard and display module are located on the bottom outside cover of each component.

译文：键盘和显示器的序号及型号位于每一个部件的底部外壳上。

In computers, this very important function of remembering or storing bits of data is performed by various types of flip–flops.

译文：在计算机中，这种记忆或存储数据位的非常重要的功能是由不同类型的触发器实现的。

10. 不定式结构

在英文科技产品说明书中，许多句子以不定式结构开头，是因为不定式结构语气客观、形式简单、通俗易懂，往往可以明确地表达目的和功能等。不定式结构通常充当目的状语，一般译为"如需……"，也可译为"为了……""要……"。例如：

To remove the condensation, turn on the power and wait approximately two hours before using the fax machine.

译文：为了清除冷凝物，请开启电源并等候大约两个小时，然后再使用传真机。

To exit from the power–saving mode, press the power button or use the keyboard or the mouse.

译文：如需退出省电模式，可按电源按钮，也可按键盘或鼠标。

To place a print cartridge, press the power button to turn on the printer, then open the top cover.

译文：要安装墨盒，先按电源按钮启动打印机，然后再打开墨盒顶盖。

11. if 从句

英文科技产品说明书里常用 if 引导的句子。其中，由 if 引导的条件状语从句实际上是对用户在使用产品的过程中可能遇到的某种情形的描述，而接在 if 从句之后的主句往往表达的是商家建议用户采取的应对措施。这类句子通常可译为"如果……，请……"。例如：

If the door was opened on an appliance which has a warm water connection or is heated, first leave the door ajar for several minutes and then close.

译文：如果门是在有热水连接或加热的设备上打开的，请先让门半开着几分钟，然后再关闭。

If you cannot confirm that your Nokia battery with the hologram on the label is an authentic Nokia battery, please do not use the battery.

译文：如果您无法确认自己手中贴有全息标签的诺基亚电池是真正的诺基亚原厂电池，请不要使用该电池。

If the fax machine is moved from a cold to a warm place, it is possible that condensation may form on the scanning glass, preventing proper scanning of documents for transmission.

译文：如果将传真机从寒冷的地方移至温暖的地方，传真机的扫描玻璃上可能形成冷凝物，这会妨碍在传送文件时进行正常扫描。

12. 由 For（或 In 等）引出的短语

举例如下。

For prevention of the advance of cataract.

译文：用于预防白内障的发展。

四、科技产品说明书翻译技巧

下面主要从引申译法、词类转换、句法转换来讨论科技产品说明书的翻译技巧。

（一）引申译法

当英语句子中的某个词按词典的释义直译不符合汉语修辞手法或语言规范时，则可以在不脱离该英语词本义的前提下，灵活选择恰当的汉语词语或词组译出。例如：

Install two "AA" batteries observing the correct polarity to test and fix this problem.

译文：按正确的极性安装两节"AA"电池以进行测试并解决该问题。

原文中，"fix"的本意是"修理；固定"，这里引申译为"解决，处理"。

Because of the circuitous and directional flow of waterways, barges often have an energy advantage over railways.

译文：由于河道迂回曲折且水流具有方向性，因而水运较之铁路运输，常常具有节能优势。

原文中的"railways"应译为"铁路运输"；"barges"的本意为"驳船"，这里也应扩大其语义外延，译为"水运"，这样才能体现作者的原意，也才符合汉语的表达习惯。

以下为词义引申的四个分类。

1. 技术性引申

技术性引申的目的主要是使译文中涉及科学技术概念的词语符合技术语言规范。例如：

After the spring has been closed to its solid height, the compressive force is removed.

译文：弹簧被压缩到接近压并高度之后，就没有压力了。

The adjustment screw has stops at both sides.

译文：调整螺钉的两端设有定位块。

The probe was on the course for Saturn.

译文：探测器在去往土星的路程上。

2. 修辞性引申

修辞性引申的目的是使译文语言流畅，文句通顺，符合汉语的表达习惯。例如：

Hand picks have been replaced by pneumatic picks and electric drills.

译文：手镐已被风镐和电钻取而代之。

Computers come in a wide variety of sizes and capabilities.

译文：计算机大小不一，能力各异。

The antivenin will not help, but a blood transfusion will. The treatment works.

译文：抗蛇毒血清无济于事，但输血有用。这个疗法很有效。

The best solution that science and medicine have come up with so far is the sleeping pill, which is another mixed blessing.

译文：迄今为止，科学和医学对此提出的最佳解决方法是安眠药片，但这也是既有利，又有弊。

3. 具体化引申

把原文语句中含义较概括、抽象、笼统的词引申为意思较为具体的词，尤其是将不定代词进行具体化引申，避免译文概念不清或不符合汉语表达习惯的情况出现。例如：

While this restriction on the size of the circuit holds, the law is valid.

译文：只要电路尺寸符合上述限制，这条定律就能适用于该电路。

This suggests that matter can be converted into energy and vice versa.

译文：这就是说物质可以转化为能量，能量也可以转化为物质。

Many health-conscious women increase their risk by rejecting red meat, which contains the most easily absorbed form of iron.

译文：许多具有保健意识的女性会由于她们拒绝食用红肉，而增加其（缺铁的）风险，因为这些肉所含的铁质最易被吸收。

The data types of arrays and records are native to many programming languages.

译文：数组和记录的数据类型是许多编程语言所固有的。

4. 抽象化引申

有些词在英语中比较具体、形象，如果在译文中不需要强调它的具体名称或具体说明，汉译时，则可以把它抽象化或概括化，用比较抽象、概括的语言来表达。例如：

Quantum chemistry is still in its infancy.

译文：量子化学仍处于发展初期。

Chemical control will do most of things in pest control.

译文：化学防治能在病虫害防治中起主要作用。

（二）词类转换

在科技产品说明书的翻译中，常常需要将英语句子中属于某种词类的词，译成另一种词类的汉语词，以适应汉语的表达习惯或达到某种修辞目的。例如：

When it goes wrong, the performance test of BAND SW（band switch）have priority.

译文：在电视机信号不正常的情况下，频段开关性能测试都要优先。

原文中的名词"priority"转译为动词"优先"，即把它原来的名词性质转换成了动词性质。

The shadow cast by an object is long or short according as the sun is high up in the heaven or near the horizon.

译文：物体投影的长短取决于太阳是高挂于天空还是靠近地平线。

原文中的副词短语"according as"转译为动词"取决于"，形容词短语"high up"转译为动词"高挂"，介词"near"也转译为动词"靠近"。

Even the protective environment is no insurance against death from lack oxygen.

译文：即使有防护设施，也不能保证不发生因缺氧而死亡的情况。

原文中英语的谓语动词只有"is"，但是被省译了，译文中增译了动词"有"，并且名词"insurance"、介词"against"、名词"death"和名词"lack"均转译为汉语的动词。

以下为词类转换的四个分类。

1. 转译为动词

（1）名词转译为动词。例如：

These depressing pumps ensure contamination-free transfer of abrasive and aggressive fluids such as acids, dyes and alcohol among others.

译文：在输送酸、染料、醇及其他摩擦力大、腐蚀性强的流体时，这类压缩泵能够保证输送无污染。（名词"transfer"转译为动词。）

Despite all the improvements, rubber still has a number of limitations.

译文：尽管改进了很多，但合成橡胶仍有一些缺陷。（名词"improvements"转译为动词。）

High precision implies a high degree of exactness but with no implication as to accuracy.

译文：高精度意味着高度精确，但并不表明具有准确性。（名词"implication"转译为动词。）

（2）形容词转译为动词。一些表示心理活动、心理状态的形容词作表语时，通常可以转译为动词。有些具有动词意义的形容词也可以转译为汉语的动词。例如：

If extremely low – cost power were ever to become available from large nuclear power plants, electrolytic hydrogen would become competitive.

译文：如果能从大型核电站获得成本极低的电力，电解氢就会具有更大的竞争力。（形容词"available"转译为动词。）

The circuits are connected in parallel in the interest of a small resistance.

译文：将电路并联是为了减小电阻。（形容词"small"转译为动词。）

Once inside the body, the vaccine separates from the gold particles and becomes active.

译文：一旦进入体内，疫苗立即与微金粒分离并"激活"。（形容词"active"转译为动词。）

（3）副词转译为动词。英语中有很多副词在古英语中曾是动词，在翻译时常常可以将副词转译为汉语的动词，尤其当它们在原文句子中作表语或状语时。例如：

If one generator is out of order, the other will produce electricity instead.

译文：如果一台发电机发生故障，另一台便代替它发电。（副词"instead"转译为动词。）

Such service will certainly go far in extending the life cycle of older equipment.

译文：这种保养肯定会在延长陈旧设备的使用周期中大有作为。（副词"certainly"转译为动词。）

（4）介词转译为动词。英语中有很多介词在古英语中曾是动词，在翻译时常常可以将介词转译为汉语的动词。例如：

In any machine, input work equals output work plus work done against friction.

译文：任何机器的输入功都等于输出功加上克服摩擦所做的功。（介词"against"转译为动词。）

This type of film develops in twenty minutes.

译文：冲洗此类胶片需要20分钟。（介词"in"转译为动词。）

An analog computer manipulates data by analog means.

译文：模拟计算机采用模拟方式处理数据。（介词"by"转译为动词。）

2. 转译为名词

英语中的动词、代词和形容词等也可以转译为汉语的名词。

（1）动词转译为名词。例如：

Boiling point is defined as the temperature at which the vapor pressure is equal to that of the atmosphere.

译文：沸点的定义就是蒸汽压等于大气压时的温度。（动词"define"转译为名词。）

Tests showed that the cooling air must flow at a rate of at least 17m/s.

译文：实验表明，冷却空气的流速至少应为 17m/s。（动词"flow"转译为名词。）

Black holes act like huge drains in the universe.

译文：黑洞的作用像宇宙中巨大的吸管。（动词"act"转译为名词。）

（2）代词转译为名词。所谓代词转译为名词，实际上就是将代词所代替的名词翻译出来，我们也可称之为"还原"。例如：

The radioactivity of the new element is several million times stronger than that of uranium.

译文：这种新元素的放射性比铀的放射性强几百万倍。

This means the permittivity of oil is greater than that of air.

译文：这就意味着油的介电常数大于空气的介电常数。

The most effective measures for deterring gene piracy will be those that prevent it altogether.

译文：制止基因盗窃的最有效的措施应该是能彻底防范盗窃的措施。

（3）形容词转译为名词。例如：

Television is different from radio in that it sends and receives a picture.

译文：电视和无线电的区别在于电视发送和接收图像。（形容词"different"转译为名词。）

A body that recovers completely and resumes its original dimensions is said to be perfectly elastic.

译文：能完全恢复并复原到原始尺寸的物体称为完全弹性体。（形容词"elastic"转译为名词。）

All metals tend to be ductile.

译文：所有金属都有延展性。（形容词"ductile"转译为名词。）

About 20 kilometers thick, this giant umbrella is made up of a layer of ozone gas.

译文：地球的这一巨型保护伞由一层臭氧组成，其厚度约为 20 公里。（形容词"thick"转译为名词。）

（4）副词转译为名词。除动词、代词和形容词可以转译为名词外，有时副词、介词甚至连词也可以转译为名词。例如：

The device is shown schematically in Fig. 2.

译文：图 2 是这种装置的简图。（副词"schematically"转译为名词。）

These parts must be proportionally correct.

译文：这些零件的比例必须准确无误。（副词"proportionally"转择为名词。）

3. 转译为形容词

（1）名词转译为形容词。例如：

Gene mutation is of great importance in breeding new varieties.

译文：在新品种培育方面，基因突变是非常重要的。（名词"importance"转译为形容词。）

Much less is connected with the separation of genera, and there is considerable uniformity of opinion as to the delimitation of families.

译文：这与属的划分关系不大，而在科的划分上观点是相当一致的。（名词"uniformity"转译为形容词。）

Peptides can act directly on the brain to change aspects of mental activity.

译文：肽能够直接作用于大脑，改变各种脑力活动。（名词"aspects"转译为形容词。）

（2）副词转译为形容词。当英语的动词转译为汉语的名词时，修饰该英语动词的副词往往随之转译为汉语的形容词。例如：

Variation is common to all plants whether they reproduce asexually or sexually.

译文：变异对所有的植物，无论是无性繁殖还是有性繁殖都是常见的。［动词"reproduce"转译为名词"繁殖"，副词"asexually"和"sexually"则分别转择为形容词"无性（的）"和"有性（的）"。］

This communication system is chiefly characterized by its simplicity of operation.

译文：这种通信系统的主要特点是操作简单。（动词"characterized"转译为名词"特点"，副词"chiefly"则转译为形容词"主要"。）

（3）动词转译为形容词。例如：

Light waves differ in frequency just as sound waves do.

译文：同声波一样，光波也有不同的频率。（动词"differ"转译为形容词。）

The range of the spectrum in which heat is radiated mostly lies within the infrared.

译文：辐射热的光谱段大部分位于红外区。（动词"radiated"转译为形容词。）

4. 转译为副词

（1）形容词转译为副词。英语中能转译为汉语副词的主要是形容词。形容词转译为副词有以下三种情况。

①当英语的名词转译为汉语的动词时，修饰该英语名词的形容词就相应地转译为汉语的副词。例如：

In case of use without conditioning the electrode, frequent calibrations are required.

译文：如果在使用前没有调节电极，则需要经常校定。（名词"calibrations"转译为动词"校定"，形容词"frequent"则转译为副词"经常"。）

The language allows a concise expression of arithmetic and logic processes.

译文：这种语言能简要地表达算术和逻辑过程。（名词"expression"转译为动词"表达"，形容词"concise"则转译为副词"简要地"。）

A further word of caution regarding the selection of standard sizes of materials is necessary.

译文：必须进一步提醒关于选择材料的标准规格之事宜。（名词词组"word of caution"转译为动词"提醒"，形容词"further"则转译为副词"进一步"。）

②在系动词＋表语的句型结构中，作表语的名词转译为汉语的形容词时，修饰该英语名词的形容词就相应地转译为汉语的副词。例如：

This experiment is an absolute necessity in determining the solubility.

译文：对确定溶解度来说，这种试验是绝对必要的。（名词"necessity"转译为形容词"必要的"，形容词"absolute"则转译为副词"绝对"。）

These characteristics of nonmetal are of great importance.

译文：非金属的这些特性是非常重要的。（名词"importance"转译为形容词"重要的"，形容词"great"则转译为副词"非常"。）

③除以上两种形式外，其他形式的形容词也可转译为副词。例如：

In actual tests, this point is difficult to obtain.

译文：在实际的测试中，很难测到这个点。（形容词"difficult"转译为副词。）

There is superficial similarity between the two devices.

译文：这两个装置在表面上有相似之处。（形容词"superficial"转译为副词。）

（2）动词转译为副词。当英语句子中的谓语动词后面的不定式短语或分词转译为汉语句子中的谓语动词时，原来的谓语动词就相应地转译为汉语的副词。例如：

Rapid evaporation tends to make the steam wet.

译文：快速蒸发往往使蒸汽的湿度加大。（动词"tends"转译为副词。）

The molecules continue to stay close together, but do not continue to retain a regular fixed arrangement.

译文：分子仍然紧密地聚集在一起，但不再继续保持有规则的固定排列形式。（动词"continue"转译为副词。）

（3）名词转译为副词。英语中一些具有副词含义的名词有时也可转译为副词。例如：

Quasi-stars were discovered in 1963 as a result of an effort to overcome the shortcomings of radio telescopes.

译文：类星体是1963年发现的，是人们努力克服射电望远镜的缺点所取得的一项成果。（名词"effort"转译为副词。）

He had the fortune to be able to make the operations that are taking place closer to where the data is actually store.

译文：他幸运地能在更靠近数据实际存储的地方进行操作。（名词"fortune"转译为副词。）

（三）句法转换

翻译作为一种逻辑活动，译文应该是逻辑活动的产品，即要寻找事物的逻辑关系、内在联系，要体现原文的精神实质，达到准确、通顺的要求，也就是说在必要的时候要"得意忘形"。以下列举几种句法转换的方法。

1. 按逻辑关系

按逻辑关系组织译文是科技产品说明书翻译中非常重要的技巧之一。例如：

Fatigue failure of structural components of an aircraft of fail – safe design is quite acceptable, provided it does not occur often enough to endanger the aircraft, reduce its service life, or reduce its utilization and economy by excessive maintenance.

译文：按照故障安全设计的飞机结构部件，只要其疲劳断裂不至于频繁发生以危及飞机安全，缩短其使用寿命，或因维修过多减小其利用率和经济效益，那么偶尔的疲劳断裂是完全允许的。

经过原文和译文的对比可看出，信息中心位置发生了变化，译文中移至句末。而且，译文没有拘泥于原文字面结构，根据逻辑内涵添加了"偶尔的"这个词，使得阐述明晰化。

2. 按时间顺序

在科技产品说明书的翻译中，有时需要根据事件发生的先后顺序组织译文。例如：

When a flash of light falls on the retina, the impression of the light in the brain persists for some time—about 1/10th of a second—after the actual light has ceased.

译文：光线射到视网膜上，而在光实际上已经熄灭之后，脑子里光的印象还会留存一会儿——大约为十分之一秒。

经过分析可发现，原文包含了发生在不同时间的几个连续动作，即光线射到视网膜上、光熄灭和光的印象在人脑内留存约十分之一秒，因此可以按照其先后顺序表达含义。

3. 按肯定、否定

在科技产品说明书的翻译中，如果将肯定、否定关系颠倒，那意思将大相径庭。例如：

The electric locomotive is not dependent, like its steam counterpart, on the competence of driving and firing or the quality of the fuel burned.

译文：与蒸汽机车不同的是，电力机车不依赖驱动和点火能力，也不依赖所用燃料的质量。

插入成分"like its steam counterpart"的实际含义是"while the steam locomotive depends on the competence of driving and firing, and the quality of the fuel burned"，而这层意思是可以通过常识判断出的。

五、课后练习

（一）思考题

1. 科技产品说明书有哪些特点？

2. 翻译科技产品说明书有哪些技巧?

(二) 翻译练习

1. 中译英

(1) 请将下面蓝牙键盘说明书译成英文。

连接蓝牙键盘

步骤 1:打开键盘。蓝牙 LED 指示灯闪烁 5 秒钟,然后再次熄灭。

步骤 2:按下链接/连接按钮。蓝牙 LED 指示灯闪烁,表明设备已准备好进行连接。键盘现已准备就绪,可以连接到 Samsung Galaxy Tab 10.1。

步骤 3:打开并解锁 Galaxy Tab。通过应用菜单打开设置菜单。

步骤 4:选择"蓝牙"。如果蓝牙关闭,则将其激活。如果启用了蓝牙,菜单将显示"蓝牙键盘!"。选择此设备。

步骤 5:现在,Galaxy Tab 显示要连接的代码。通过键盘插入代码,然后按回车键确认。现在将对连接进行安全编码。

步骤 6:无线键盘现在已成功连接到 Samsung Galaxy Tab 10.1。一旦成功连接,键盘将保存所有连接数据,直到连接到另一个设备。

注意:键盘也可以用于 iPad、iPhone 和 iPod Touch。必须安装 iOS 4.0 或更高版本。对于其他支持蓝牙的设备,请在尝试连接到键盘之前验证蓝牙标准和兼容性。

给集成电池充电

电池电量不足时,"电源"指示灯将开始闪烁。现在该给键盘充电了。

步骤 1:将带有迷你 USB 插头的 USB 充电电缆连接到键盘的充电接口。

步骤 2:将 USB – A 连接到电源适配器或计算机的 USB 接口。

步骤 3:在充电过程中,充电完成时,充电状态 LED 指示灯将点亮,然后熄灭。

节能睡眠模式

如果键盘在 10 分钟内处于非活动状态,它将进入睡眠模式。要从睡眠模式激活,必须按任意键并等待 3 秒钟。

安全注意事项

请避免以下情况:

远离尖锐的物体、油、化学药品或任何其他有机液体。

请勿在键盘上方放置重物。

避免明火和高温。

请勿置于直射阳光下。

(2) 请将下面 ZigBee 恒流调色温电源说明书译成英文。

操作

- 按照接线图将设备正确接线并上电。
- 此 ZigBee 设备是一种无线接收器,可与各种兼容 ZigBee 的系统通信。此接收器可接收与之兼容的 ZigBee 系统所发射的无线射频信号，并被该信号控制。

加入一个 ZigBee 网络

第 1 步：如果此设备已经加入一个 ZigBee 网络，请将此设备退出该 ZigBee 网络，否则入网可能失败。有关更多信息，请参阅"手动恢复出厂设置"部分。

第 2 步：从你的 ZigBee 网关选择添加 LED 灯，进行入网模式，请参阅 ZigBee 网关说明书。

第 3 步：将此设备重新上电一次，设备将进入入网模式（所接 LED 灯慢闪 2 次）。入网模式持续 15 秒，如果超时，请重复此步操作。

第 4 步：所接 LED 灯会闪烁 5 次，然后常亮，此设备将会出现在你的网关的控制界面上，你可以通过网关控制界面控制此设备。

手动恢复出厂设置

第 1 步：如果无法操作"Prog"按键，连续快速短按"Prog"按键 5 次，或者将此设备连续重新上电 5 次。

第 2 步：设备所接的 LED 灯会闪烁 3 次以指示设备成功恢复出厂设置。

注意：

- 如果设备已经在出厂设置状态，当再次恢复出厂设置时，设备将不会有指示。
- 设备恢复出厂设置后，所有配置参数都会被恢复为出厂默认状态。

安全 & 警告

安装设备时请不要通电。

请不要将设备暴露在潮湿环境下。

（3）请将下面多士炉使用手册译成英文。

使用多士炉

- 确保控制杆（B）处于提起状态。将烘烤程度选择旋钮（D）旋转到你所需的设置。
- 将机器接通电源。
- 将切片面包放入烘烤槽（A），然后往下按下控制杆（B）直到锁定。
  注意：如果机器没有接通电源，那么控制杆（B）将不会锁定。
- 当烘烤完成后，控制杆（B）会自动弹起，里面的切片面包也会同时弹起。
- 如果没有达到所需的烘烤程度，可通过旋转烘烤程度选择旋钮（D）增加烘烤程度。
- 通过按压停止/取消按钮（I）可随时停止烘烤。切勿用将控制杆（B）往上提的方式来停止烘烤。
  注意：机器在运行时烘烤槽（A）会变得很热，切勿触摸。

解冻功能

冰的切片面包可以通过按一下解冻按钮（F）进行烘烤，同时需要将控制杆（B）往下按直到锁定。为了达到所需的烘烤程度，烘烤的时间会长一些。当解冻功能启动时，解冻指示灯亮起。

再加热功能

如果烘烤的面包不够热，那么按下控制杆（B）直到锁定，然后再按一下再加热按钮（G）即可进行再加热。注意这只能改变再加热的时间，并不能改变烘烤程度。再加热功能通过按一下停止/取消按钮（I）可随时停止。当再加热功能启动时，再加热指示灯亮起。

烤面包圈功能

烤面包圈功能可以烘烤面包、面包圈、松饼等。这也就是只能一面烘烤（内部那一面），另外的一面只能加热（外部那一面）。烤面包圈及松饼在烘烤前需要对半切。按下控制杆（B）直到锁定，然后再按一下烤面包圈按钮（H）即可进行烘烤。当烤面包圈功能启动时，烤面包圈指示灯亮起。

预防措施

不要在没有面包的情况下使用本机器（第一次使用除外）。

不要烘烤太薄或碎的切片面包。

不要烘烤那些易散落的食物，以免增加清洁难度，而且会引起烟火。

定期清洁残屑盘：残屑会引起冒烟或着火。

不能将太大的食物硬塞入烘烤槽中。

不要将刀叉或其他器皿插入烘烤槽来移动面包，这会损坏机器，且很容易触电。

如果机器被堵塞，那么将电源插头拔下，待机器完全冷却后，把机器倒放，并轻轻地摇晃，取出里面的食物。

清洁及保养

在进行清洁之前，请先拔掉电源插头，并待机器冷却。

机器的外部需要用一块柔软的布进行清洁。不要用带有腐蚀性的洗涤剂进行清洁，这会损坏机器表面。

切勿浸入水中。

每次使用完请拔下电源插头，并把残屑盘清理干净。

不要用锋利或金属器皿接触机器，特别是烘烤槽。这会有触电危险。

2．英译中

（1）请将下面除湿机说明书译成中文。

Important Warnings

Danger!

This is an electrical appliance, therefore the following safety warnings must be strictly obeyed:

Do not touch the appliance with wet hands.

Do not touch the appliance plugwith wet hands.

Make sure the mains socket used is always accessible, and unplug the electrical plug when necessary.

Unplug the power plug directly. Do not pull the power cord to avoid damage.

To completely disconnect the electrical power, please pull theelectrical plug from the socket.

If the appliance malfunctions, do not attempt to repair it. Turn the appliance off, unplug the plug from the mains socket and contact a Service Center.

Do not move the appliance by pulling the power cable.

It is dangerous to modify or alter the characteristics of the appliance in any way.

If the power cable is damaged, it must be replaced by the manufacturer or an authorized technical service center in order to avoid all risk.

Do not use power strips.

The appliance must be installed in accordance with the national electrical distribution rules.

The power socket must be grounded, and the electrician will check whether your circuit is qualified.

This appliance is not intended for commercial use and is for family use only.

General safeguards

Do not install the appliance in rooms containing gas, oil or sulphur. Do not install near sources of heat.

Keep the appliance at least 61cm away from flammable substances (e.g. alcohol) or pressurized containers (e.g. aerosol cans).

Do not restheavy or hot objects on top of the appliance.

Always transport the appliance upright or resting on one side. Remember to drain the water tank before moving the appliance. Wait at least 6 hour after transporting the appliance before starting it.

The materials used for packaging can be recycled. You are therefore recommended to dispose of them in special waste collection containers.

Do not use the appliance outdoors.

Do not use the appliance in laundry rooms.

(2) 请将下面手动塞拉门安装调试说明书译成中文。

1. Mounting the positioning seat

Mount the positioning seat and shims onto the rear door frame with 2 M5 × 12 hexagon socket head cap screws, spring washers and flat washers (after the end of installation and adjustment of the whole door system, commission the positioning seat by tapping threaded holes).

2. Mounting the lower slide way

Mount the lower slide way onto the bottom of lower door frame with 3 M10 × 22 hexagon screws, spring washers and big washers, with special attention paid to the installation dimension of (72 ± 1) mm during the mounting.

3. Mounting the bottom rubber stopper

Put the rubber stopper into the tapped thread hole, and fix it with nut (the rubber stopper should be mounted after the end of the installation and adjustment of the whole door system).

4. Mounting the drive mechanism

Connect the carrier to the car body with 4 M10 × 30 hexagon bolts, spring washers and big washers. In the initial mounting, 4 adjusting shims should be put in advance between the carrier and body. When the drive mechanism is mounted, the nylon roller on the door carrying frame should be fit into the upper slide way.

5. Mounting the lower support and the door leaf

5.1 Mounting the lower support

Mount the lower support onto the door panel with 5 hexagon socket head cap screws (applied with appropriate amount of thread lubricant), 8 spring washers and 8 flat washers. Initially adjust it at the middle position of the long hole.

5.2 Mounting the door leaf

Remove the protective hood at the top of door leaf.

Move the door carrying frame to the middle of the door frame, then lift the door leaf and align the two nylon rollers on the leaf lower support (part 23) with the slide channel of the lower slide way, then approach the top of door leaf to the car body until it rests on the door carrying frame (part 24), embed the two positioning blocks with fitting dimension 44mm on the door carrying frame (part 24) into the positioning groove of the aluminum section at the top of door leaf, make coarse adjustment of the two eccentric shafts to the vertical middle position, align the 8 long holes on the door carrying frame with the M8 threaded holes on the top of door, and then tighten them with 8 spring washers, flat washers and M8 × 30 hexagon socket head cap screws applied with appropriate quantity of thread lubricant.

（3） 请将下面吸尘器说明书译成中文。

Assembling and Charging Vacuum Cleaner

1. Insert foot into the body of the vacuum cleaner until you hear a "click".

2. Insert charging adapter plug into the charging port on the back of the machine.

3. Plug the adapter into the wall outlet. Charge the machine completely for at least 4 hours prior to first use.

LED status

During charging

Battery status	Indicator light status
Charging	Red light flashes once every second
Fully charged	Green light lights up once every 10 minutes, and then flashes once every 1 minute
Charger, battery or motor error	The red and green lights flash at the same time. Please contact the Consumer Service Department

Emptying the dirt tank

1. Make sure the vacuum cleaner is off. Remove the hand vacuum cleaner by pressing the release button on its top handle.

2. Hold the hand vacuum cleaner vertically and press the release button on the front to release the dirt tank.

3. Grasp filter tabs and pull up to remove filter assembly, and empty dirt into waste container.

4. Replace filter assembly in the dirt tank, then snap the dirt tank back into place on the hand vacuum cleaner.

WARNING: To reduce the risk of fire, electric shock or injury, turn power off and unplug the plug from the electrical outlet before performing maintenance or troubleshooting.

Cleaning the filter

1. Turn power off and remove the dirt tank as directed in the "Emptying the dirt tank" section.

2. Grasp filter tabs and lift up to remove the washable filter assembly from the dirt tank. Grip the filter assembly, turn counter-clockwise and pull down to remove filter from filter screen.

3. Tap firmly against the inside of a waste container to remove any visible dirt.

4. Rinse the filter and screen in warm water to clean. Let them dry completely.

5. Replace filter into the filter screen and turn clockwise to lock into place. Reattach filter assembly into the dirt tank, and attach the dirt tank to the hand vacuum cleaner until it snaps securely in place.

学思践悟

第八章 科技会展文案

一、科技会展文案文本介绍

(一) 科技会展文案概述

科技会展文案是指因会展活动需要而产生的并在会展活动进行与发展中产生的,以文字、图片为主,记载或传播各种信息的文书材料。会展文案写作代表机构和组织的集体意志,并以相关机构和组织的名义发布,诸如会展合同、会展法规之类的文件一旦签署或发布,就具有法律效力,因此其写作者必须具备与会参展、参观的合法资格,具备必要的资质和法律承受能力。

(二) 科技会展方案的分类

滕超在《会展翻译研究与实践》一书中提到,影响国际会展组织者和海外参与者之间语言交流的文案有以下几类。

1. 会展策划阶段的文案

会展策划阶段的文案即会展前文案,主要是指申请举办或注册国际会展项目的正式报告,也是会展组织者在立项策划书及可行性研究报告的基础上完成的结论性成果。特别是具有重大国际影响力的世界性会展活动,都要求东道主城市提交申请报告,由相关国际组织的权力机构审议通过,方能取得举办相应会展活动的资格。其重要性由此可见一斑。至于国际组织定期举行的会议,则通常由负责管理的领导机构主办(如联合国大会)或由成员组织轮流举办(如 APEC 会议),故多数情况下组织者无须事先提出申请。

2. 会展运作阶段的文案

会展运作阶段的文案即会展中文案,主要包括会展规章类文案、会展合同类文案和会展指南类文案。

3. 会展测评阶段的文案

会展测评阶段的文案即会展后文案,包括各种形式的会展调研、评估、总结报告。

4. 贯穿会展全过程的事务类文案

这类文案涉及宣传推广、公共礼仪等领域。

除上述文案外,广义的会展翻译研究对象还包括其他会展参与者之间的语际交流,参与者的交流方式并不止于书面,会展口译也是需要审慎探索的领域。但会展组织者作为项目主导,其有针对性的书面信息发送与接收,不仅是会展活动中所独有的,而且直接关系到会展各个环节的顺利实施,对活动的成功举办最具决定性意义。

（三）科技会展文案文本分析

1. 会展指南类文案文本分析

科技会展文案包括会展申请类报告、会展规章与合同、会展指南类文案、会展评估类文案、会展事务类文案等。科技会展文案文本分析对译者正确选择翻译策略起着至关重要的作用。

（1）会展指南类文案的主题、内容如下。

①会展的背景信息。

　a. 名称、主题、宗旨、历史等。

　b. 举办时间：分项列明开闭幕时间、布展和撤展时间、观展时间、员工作息时间等。

　c. 举办地点：要求写明具体的城市和场馆。

　d. 组织机构：包括主办、承办、协办、支持及赞助单位。

　e. 指定承建商、代理商、餐饮住宿等。

②场馆的基本情况。

　a. 参展规则（包括资格、程序、费用等）。

　b. 展位搭建指南。

　c. 展品运输指南。

　d. 其他服务指南。

　e. 相关规章条例。

实际上，招商文案也有相当篇幅用于描述会展活动的基本情况，但其主体旨在阐明招商项目及其申请规则（包括资格、程序、费用等）。

（2）会展指南类方案的篇章结构主要涵盖以下四个环节。

①标题。标题应当注明会展名称及"参展说明书"、"展览商服务手册"或"展览手册"等文案类型。后者近似于英语平行文本的标题"Exhibitor Manual"。但他国文字译入英语的参展说明书也有使用所有格的。例如，2008 年在西班牙马德里举行的世界石油大会，其参展说明书即被译为"Exhibitor's Manual"。

②前言。前言应当首先对参展商参加本次会展表示欢迎和感谢，然后说明编撰指南的宗旨及其内容，提醒参展商遵照执行其中的相关规定。该部分文字通常要求言简意赅。例如，《2009 年美国水产业展览手册》（*Exhibitor Manual of Aquaculture America* 2009）的前言：

Thank you for participating in AQUACULTURE AMERICA 2009! This manual will help you prepare everything you need for your booth so that when the exhibit doors open, you are ready to do business!

③正文。正文既可采取章节或序号形式，也可通过大小标题字体、字号的变化表示层次。

④相关图示和表格。会展指南类方案中的图示和表格包括辅助说明性和实用性两类。前者多穿插于正文的关联内容之中，后者则作为正文或相关章节末尾的附录。

与之相似的是各类招商说明书的篇章结构。但如果文案采用公告形式，则无前言要素，需要另添加落款及日期；如果文案采用信函形式，则正文前还应注明称谓。

2. 会展评估类文案文本分析

(1) 会展评估类方案的主题、内容如下。

①会展评估类文案的主题牵涉广泛，大致可划分为三类：会展工作测评报告、会展质量测评报告和会展效果测评报告。会展工作测评报告主要反映工作质量、效率和成本效益。会展质量测评报告通过参展商数量、质量及其参展时间，观展者数量、质量及其观展时间，人流密度指数等方面数据的调研，评估项目质量。会展效果测评报告针对的则是项目成果。

②但无论报告的主题如何变化，此类文案的内容大致应当包括以下环节：测评的背景和目的、测评方法、测评结果、结论和建议。

(2) 会展评估类文案的篇章结构包含标题、序言、目录、正文及署名等要素。

①标题。会展评估类文案的标题通常由会展名称和文案类型构成。其中，会展名称可以使用简称，文案类型则包括调研报告（survey report）、评估报告（evaluation/appraisal report）、总结报告（final report）等。

②序言。会展评估类文案的序言主要介绍会展项目背景、会展项目组织者、会展项目审批机构等基本信息，此外通常还包含活动的各项总体指标，以体现活动取得的主要成绩。如果项目是定期举办的会展活动，那么大部分相关内容应当以现在时态翻译。需要译者特别关注的是，完整段落表述的基本信息与会展公告或指南类文案简介部分经常采用的、类似表格的非完整句不同，句型选择至关重要。

③目录。会展评估类文案中紧接序言的结构要素是目录，篇幅较长的文案配置目录，方便读者阅览。编撰目录的要求在于清晰，并与正文主体保持一致。

④正文。会展评估类文案的正文，其内容极具多样性的特征，但就语篇风格而言，最值得关注的无疑是大量堆砌的数据指标和引述。数据指标是会展评估类文案立论的基础，其重要性不容忽视。引述会展参与者的观点、意见及评论等，作为撰写者主观价值判断的客观基础，是会展评估类文案为提升语篇客观性而普遍采用的策略。

⑤署名。公开发表的会展评估类文案，署名部分有的可能较为复杂，不但涉及文稿撰写者、版面设计者和出版商，还需要注明享有独立著作权的摄影人。

3. 会展事务类文案文本分析

会展事务类文案是筹办大型活动不可或缺的交流手段。

(1) 会展事务类文案的主题、内容如下。

①会展致辞：致辞即在社交活动的特定场合，特别是迎送及宴会之际，主宾双方就活动或现场相关的话题发表的演讲，具体到会展而言，常见的致辞类型包括开幕词（opening speech）、闭幕词（closing speech）、祝酒词（toast）、欢迎词、答谢词、欢送词、告别词等。

②会展书信：除以招展、招商为目标撰写的邀请函外，还包括申请函、支持函、担保函等类型。

③特殊会展表格：事务礼仪性会展文案中有些虽冠以书函的名称，却并不具备书函的格式，反而采用了类似表格的固定模板。其中，较为典型且常用的就是承诺书和委托书，与申请表等并无本质差别。

(2) 不同会展事务类文案的篇章结构介绍如下。

①会展致辞由称呼（salutation）、开篇（opening sentences）、正文（body）、结束语（closing sentences）等构成。有的致辞者在讲话结束后，还将邀请其他致辞者讲话。祝酒词

结束后，则将提议宾主举杯共饮。

　　a. 称呼。致辞伊始，恰当地称呼主人/主要人物及全体听众，是十分必要的。称呼的基本原则是必须遵循社交规范。英美文化的传统习惯是先称呼对方的主要出席者，然后再称呼其他所有出席者。

　　b. 开篇。称呼之后，英文会展致辞常用欢迎、感谢、祝贺或其他致意之词来开篇。

　　c. 正文。会展致辞正文的撰写应当着眼于活动场合的独特性质，要求结构严谨、层次清晰，强调段落间的逻辑关系。

　　d. 结束语。致辞结束时，一般应当表达演讲者真诚的谢意、美好的祝愿和殷切的希望等。

　　②会展书信的篇章结构包括标题（title）、称呼（salutation）、正文（body）、署名（signature）和日期（date）五项必备要素及附件（enclosure）等可选要素。

　　a. 标题。中文会展书信通常带有标题，标题总是居中书写。例如，会议邀请函的标题，可能就写"邀请函"，也可能由会议名称和"邀请函"组成。其他常见标题类型还包括申请函、担保书、支持信等。英文商业书信习惯在称呼下一行中间，以标题形式注明书信的要旨，方便收信人迅速了解所述内容。此类英文标题前还可加上"Subj."（即 Subject 的缩写，也可写全称）或"Re："（即 With reference to 的缩写），且下方常加横线使之更为凸显。

　　b. 称呼。顶格书写的称呼部分用于表明收信人的名称（单位）或姓名（个人）。由个人收信的，中文会展书信中常前冠"尊敬的"等敬语词，以示尊重。然而，英文书信中直接将中文敬语词译成"Respected"，并置于收信人姓名左侧，这是不符合英文书信文体特征的。事实上，英文会展书信的称呼部分总是以"Dear"起始。

　　c. 正文。正文构成会展书信的主体内容。不同功能的事务书信，内容大相径庭，但有时会在正文首尾分别包含应酬语和祝颂语。开始应酬语（greetings），一般即"您好"，多见于以个体为收信人的会展邀请函。该部分文字既无实质性信息，在英语中也没有相对应的结构要件，故英译时可以完全移除。结束祝颂语（compliments），一般即"此致敬礼"或"顺颂时祺"等惯用语。通常，正文下方空两格写"此致"或"顺颂"，随后另起一行顶格写"敬礼"或"时祺"。该部分文字同样无实质性信息，在英语中也没有相对应的结构要件，故英译时可以完全移除。

　　d. 署名。中文会展书信的署名部分位于书信右下方，以单位为收信人的书信可署单位名称并盖章，但更多的是由发文单位负责相关事务者亲自署名，并在姓名前写明职务。这与英语书信结尾处先签名后注明职务的次序有所不同，翻译时应当特别注意格式变换。英文会展书信在尾部另行附加结束语"Yours faithfully/truly /sincerely/cordially/respectfully"，然后署名（写信人的姓名和职务），正式英文事务书信的署名部分略低于结束语一两行。建议写信人采用与其姓名有所联系的亲笔签名，并在其下方打印自己的姓名及职务。

　　e. 日期。中文会展书信的日期部分位于署名下方。英文会展书信总在信头位置注明日期。正式英文书信的日期要求年、月、日齐备，且不应缩写英文月份名称，还习惯使用简洁的基数词表示"日"。另外，英式日期写法按"日　月　年"的次序排列，美式日期写法按"月　日　年"的次序排列。

　　③特殊会展表格由标题（title）、特别提醒（alerts）、联系信息（contact information）、参展商信息（exhibitor's information）、签名栏（signature）等构成。

　　a. 标题。通常，中文特殊会展表格的标题由"内容＋类型"的方式确定。至于表格前的编号，仅是组织者考虑到大型会展要求填写的表格繁多，为方便参与者索引查询而采取的策略。中文会展申请表的标题由"申请内容＋'申请表'"的方式确定。中文会展承诺书的标题由"承诺内容＋'承诺书'"的方式确定。中文会展委托书的标题由"委托内容＋'委托书'"的方式确定。

b. 特别提醒。特殊会展表格的特别提醒事项专指填表对象、交表日期等需要特别强调的关键因素。

c. 联系信息。联系信息主要是指会展组织者指定接受表格的部门及其联系方式，包括地址、电话、传真、电子邮箱等。

d. 参展商信息。尽管很多会展表格正文陈述参展商信息，也有会展表格在头部单列参展商信息的。

e. 签名栏。所谓签名，包括当事人采用的任何符号，旨在体现其证明相关文件真实性的意图。而签名栏则是特殊会展表格中供当事人及见证人（如果有）签名处。译者必须牢记的是，英美国家多要求签名不应当出现在独立的文页上。换言之，签名栏所处文页，必须包含特殊会展表格的部分内容。

二、科技会展文案范文展示

中国北京国际科技产业博览会
China Beijing International High–tech Expo

中国北京国际科技产业博览会（China Beijing International High–tech Expo，简称科博会）是由中国科技部、商务部、教育部、工业和信息化部、中国国际贸易促进委员会、国家知识产权局及北京市人民政府等多个国家政府机构于每年5月在北京举办的大型国际科技交流合作盛会。

The China Beijing International High–tech Expo (CHITEC) is a large–scale national–level international science and technology exchange and cooperation event held in Beijing every May that is sponsored by a number of national government agencies, including China's Ministry of Science and Technology, the Ministry of Commerce, the Ministry of Education, the Ministry of Industry and Information Technology, the China Council for the Promotion of International Trade (CCPIT), the China National Intellectual Property Administration, as well as the People's Government of Beijing Municipality.

科博会创办于1998年，当时定名为"中国北京高新技术产业国际周"，从2002年第五届起正式更名为科博会。其目的是促进科技产业的商业化、市场化运营及国际化。20年来，科博会已发展成为中国展示国家前沿技术、传播前沿理念、发布产业政策、促进国际经济技术合作的高度专业化和国际化的科技贸易盛会。

CHITEC was initiated in 1998 as China Beijing High–tech Industry International Week. It was given its current name in 2002 at the fifth edition of the event. The aim of the event is to promote the commercialization, market–oriented operation, and internationalization of the technology industry. Over the past two decades, CHITEC has grown into a highly–specialized and international technology and trade event at which China can showcase state – of – the – art technologies, spread cutting – edge ideas, release industrial policies, as well as promote international economic and technical cooperation.

科博会以展览、论坛和商务会谈的形式运作。展览主要突出与高新技术相关的产业，包括电子和通信技术、计算机网络、能源和环境保护、生物医药及汽车技术。论坛关注鼓舞人心的话题，并开展了一些涉及自主创新、能源战略、创新服务业、新技术和文化创意产业及循环经济的品牌活动。商务会谈涉及国际投资项目、国内企业海外投资、省级代表团等相关活动。

CHITEC is operated in the form of exhibitions, forums, and business talks. The exhibitions highlight industries related to new and high technologies, including electronic and communications technologies, computer networks, energy and environmental protection, biomedicine, as well as auto technology. The forums focus on inspiring topics and carry out a number of brand – named activities involving independent innovation, energy strategies, innovative service industries, new technologies and the cultural creative industry, as well as the circular economy. Business talks involve activities related to international investment projects, overseas investment by Chinese enterprises, and provincial delegations.

在第二十二届科博会期间，来自两个国际组织、33个国家和地区的57个代表团参加了多项活动。来自全国21个省区市的代表参加了展览，主要展览接待了近65 000人次。共有247名国内外知名人士在研讨会和论坛上发表了演讲，6800名专业人士到会交流。14个项目推广活动共吸引国内外客商3100余人。

据统计，达成交易、合作协议、项目共88个，总金额为286亿元人民币。

第二十三届中国北京国际科技产业博览会将于2020年9月17日至20日举行。

During the 22nd CHITEC, 57 delegates from two international organizations, as well as 33 countries and regions took part in a number of activities. Delegates from 21 provinces, regions and cities participated in exhibitions, with the main exhibition receiving nearly 65,000 visits. A total of 247 public figures both at home and abroad delivered speeches at the symposium and forums, while 6,800 professionals attended meetings and conducted exchanges. A total of 14 project promotion activities attracted over 3,100 merchants from home and abroad.

According to statistics, the number of deals, cooperation agreements, and projects reached was 88, with an amount totaling 28.60 billion yuan.

The 23rd China Beijing International High – tech Expo will be held from Sept 17 to 20, 2020.

联系方式：
展会筹备组：
电话：86 – 10 – 8070377, 88070392
传真：86 – 10 – 68062325
电子邮件：zhangzhiliang@ ccpitbj. org
展览业务：
电话：86 – 10 – 68062929, 68063939, 68065959
传真：86 – 10 – 68066929

电子邮件：wangpeng@ ccpitbj. org，haocheng@ ccpitbj. org
海外接待组：
电话：86 – 10 – 88070373，88070332
传真：86 – 10 – 68059013
电子邮件：wulanlan@ ccpitbj. org，chengquan@ ccpitbj. org
官方网站：
http：//www. chitec. cn（中文网站）
http：//english. chitec. cn（英文网站）
Contacts：
Plan team：
Tel：86 – 10 – 8070377，88070392
Fax：86 – 10 – 68062325
Email：zhangzhiliang@ ccpitbj. org
Exhibition team：
Tel：86 – 10 – 68062929，68063939，68065959
Fax：86 – 10 – 68066929
Email：wangpeng@ ccpitbj. org，haocheng@ ccpitbj. org
Overseas reception team：
Tel：86 – 10 – 88070373，88070332
Fax：86 – 10 – 68059013
Email：wulanlan@ ccpitbj. org，chengquan@ ccpitbj. org
Official websites：
http：//www. chitec. cn（Chinese）
http：//english. chitec. cn（English）

三、科技会展文案常用词汇和句型

（一）科技会展文案常用词汇

1. 举办日期类词汇
registration（open/opening）hours/times 注册（开放）时间
exhibitor move – in hours/times 参展商进场布展时间
exhibitor move – out hours/times 参展商撤展出场时间
exhibition/show hours/times 展览时间
event duration 活动持续时间
2. 证件服务有关词汇
admission ticket 入场券
definition clause 定义条款

applicant 申请者
exhibition pass 展区出入证
attendee 出席者，在场者
exhibition/show badge 展览证
badge 胸章
exhibitor (ID) badge 参展商证
badge service 证件服务
registration area 登记处
conference badge 会议证
3. 展位搭建服务有关词汇
booth/stand 展位
booth/stand location 展位地点
information counter 问询台
back drape 背景帷幕
ancillary facility 辅助设施
installation & dismantling 安装和拆除
4. 展品运输服务词汇
material handling 材料装卸搬运
exhibit shipment 展品运输
truck dock 货车装卸区
5. 展品描述词汇
attractive design 款式新颖
complete in specifications 规格齐全
color brilliancy 色泽光润
deft design 设计精巧
dependable performance 性能可靠
durable in use 经久耐用
easy and simple to handle 操作简便
reliable reputation 信誉可靠
selling well all over the world 畅销世界各地
the king of quality 质量最佳
wide varieties 种类繁多
to win warm praise from customers 深受欢迎
world-wide renowned 闻名世界
6. 展位描述词汇
backwall booth 靠墙展位
corner booth 角落展位
double-decker booth 双层展位
island booth 岛形展位

multiple-story exhibit　多层展台
transient space　临时摊位

7. 展馆标志词汇

facility　展览馆，展览设施
fire exit　展馆内的紧急出口
floor load　展馆地面最大承重量
floor plan　展馆平面图
floor port　展馆地面接口

8. 会展有关人员词汇

attendance　参观人数
consignee　（展品）收货人
dealer　经销商
dispatcher　调度员
each party's authority　当事人的合法依据
exhibit designer/producer　展台设计师/搭建商
exposition manager　展厅经理
sponsor　赞助商，赞助者
international sales agent　国际销售代理
merchant　商人
organizer　组织者
participant　与会者
professional visitors　业内人士
registrant　登记者
retailer　零售商
wholesaler　批发商

9. 会展协定/协议有关词汇

basic conditions　基本条款
bilateral agreement　双边协定
concluding sentence　结尾语
contract for technology transfer and importation of equipment and materials　技术转让和设备、材料进口合同
contract for work　工程合同
contractual language　协议文字
effect upon termination of the agreement　协议终止后的效力
entire agreement clause　完整条款
general terms and conditions　一般条款
gentlemen's agreement　君子协定
inter-governmental trade agreement　政府间贸易协定
multilateral agreement　多边协定

non – governmental trade agreement　民间贸易协定
patent licence agreement　专利许可协议
payment agreement　支付协定
protocol　协议，议定书
settlement of dispute　解决争端
technical assistance agreement for plant construction　工厂建设技术援助协议
technical assistance agreement　技术援助协议
technical cooperation agreement　技术合作协议
technical documentation　技术资料
technical information agreement　技术信息协议
technical modifications and improvements　技术修改和改进
technology transfer agreement　技术转让协议
term of the agreement　协议期限
counter – sign　会签
trade agreement　贸易协定
trade and payment agreement　贸易与支付协定
verbal agreement　口头协议
written agreement　书面协议
agreement of intent　意向协议

10. 会展合同有关词汇

amendment of contract　修改合同
amendment　修改
arbitration　仲裁
assignment and amendment of the agreement　协议的转让和修改
signing parties　签约当事人，签约方
attachment/appendix　附录
blank form　空白格式
breach of contract　违反合同
cancellation of contract　撤销合同，解除合同
consultation　咨询
contract terms　合同条款
copy　副本
definition　定义
delivery of technical documentation　技术资料的交付
escape clause　免责条款
exchange of letter　换文
expiration of contract　合同到期
failure to exercise right　未行使权力
governing law　适用的法律

industrial property rights and know-how　工业产权和专有技术
interpretation of contract　合同的解释
jurisdiction　管辖权
letter of intent　意向书
liability　责任
licence agreement　许可协议
manufacture of contractual products　合同产品的制造
memorandum　备忘录
model contract　格式合同，标准合同
notices　通知

（二）科技会展文案常用句型

1. 使用被动语态

使用被动语态可以让科技会展文案的内容显得更加客观、直接、简洁、准确，避免拐弯抹角和累赘啰唆。被动语态还能缓和语气，使行文衔接连贯、简短有力，并使有关成分相互作用、相互关联和相互构建。作为一种信息重组的语用手段，被动语态不仅有内在的逻辑因果关系，而且有话题的确立功能、连接功能、转换功能、焦点对比功能。因此，在英文科技会展文案中，常用被动语态实现陈述的开门见山和表达的直陈事理。同时，被动语态由于其结构的特点，将主要信息前置，陈述要点放在句首的主语位置，可以突出重点或起到强调的作用。例如：

CHITEC was initiated in 1998 as China Beijing High-tech Industry International Week.

译文：科博会创办于1998年，当时定名为"中国北京高新技术产业国际周"。

2. 使用一般现在时和现在进行时

现在时态中的一般现在时和现在进行时可以表明事件（活动）正在或即将进行。通过这两种时态的运用，可以使读者产生强烈的现时感，使谈论的内容具有可靠性和客观性。例如：

This product is now in great demand and we have on hand many inquiries from other countries.

译文：这种产品现在需求量很大，我们手头上有来自其他国家的很多询盘。

It is our permanent principle that contracts are honored and commercial integrity is maintained.

译文：重合同、守信用是我们的一贯原则。

四、科技会展文案翻译技巧

科技会展文案既具有会展文案的共性，也无法脱离科技文献本身的特殊性，即要考虑科技文体特殊的知识性、专业性，结合科技英语翻译的方法与策略理解科技会展文案的翻译。这就要求译者不能孤立地理解所译的文本语言，而是要理解文本的全部交际内涵与外延。科技会展文案的翻译不是单纯的信息接收，而是要构建源语文本与目的语文本之间的联系路

径,也就是以源语文本的信息接收为基础,联系目的语文本所处的情境,比照目的语平行文本特征,选择正确的翻译策略和技巧,解决会展文本翻译中出现的跨文化交际的实际语言问题。

(一) 科技会展文案翻译的基本要求

1. 语言简洁

科技会展文案的主要目的就是向外界推广科技会展活动本身,或者对外进行企业宣传。无论是中文科技会展文案还是英文科技会展文案,其语言必然是简洁明了的,只有这样,才能够吸引读者,进行成功推介。因此,在科技会展文案翻译过程中,我们也应该继续秉承科技会展文案的这一特点,用语不能够太过复杂,少用生僻字眼,应该选择简单、易懂的词汇进行翻译。

2. 信息全面

科技会展文案作为一种商务文体,在对其进行翻译时,不求信息的完全对等,但是译文必须能够全面反映出原文的信息点,务必使译文语义和作者思想保持一致,译者不可以随便更改或发挥,甚至"强加于人"。无论译者是否认同原文的观点,都要忠实再现其思想。

3. 注重中西文化差异

翻译是一种跨语言、跨文化的交际活动,社会文化不同,思维、观念也不同。如果翻译时语用失误,科技会展文案就有可能出现译文与原文相悖的情况。所以,科技会展文案翻译不仅要考虑语言特点,更要注重科技会展文案的文化内涵。例如,某企业生产的"玉兔"牌产品商标,不宜译作"Jade Rabbit",而应译作"Moon Rabbit",因为玉兔是我国神话中陪伴嫦娥生活在月宫桂花树下的兔子,已经成为月亮的代称。"Moon Rabbit"这一译名既体现了中国的古老文化,又避免了误解,不会让人以为是玉做的兔子。

(二) 纽马克的翻译思想在科技会展文案翻译中的应用

科技会展文案翻译作为跨文化交际语境下的特定行为,尤其需要译者明确翻译目的,掌握文本功能,选择相应的翻译策略。

1. 信息功能,清晰与简洁兼收

在纽马克的学说中,语言描述功能(representative function)即信息功能(informative function),其核心存在于言外现实情景(external situation/reality outside language),即文本所处的语域或所涉及的应用情境。就科技会展文案翻译而言,语言描述功能主要表现为告知与指令两种形式。前者用于告知读者不了解的事实或状态,而后者指令的则是正确的行为方式,较为典型的如会展规章与合同。译者在翻译过程中,如能充分考虑到文本的特征和功能,许多信息资料在翻译过程中得以进一步整理,译者翻译的文本就可做到清晰准确。例如,世博会宣传手册中有"城市人"这一新术语,界定如下:

Urbanized people: The soul of a city is in its people, who constitute its body likes cell. It is people who give the city its culture, character and innovative power. With more and more people becoming urbanized, the city's population grows ever larger and more diversified. Meanwhile, people outside the city also grow under the influence of urbanization, accordingly. The city should function as the motive power for upgrading the quality of people's life and as a main source of human innovation and creation.

译文：城市人：人是城市的细胞，又是城市的灵魂。人赋予城市文化、性格和创新力。随着越来越多的人成为"城市人"，城市人口数量与日俱增，类型也更具多样性。在城市化的影响下，城外的人口数量也随之增长。城市要成为提高人类生存质量的原动力，也应该成为人类创新和创造的温床。

原文案信息属于信息类文本，原文案中的"Urbanized people"一词不仅对源语信息接收者是全新的概念，对目的语读者也是如此。该译文的译者正确地理解"Urbanized people"和"the influence of urbanization"这些核心词，利用原文中对新术语所给出的界定"give the city its culture, character and innovative power""more diversified""the motive power for upgrading the quality of people's life""a main source of human innovation and creation"提升"Urbanized people"这一新概念的清晰度，在译文中明示读者：该文案中的"城市人"与"城里人""城市居民"等不是同一概念。"城市人"指的并非进入城市生活的人，实则泛指受"城市化"影响的所有人。对于"城市人"这个全新的概念，译者的理解没有仅着眼于字面意思，而是依据它的内涵进行实质性翻译。

2. 表达功能，逻辑和美学并蓄

纽马克认为，语言表达功能（expressive function）的核心在于说话人或撰写者的思想，通常运用具有美学效果的修辞手法体现，其作用在于表达语言使用者的情感。科技会展文案是典型的表现语言表达功能的语篇形式，同时也显示出信息发送者的情感和审美情趣，需要译者特别谨慎处理。例如：

中国2010年上海世博会将是一曲以"创新"和"融合"为主旋律的交响乐：创新是世博会亘古不变的灵魂；跨文化的碰撞和融合，则是世博会一如既往的使命。"以人为本、科技创新、文化多元、合作共赢、面向未来"——上海世博会将在新的时代背景下继续弘扬"创新"和"融合"的主旋律，创作一曲人类新世纪的美妙乐章。

译文：Expo 2010 Shanghai China will be centered on innovation and interaction. Innovation is the soul while cultural interaction is an important mission of the world Exposition. In the new era, Expo 2010 Shanghai China will contribute to human–centered development, scientific and technological innovation, cultural diversity, win–win cooperation for a better future, thus composing a melody with the key notes of highlighting innovation and interaction in the new century.

该例选自《中国2010年上海世博会注册报告》序言部分，文本整体上表现出显著的文学色彩。译文不仅译出原文所给出的所有信息，也显示出原文音美、意美的特点。

3. 诉求功能，形式与内容共存

纽马克认为，语言诉求的对象是读者或被打招呼的人。这种诉求功能既可直接表现为标志性的语言形式，如祈使句（imperative sentence）或修辞性的疑问句（rhetorical question），也能借助指向描述或表达功能的语言或风格手段间接地加以实现。诉求的文本包括说明书、宣传资料、申请书、案情资料、通俗读物等。这些文本形式在科技会展文案中也很普遍。例如：

Permission for use

If the pavilion has passed the quality inspections and tests on completion carried out by relevant authorities and any contractor involved in the construction of this pavilion, the Official Participant shall, within 5 days after obtaining the Acceptance Certificate for Civil Engineering after Test on Completion, submit an Application for Permit to Use a Pavilion to the Technical Office of the Organizer. The Official Participant may inquire the Technical Office of the Organizer about the documents to be submitted as required for the purpose of application.

译文：使用许可

经政府相关部门和参与建设的各单位进行竣工质量检查、验收合格的展馆，官方参展者应在获得《建筑工程竣工验收备案证书》之后5日内向组织者技术办公室提交《展馆使用许可申请书》。所需提交资料的内容，官方参展者可向组织者技术办公室咨询。

该例是典型的以实现诉求功能为目标的会展文本。对于这一文本的翻译，在内容上要忠实传递源语信息，在形式上则要遵循目的语的表达习惯，使目的语读者产生与源语信息接收者类似的效果。

(三) 科技会展文案翻译手段

在具体的翻译过程中，译者应该根据具体情况采用增译、编译、改译、缩译等翻译手段对原文进行处理，目的是要跳出原文的束缚，以目的语读者的立场来行文。

1. 增译

这种译法要符合目的语的表达方式。注重意合是汉语句式的特点，其逻辑关系通常隐藏在上下文和语境当中，英译时可以借助上下文和语境的补充意义来理解文本，从而弥补汉语在逻辑关系和语句连贯方面的不足。而英语文本的逻辑关系较为清晰，汉译时需要通过词法和句式结构准确判断出其逻辑关系。这就要求译者在翻译过程中注意两者之间的转换，实现形合与意合的对等。例如：

俄罗斯科技部与工贸部联合提议，为培养掌握智能技术、信息技术、计算机建模、机器人技术和附加技术的专业人才，建立系统的培训机制。

译文：Russian Ministry of Science and Technology and Ministry of Industry and Trade jointly proposed to establish a systematic training mechanism in order to cultivate professionals who master intelligence technology, information technology, computer modeling, the robot technology, and additional technology.

汉语注重一种内在、隐含关系。原文为多个短句，它们之间只有意义上的联系，没有形式上的关联。英语是重形合的语言，在构词、构句上都倾向考量形式因素。因此，译文需要添加连接词 in order to 和关系代词 who，力求在形合时起到衔接作用，组合成为一个形式与意义并重的整体。

2. 编译

编译即先编辑，将原文内容条理化，使之更完美、更精致，再翻译。编译包括编选（从原文中选取一部分加以整理）、编排（按一定的顺序将原文内容重新排列先后）、编写（将原文提供的材料加以整理，译成目的语文字）等。例如，下列广交会中对广州区位优势的介绍：

广东是中国第一经济大省，在电子信息、电器机械、石油化工、森工造纸、医药、汽车、纺织服装、食品饮料、建筑材料、家具、旅游、印刷、玩具、物流、畜牧、礼品、自行车、模具、包装、体育用品、花卉园艺、制鞋、照明、文化用品及教育等多个工业行业在全国甚至在国际上具有很强的竞争力。广州连续多年工业总产值稳居全国十大城市前茅，是广东乃至华南地区工业门类最齐全的城市。广州作为中心城市的辐射作用，已超越华南地区，正向粤港澳经济圈乃至整个东南亚地区稳步拓展。

译文：As a hugely successful economic venue, Guangdong Province outperforms the other parts of China in many fields, namely, manufacturing, logistics, gardening and tourism, among others. For many consecutive years, the total industrial output value of Guangzhou has ranked among the top ten cities in China, and it is the city with the most complete industrial categories in Guangdong and even South China. As a central city, Guangzhou has surpassed South China and is steadily expanding to Guangdong – Hong Kong – Macao economic circle and even the whole Southeast Asia.

原文中列举了广东省很多极具竞争力的行业，但在翻译时并没有逐个翻译，而是进行了相应的编排，这样就避免了译文烦冗拖沓。同时，如果直译出来，译文很可能会出现头重脚轻的现象，并且无法突出信息重点。

3. 缩译

缩译中的"缩"就是压缩原文中的内容，用简洁凝练的语言表达出原文的核心信息，使篇幅变小。例如，深圳高交会的宣传：

经过8年的发展，高交会以其"国家级、国际性、高水平、大规模、讲实效、专业化、不落幕"的特点，已成为目前中国规模最大、最具影响力的科技类会展，有"中国科技第一展"之称。

译文：CHTF is an international hi – tech trade fair with a history of eight years. It has become the largest and mostinfluential science and technology exhibition in China, and is known as "China's first science and technology exhibition".

"国家级、国际性、高水平、大规模、讲实效、专业化、不落幕"这些并列词与后面"已成为目前中国规模最大、最具影响力的科技类会展，有'中国科技第一展'之称"意思上重叠。因此，这些并列词可省略不译。

4. 改译

改译是指使原文发生明显的变化，改变内容或形式。改译包括改变（改掉原文中的内容或形式，换成适合目的语读者的内容或形式）、改编（根据原文内容采用另一种体裁重写）和改造（修改整个原文，以便译文适合新的要求）等。例如：

"知人善任，唯才是用"是我集团始终坚持人才为公司之本的观念，努力健全员工招聘、培训、使用、激励和淘汰机制，注重尊重人才、重用人才、用好人才的良好风气。公司荟萃了一批具有高知识水准和良好行业素质的专业人才，他们已成为公司最宝贵的财富，为公司在国际国内激烈的竞争中保持卓越的业绩打下了坚实的基础。

译文：All Talents Are Welcome Here. The Group insists on the idea that talents are the foundation of the Group. We bend on the careful work of recruiting, training, employing, encouraging and unemployment system. We respect real talents and assign talents important posts. The Group has a group of technicians with high knowledge and good professional morality who are the wealth of the group. Their excellent work enables us to achieve a better result in the fierce competition at home and abroad.

原文的中心点是"唯才是用","尊重人才、重用人才、用好人才"的意思都是唯才是用,故在英译时尽量将该主题思想表达清楚,无须逐字一模一样地进行翻译。

五、课后练习

(一) 思考题

简述科技会展文案翻译的基本要求。

(二) 翻译练习

1. 中译英练习一

第二十四届中国北京国际科技产业博览会

中国北京国际科技产业博览会(简称科博会)是经国务院批准,由中国科技部、国家知识产权局、中国贸促会和北京市人民政府等多个国家政府机构共同主办,北京市贸促会承办的大型国家级国际科技交流合作盛会。

展览展示

主展场设在中国国际展览中心(静安庄馆),拥有 50 000 平方米,设置 13 个专题展区:国际展区、前沿科技热点展区、科技冬奥展区、北京高精尖产业创新成果展区、人工智能展区、机器人展区、金融科技展区、智慧教育展区、首都文化科技融合发展成果展区、科技产业功能区创新成果展区、首都青年科技创新创业成果展区、首都汽车科技展区、省区市科技创新成果展区。

网上展示推介将利用科博会官网提供优秀企业及产品网上推介、招商合作项目网上对接等。

参展指南

一、时间、地点

展会时间:2021 年 9 月 16—19 日

展会地点:中国国际展览中心

主办单位:北京市人民政府、中华人民共和国科技部、国家知识产权局

展会规模:50 000m²

二、相关活动

开幕式暨主题报告会;党和国家领导人参观展览专场;产品发布/推介会;项目发布与采购专场;优秀项目评选活动;国际电子技术论坛;网络信息技术研讨会;教育装备与学校对接会;机器人技术论坛;3D 打印大会;物联网技术与应用论坛;国际金融论坛;VR/AR 大型体验活动。

三、科博会优势

经过二十三年的培育和发展,科博会已经成为一个具有广泛国际影响力的综合性科技盛会,成为中国与世界各国进行科技交流合作的重要平台,成为我国科技经贸领域最具代表性和权威性的重大国际博览会之一。在展示国内外最新科技成果、传播前沿思想理念、促进科

技交流合作等方面发挥了积极作用。

1. 顶级的年度例展：经国务院批准，由八部委联合主办，党中央、全国人大、国务院、全国政协、中央军委、有关部委领导多次莅临科博会参观。

2. 综合性科技盛会：第二十三届科博会举办了 15 场推介洽谈、12 场专业论坛、多场说明会。这些内容丰富而理性务实的招商推介活动和寻求产业合作的专业活动，成果显著，促成签署科技合作、技术成果交易项目 312 个，协议总金额 960 亿元人民币。

3. 一流的商业平台：第二十三届科博会吸引国内外观众达 23 万人次，来自 9 个国际组织和 37 多个国家和地区的 80 多个境外代表团组，全国 32 个省区市、计划单列市政府代表团参加科博会，2000 余家跨国公司、国内行业领军企业、大型骨干企业集团及高成长性中小型企业参展。

四、参展区域

1. 消费电子

VR/AR 相关产品、数码产品、汽车电子产品、云计算与技术应用、智能家居、智能家电与电器、移动智能终端及周边、可穿戴式设备、LED 与照明、电子制造、品牌专区。

2. 教育装备及网络教育

教育教学数字化、信息化，网络教育机器人，互联网教育及平台，教学与校园视听广播网络系统，数字教室、数字校园，教学仪器设备，电脑、电教器材，实验设备系统，科普教育，教学软件。

3. 智能科技

智慧城市、智慧园区，智慧物流，智慧交通，智能家居，安防及监测技术，智慧照明、智能 LED 照明，云计算、云存储，传感器、识别技术，短距离通信技术与产品，管理系统软件，物联网示范应用，其他物联网元器件。

4. 电子信息与现代通信

网络信息技术及解决方案、集成电路和电子元器件、电子信息成套产品设备、现代通信设备、电力电子器件、激光和光电子器件、光机电一体化、液晶显示、因特网与电子商务。

5. 机器人科技

工业机器人、教育机器人、服务机器人、特种机器人、清洁机器人、医用机器人、检查维护保养机器人、建筑机器人、水下机器人、机器人组成商、机器人供配商。

6. 3D 打印技术

3D 打印机，3D 打印机制造设备，3D 打印技术，3D 激光设备，3D 打印控制设备，测量设备，逆向工程软件和技术，其他快速成型技术，相关零部件、辅料，3D 打印耗材。

7. 环境保护和新能源产业

环境管理、污染控制与减少、新材料与新能源、化工新材料、功能金属材料、新型建材、光电子材料、太阳能热利用、节能新技术。

8. 现代工程与先进制造技术

工程及加工机械、微型机械，现代出版和印刷设备，科学仪器和检测控制设备，医疗器械，现代城市建设与交通工程。

五、参展费用

室内展位

人民币￥1600.00元/平方米

会刊、广告费用

封面：40 000元　封底：20 000元　封二：18 000元　封三：16 000元

彩色整版：6000元

桁架广告：350元/平方米　吊带：20 000元/展期　墙体广告：260元/平方米

通道地标广告：3000元/6个　参观券：10 000元/万张　参观证：20 000元/展期　礼品袋：15 000元/千个

六、参展程序

1. 按要求填写好"参展申请表"并交回展览主办单位。通过传真方式报名参展也可接受。请注意截止日期为2021年9月10日。

2. 收到"参展申请表"后，展览主办单位将向参展公司寄发正式合同一式两份，以待会签。

3. 参展公司需要按主办单位发出的形式发票的要求，通过银行电汇展台租金的50%作为预订金（人民币）或一次付清全部款项，以落实展台位置。展台租金的余额部分应不迟于2021年9月10日汇付。

4. 在确认展台后，主办单位将向参展公司寄发《参展商手册》，手册包括展品运输、展台设计搭建、旅行及住宿安排、物品租用和服务员、广告及签证申请等有关信息。参展商必须按要求填写好手册中的有关表格，并于截止日期前交回主办单位。

5. 只有收到展台预订金后，才能落实所预订的展位。展位分配按"先预订交费，先落实确认"的原则售完为止。

组委会办公室：

展览业务

李文静：15652230005

电话：010-53515097

QQ：2638342269

邮箱：ddgjexpo@sina.cn

http：//www.vanzol.com/chitec/

2. 中译英练习二

第131届中国进出口商品交易会

第131届中国进出口商品交易会（广交会）将于4月15—24日在网上举办，展期10天。本届广交会以联通国内国际双循环为主题，展览内容包括线上展示平台、供采对接服务、跨境电商专区三部分，在官网设立展商展品、全球供采对接、新品发布、展商连线、虚拟展馆、新闻与活动、大会服务等栏目，按照16大类商品设置50个展区，境内外参展企业2.5万多家，并继续设立"乡村振兴"专区，供所有脱贫地区参展企业集中展示。

本届广交会围绕提升贸易对接成效和用户体验，持续优化提升线上平台功能，多措并举便利展客商互动交流和贸易成交。欢迎中外企业和客商积极参加，共享商机，共谋发展。

学思践悟

第九章 科技文章

一、科技文章文本介绍

(一) 科技文章概述

科技文章一般指自然科学和工程技术方面的科学著作、论文、教科书和科技报告等。常见的科技文章包括科技论文、科研报告、科普文章、科技实用手段（如仪器、机械等）的结构描述和操作描述。这类科技作品的特点是对科学问题论述完整、清楚、准确、系统，而且逻辑关系严谨，推导说理明晰。它在语言运用上要求多用词义明确的科技术语，多用符合书面语规范的主从复合句、长句及各类短语等，不用或少用带有感情色彩的词语。

(二) 科技文章的特点

1. 词汇特点

(1) 英文科技文章涉及许多专业术语。尽管随着科技知识的普及，有些词语已为大众所熟知并成为普通语汇的一部分，但绝大多数词语仍仅见于科技文章之中。这些词语大多源自希腊语和拉丁语，词义单一，能避免产生歧义，符合科技英语力求准确的要求。专业术语可以分为两大类。一类是纯专业术语，即该词汇的意义唯一。例如，"grade contour"只能翻译成"坡度等高线"，同样"坡度等高线"要翻译成英语也只能使用"grade contour"。还有一类是半专业词汇，即在不同的学科领域中，该词汇的意义会发生改变。例如，"cell"在生物学中指的是"细胞"，在物理学中又可指"电池"。

(2) 在英文科技文章中还会大量使用缩略词和复合词。大量使用缩略词与复合词，把名词和名词（或名词和形容词）直接拼接叠加来合成新词的方法可以避免不断创造新词，同时还使语言变得精练。复合词从过去的双词组合发展到多词组合。了解英文科技文章中常用的一些复合词与缩略词会大大提高翻译效率。例如，"IABSE"就是"International Association for Bridge and Structural Engineering"（国际桥梁及结构工程协会）的首字母缩写，"CAD"是"Computer Aided Design"（计算机辅助设计）的首字母缩写。

2. 句法特点

(1) 根据上海交通大学的计算机语料统计，科技英语句的平均长度为21.4个单词，不超过7个单词的短句仅占8.77%，超过40个单词的长句占6.3%。为了表述复杂概念，使之逻辑严密、结构紧凑，一般来讲，英文科技文章中的句子长度较长，结构较复杂，信息量较大。

(2) 为使陈述显得客观，英文科技文章大量使用被动语态。其一，科技文章反映的是客观事实及据此做出的科学推论，因此语言运用要体现客观性和普遍性，避免主观臆断、以偏概全。其二，科学研究的对象是客观世界，科技文章描述的是科学研究的对象、手段、过程、结果等各个方面，揭示客观世界的规律，使用被动句则可以突出客观世界这一科学研究的主体。其三，人是科学研究的从事者，是客观世界的施动者，但在很多时候，并不知道或没有必要知道施动者是谁，这时被动句就可派上大用场。

(3)《当代英语语法》（*A Grammar of Contemporary*）在论述科技英语时提出，大量使用名词化结构是科技英语的特点之一。因为科技文体要求行文简洁、表达客观、内容确切、信息量大、强调存在的事实，而非某一行为。

3. 语篇特点

(1) 科技文章的结构逻辑性强，层次清楚。科技文章的段落逻辑较简单，一般是总—分结构（少数采用分—总结构）。在段落的开头就摆出总结性语句，后面就一步一步地进行叙述，一般不存在插叙、倒叙。为了使科技文章的结构更加清楚明白，经常借助子标题及编号等形式。科技文章中还会出现表格、数学公式等。

(2) 科技文章的语言具有规范性和逻辑性。在科技文章中，由于其阐述的是科学问题或观点，因此在语言上表现出规范性和逻辑性，对学术性的用语和语言文字都有更高的要求。在科技文章中，要求语言精练，无论是在结构上还是在逻辑上都要严密。科技文章在写作过程中一般是通过提出论题、收集材料、推断结论等完成文章的全过程的，而在这个过程中，最基本的原则就是要保持逻辑学的规律。因此，在翻译过程中，也应该注重科技文章语言的逻辑性和规范性，保证翻译的最终结果也具备逻辑性和规范性。

(3) 科技文章的语言具有准确性和专业性。在撰写科技文章时，有时会采用缩略词，尤其是对于二次出现的词语或概念，一般会使用首字母的缩略词来代替，这就要求翻译人员应该联系文章的上下文进行翻译。在科技文章中，常常会采用一些专业术语，一些不常用的词语也会出现，所有的词语和句子都不应有任何歧义。因此，在科技文章中对每一个词汇和句子的准确性要求都非常高。在翻译科技文章时也要保证翻译的内容不会出现任何歧义。因此，对翻译人员的专业性提出了更高的要求，要求翻译人员对一些常用语及专业术语熟悉掌握。

二、科技文章范文展示

范文一

Science and Technology: the Life or Death Question for the Profession

In 1965, three years before he co-founded Intel, Gordon Moore predicted that the processing power of computers would double every two years or so. Sceptics at the time said that "Moore's Law" would hold true for a couple of years, but 51 years later it still cannot be overturned. By 2020, it is predicted that an average desktop machine will have roughly the same processing power as a single human brain and by 2050 a similar machine will exceed the processing power of all human brains on the planet.

This exponential growth in science and technology is impacting on the professions' career and bringing their future existence into question. It is time the professions reviewed some of their working practices. Avoid "technical myopia" (the tendency to underestimate the potential of tomorrow's applications by evaluating them in terms of today's technologies), because professionals need to understand that machines are becoming increasingly capable. More and more tasks that once required human beings are being performed more productively, cheaply, easily, quickly, and to a higher standard by a range of systems.

Blurring boundaries

And as machines become more intelligent, the boundaries between what they do and humans do are beginning to blur. Machines can now discern patterns, identify trends and make accurate predictions through the use of Big Data. They can perform tasks previously thought to require human intelligence (think IBM's Watson can act as a c – suite adviser, scanning strategy documents, listening to and digesting conversations at meetings and providing analytical advice based on its insights, such as possible companies to invest in).

Robotics are so advanced now that they can interact with manual skill and dexterity in the physical world—the Google self – driving car, for instance. Then there are systems in existence that can detect and express emotions. They can look at someone's face and work out, say, whether they are happy or sad or their smile is genuine with better acuity than humans.

Letter – writing, face – to – face meetings and fixed – line telephone calls have largely been replaced by email (in 2014 more than 196 billion emails were sent each day) and video calls, while the internet has become the go – to research facility and now offers its users theplatform to generate information and make it available to others. Users now socialise and share online: YouTube has over one billion users who upload about 300 hours of video every minute, while Twitter's 288 million users send on average 500 million tweets a day.

Mass collaboration

As well as communities that have grown up via generic platforms like Facebook and LinkedIn, other online communities or platforms of experience are beginning to emerge. One example is Patients Like Me where people share medical advice and guidance based on their experiences. Perhaps the most dramatic result of connected human beings is their ability to co – operate online. Mass collaboration has resulted in Wikipedia becoming a reliable body of knowledge, Linux the most commonly used operating system and crowd – sourcing a new way of raising funds, providing solutions or helping other people.

And there is no end to all this change in sight. The commercial imperative, the drive towards technological invention, is so strong that the technology that will changeour lives in 2025 is likely not invented yet.

Technology will certainly displace many traditional professional roles in its mid – stage. But that does not mean the death knell of the professions. Rather, professionals should be helping to create new roles for themselves. For example, there will be new business models and different ways of using technology to leverage solutions. Rather than working in firms, say, professionals could become experts working for agencies that employ highly skilled individuals who make their time available. The task for the professions is to reinvent themselves and diversify.

TheSusskinds see technology as an opportunity for those professionals prepared to be flexible and open-minded. "It's up to you to embrace these possible futures, particularly those of you who are keen and competitive, and say to yourselves this is a time when we can establish new and exciting businesses and thrive."

科学技术：行业的存亡问题

1965年，戈登·摩尔（Gordon Moore）预测，计算机的处理能力将以大约每两年翻一倍的速度增长。三年后他与别人共同创立了英特尔（Intel）。当时有人怀疑"摩尔定律"可能只会在仅仅几年内有效。但在51年后的今天，这一定律仍然无法被推翻。人们预计到2020年，计算机的平均处理能力将与人脑的处理能力大致相当，而到了2050年，计算机的平均处理能力将超过地球上所有人类大脑处理能力的总和。

科学技术在呈指数型增长，影响着专业人士的工作甚至是他们未来的去留问题。是时候重新回顾专业人士的工作实践了。专业人士必须了解到机器的工作能力会不断提升，因此他们应该避免"技术短视"（技术短视是指通过评估现在的技术水平而低估该技术在未来的应用潜力的倾向）。从前需要人类操作的任务中，有越来越多的工作现在可以通过一系列系统，按照更高的标准，以更高效、更省钱、更容易、更快捷的方式完成。

模糊的界限

随着机器变得更加智能化，机器行为和人类行为的界限开始变得模糊。现在，机器可以通过使用大数据识别不同的模式、确定变化趋势并做出准确的预测。它们可以执行以前我们认为只有人类才能完成的任务。（试想一下IBM公司的沃森（Watson）认知计算系统。这一系统可以扮演高管顾问的角色，扫描战略文件，听取并理解会议记录，基于其内部信息分析来提供咨询意见。比如，它会为你寻找值得投资的公司。）

现在，机器人技术如此先进，它们甚至可以将手工技术灵活地应用于现实世界。比如，谷歌（Google）就推出了无人驾驶汽车。此外，还有一种可以检测并且表达情感的系统。它们拥有比人类更敏锐的洞察力，可以观察别人的脸色并判断出对方是高兴还是悲伤，抑或判断出对方的笑容是否发自内心。

书信往来、面对面的会议和拨打固定电话等通信方式已经在很大程度上被电子邮件（在2014年，每天发送的电子邮件超过1960亿封）和视频通话所取代。而互联网现在已经成为必备的信息检索工具，并且向它的使用者提供了生成信息并与其他人进行信息共享的平台。如今，用户都在线上进行社交与分享：YouTube每分钟会有超过10亿的用户上传近300小时的视频，而Twitter的2.88亿用户平均每天会发送5亿条动态。

大规模协作

除通过脸书（Facebook）、领英（LinkedIn）等通用平台发展起来的社区外，其他类型的线上社区或经验交流平台也正在进入人们的视线。Patients Like Me（注：一个病患交流平台）就是一个很好的例子。通过这一平台，人们基于自身的经验分享医疗建议和指导意见。

也许在人类共同协作方面，最引人注目的成果就是他们能够通过线上进行合作。大规模协作使得维基百科（Wikipedia）成为可靠的知识来源地，Linux 成为最常用的操作系统，众包成为一种募集资金、提供解决方案或帮助其他人的新方式。

而且，没有任何迹象表明这种变化将会结束。商业世界的需要（技术发明的驱动作用）十分迫切，因此那些会改变我们 2025 年生活的技术可能还未被发明出来。

技术在其发展的中期阶段肯定会取代许多传统的专业角色。但这并不意味着这些职业的消亡。相反，专业人士应该借助技术帮助自己在新的浪潮中创造新角色。例如，新的商业模式和不同的技术运用方式可以为人们提供解决方案。比如说，专业人士可以成为为机构工作的专家，而不是在公司工作。这些机构聘用的是能够充分利用时间的高技能人才。这些职业的任务是重塑自我并使之多样化。

理查德·苏士侃和丹尼尔·苏士侃（the Susskinds）认为技术的发展对专业人士来说是可以开放思想、变得更加灵活的一个机会。"是否要接受这些未来的可能性的决定权在你自己，尤其是你们中那些思维敏捷、敢于竞争的人。你要对自己说，现在正是我们可以建立一个令人振奋的商业新世界并使之发展繁荣的时刻。"

范文二

A busy brain can mean a hungry body. We often seek food after focused mental activity, like preparing for an exam or poring over spreadsheets. Researchers speculate that heavy bouts of thinking drain energy from the brain, whose capacity to store fuel is very limited. So the brain, sensing that it may soon require more calories to keep going, apparently stimulates bodily hunger, and even though there has been little in the way of physical movement or caloric expenditure, we eat. This process may partly account for the weight gain so commonly seen in college students.

Scientists at the University of Alabama at Birmingham and another institution recently experimented with exercise to counter such post-study food binges. Gary Hunter, an exercise physiologist at U.A.B., oversaw the study, which was published this month in the journal *Medicine & Science in Sports & Exercise*. Hunter notes that strenuous activity both increases the amount of blood sugar and lactate—a byproduct of intense muscle contractions—circulating in the blood and augments blood flow to the head. Because the brain uses sugar and lactate as fuel, researchers wondered if the increased flow of fuel-rich blood during exercise could feed an exhausted brain and reduce the urge to overeat.

Thirty-eight healthy college students were invited to U.A.B.'s exercise lab to determine their fitness and metabolic rates and to report what their favorite pizza was. Afterward, they sat quietly for 35 minutes before being given as much of their favorite pizza as they wanted, which established a baseline measure of self-indulgence. At a later date, the volunteers returned and spent 20 minutes tackling selections from college and graduate-school entrance exams. Hunter says this work has been used in other studies to induce mental fatigue and hunger. Next, half the students sat quietly for 15 minutes, before being given pizza. The rest of the volunteers spent those 15 minutes doing intervals on a treadmill: two minutes of hard running followed by about one minute of walking, repeated five times. This is the sort of brief but intensive routine, Hunter says, that should prompt the release of sugar and lactate into the bloodstream. These students were then allowed to gorge on pizza, too. But by and large, they did not overeat.

In fact, the researchers calculated that the exercisers consumed about 25 fewer calories than they did during their baseline session. The non-exercisers, however, consumed about 100 calories more. When the researchers factored in the calories expended on running, they determined that those students actually consumed 200 fewer total calories after their brain workouts than the resting students.

The study has limitations, of course. "We only looked at lunch," Hunter says. The researchers do not know if the runners consumed extra calories at dinner. They also cannot tell whether other types of exercise would have the same effect as running, although Hunter says they suspect that if an activity causes someone to break into a sweat, it should also increase blood sugar and lactate, feeding the brain and weakening hunger's call.

大脑忙碌可能会导致身体饥饿。我们经常在集中进行脑力活动后寻找食物，比如备考或阅读报表。研究者猜测，高强度思考会耗尽脑部能量，而脑部存储养分的能力十分有限。所以，大脑感觉自己可能很快就需要更多热量来维持运转，显然会因此刺激身体产生饥饿感。尽管几乎没有进行体力运动或出现热量消耗，我们还是会吃东西。这个过程可以部分解释大学生中常见的体重增加现象。

前不久，阿拉巴马大学伯明翰分校（University of Alabama at Birmingham）的科学家们和另一个研究机构试验通过运动对抗学习后的暴饮暴食。这项研究由阿拉巴马大学伯明翰分校的运动生理学家加里·亨特（Gary Hunter）主持，本月发表在 *Medicine & Science in Sports & Exercise* 上。亨特指出，剧烈活动会增加在血液中循环的血糖和乳酸盐（这是肌肉剧烈收缩的副产品）含量，增加脑部的血流量。大脑以糖和乳酸盐为养分，所以研究者们想知道，锻炼中增加的养分丰富的血流量是否能给疲惫的大脑提供养分，从而减轻过度进食的冲动。

38 名健康的大学生被邀请到阿拉巴马大学伯明翰分校的运动实验室，接受健康水平和代谢率检测，并报告自己最喜欢吃哪种披萨。之后，他们静坐 35 分钟，然后他们想吃多少披萨就供应多少，以确定自我放纵的基线。在后来的一个日子里，志愿者们返回这里，花费 20 分钟时间做从大学和研究生入学考试中挑选的题目。亨特说，这种方法已经在其他研究中被用于引发精神疲劳和饥饿感。接下来，其中一半学生静坐 15 分钟，然后进食披萨。剩下的志愿者用那 15 分钟时间在跑步机上做间歇运动：快跑 2 分钟，然后走大约 1 分钟，重复 5 次。亨特说，这种短暂而高强度的运动应该能促进糖和乳酸盐释放到血流中。这些学生之后也被允许尽情食用披萨。总体来说，他们没有过量进食。

实际上，研究者们计算出，这些锻炼者摄入的热量比在基线测试阶段大约少 25 卡路里。但是，那些没有锻炼的人大约多摄入了 100 卡路里。研究者们把跑步中耗费的热量计算在内后发现，那些学生实际上在脑力活动后比其他学生一共少摄入 200 卡路里。

这项研究当然具有局限性。"我们只研究了午餐。"亨特说。研究者不知道跑步者在晚餐时是否摄入了更多热量。他们也不知道其他类型的运动是否会产生和跑步一样的效果，不过亨特说，他们猜测，如果一项运动能让人出汗，那么它应该也能增加血糖和乳酸盐释放，给大脑提供养分，从而缓解饥饿感。

范文三

Display Screens

Liquid Crystal Displays (LCDs) are a familiar and ubiquitous technology. But if Harish Bhaskaran of Oxford University is right, their days may be numbered. The essential feature of LCDs is that the pixels in them switch between amorphous and crystal line phases, which changes their optical properties. In a paper in this week's *Nature*, Dr. Bhaskaran and his colleagues describe something similar in a solid material. At the least, that would stop the messy abstract – impressionist patterns which happen when an LCD is dropped too hard. At most, it might open up a new range of applications, from clothes that change colour to dimmable windscreens.

Solid phase – change materials are already used to store data in optical memory disks. They are also being considered for use in memory chips, because the switch between amorphous and crystalline states alters their electrical properties in ways that can store electronic bits of data. Dr. Bhaskaran, though, has shown that thin enough films of the right sort of material can be made to change colour, too.

This property would make them suitable both for displays that rely on reflected light (so – called electronic paper) and the older, backlit sort that rely on transmitted light. The resulting displays would be thin and could be flexible if printed on the right material—increasing the range of applications they might be used in. And they would consume little power, since energy need be used only when a pixel has to be flipped from one phase to another.

The researchers' material of choice is an alloy of germanium, antimony and tellurium. Both the crystalline and the amorphous phases of this substance are stable at any temperature a device is likely to experience, and thin films of it are more or less transparent. The power needed to effect the phase change could be fed to individual pixels by electrodes made of indium tin oxide, which is also transparent.

The colour of a pixel would depend not only on its phase, but also on its thickness, which would affect the way light waves being reflected, cancelling out some frequencies while amplifying others. (The effect is similar to the creation of colours by a thin layer of oil on a puddle.) Generally, the alloy layer needs to be thinner than 20 nanometres for that to happen.

To demonstrate their idea, the researchers sprayed films of their alloy onto pieces of silicon, quartz and plastic. They then used a device called an atomic – force microscope, which has a tip a few nanometres across, to apply appropriate electric currents in a grid pattern across the film's surface. This grid mimicked an array of pixels, creating a stable pattern. The result, as their picture of a Japanese wave shows, is a recognisable image—if not, yet, a perfect one.

Adding the indium – tin – oxide electrodes is a more complicated process, but to show it can be done in principle, Dr. Bhaskaran has made a single pixel this way. Whether his idea will get off the lab bench and into the shops remains to be seen. It is by no means the only suggestion around for a new generation of display screens. But it looks plausible.

显示屏

液晶显示屏（LCD）是一项成熟而广泛应用的技术。但如果牛津大学的 Harish Bhaskaran 的想法实现，液晶屏的好景也就没几天了。LCD 的基本特征是其像素在非晶相和晶相间转换，因此改变其光学特性。在本周的《自然》杂志上，Bhaskaran 博士及其同事陈述了一种类似的固体材料。至少该材料可以防止当 LCD 猛烈坠地后产生麻烦的抽象印象派图案。最理想情况下，它会有新的应用范围，从变色布料到可调光挡风玻璃。

固态相变材料已经用于在光存储盘中存储数据。固态相变材料也可能被用于存储芯片，因为其非晶态和晶态之间的转换也改变了其电学特性，因此它们能存储电子数据位。然而，Bhaskaran 博士展示了一种特定材料膜，在足够薄的情况下也可以改变颜色。

这一性能将使其适合于反射光显示（所谓电子纸）和早期的依赖透射光的背光显示。所得显示屏不仅薄，而且铺在合适材料上时非常灵活，增大了其可能的应用范围。而且它消耗的电能很少，因为这种材料只需要在像素变相的时候消耗能量。

研究人员选择的材料是锗、锑和碲合金。这一物质的晶相和非晶相在设备可能经历的任何温度下都很稳定，其薄膜也几乎透明。要激发相变所需的电能可以通过透明的氧化铟锡电极馈送到各个像素。

单个像素的颜色不仅取决于其相态，也同其厚度有关。厚度会影响光波的反射方式，削弱部分频率，加强其他频率。（此效应类似于池塘表面薄油层产生色彩。）一般来说，合金层需要薄于 20 纳米才会发生这一现象。

为了证明其观点，研究人员将合金薄膜喷在硅、石英和塑料上。然后，他们使用所谓原子力显微镜的设备在薄膜表面的网格上施加合适的电流。原子力显微镜的尖端仅有几纳米大小。网格模仿像素排列，产生稳定的模式。正如他们拍摄的日本海浪照片所显示的那样，成片是一个可识别的图像，即便不可识别，也是非常完美的。

加入氧化铟锡电极是更为复杂的过程，但为了表明这在原则上是可以做到的，Bhaskaran 博士用这种方式制备了单个像素。他的想法能否从实验室走进商用还是未知。这绝不是新一代显示屏的唯一方案，但颇得看好。

范文四

Cars could soon be communicating with each other using 5G to make drivers aware of upcoming hazards, scientists claim. The ultra-fast mobile internet would allow for rapid information transmission and could make drivers aware of black ice, pot holes or other dangers up ahead. Several car manufacturers are already integrating 5G into their vehicles, including as a tool to help usher in the generation of self-driving vehicles.

Experts at Glasgow Caledonian University (GCU) believe the high-speed connection will also improve the reliability and capability of automated vehicles to the point where they will be safer than the manual cars being driven today. They predict the number of road traffic accidents—which according to the World Health Organization account for more than 1.3 million deaths and up to 50 million people injured worldwide every year—will drop drastically as a result.

Dr. Dimitrios Liarokapis, a member of the research group, said: "To have a better idea of what the future will look like, think of having Tesla-like cars that not only use sensors to scan what's around them, they can also talk to each other and exchange safety – related information about their surroundings over an area that covers several square miles.

"I'm sure anyone who has had a bad experience on frozen roads would have benefited from knowing about the dangerous conditions in advance, so they could have adjusted their speed or, if possible, even avoided that route altogether. The same could be said of pot holes.

"With the help of 5G, a vehicle – generated early warning system that alerts drivers is feasible within the next few years. Cars that are close enough to the danger area will transmit warning messages to other cars around them using short – range communication technologies, but also to cars further away using 5G, fast and reliably.

"Then those cars will send the same information to cars near them and so on, forming a joined – up, multi – vehicle communication chain that stretches far and wide.

"5G is an exciting mobile technology, which will give a massive boost to smart cities and autonomous vehicles among many other things."

Automotive giant Ford is already working on connected cars. Earlier this year it revealed its intention to fit 80 percent of its 2020 vehicles with technology that warns drivers about upcoming road accidents, bad weather and traffic jams. The system pools data from other connected road users, e-mergency services and the authorities and beams it from the cloud directly to the car. Alerts pop up on the car's dashboard display warning the driver about what lies around the corner.

科学家称，汽车很快就可以通过5G网络互相"对话"，让司机意识到前方的危险。超高速的移动互联网将使信息可以快速传输，并让司机意识到前方的黑冰、坑洞或其他危险。几家汽车制造商已经将5G网络技术融入汽车生产，包括以此来帮助引领自动驾驶汽车时代的到来。

格拉斯哥卡利多尼亚大学的专家认为，高速互联网还将提高自动化车辆的可靠性和性能，使其比目前使用的手动驾驶汽车更安全。他们预测，道路交通事故的数量将因此大幅下降。据世界卫生组织统计，全球每年因道路交通事故死亡的人数超过130万，受伤人数高达5000万。

研究小组的Dimitrios Liarokapis博士说："为了更好地了解未来会是什么样子，想想你有特斯拉那样的车，它不仅使用传感器来扫描周围，车辆之间还可以相互交流，并且交换附近几平方英里内的安全信息。

"我相信任何在冰冻道路上有过糟糕经历的人都会从提前了解危险情况中受益，这样他们就可以调整车速，如果可能的话，甚至完全避开这条路线。路遇坑洞也是如此。

"在5G网络的帮助下，未来几年内，可以向司机发出警报的车辆生成预警系统将会出现。距离危险区域足够近的汽车将使用短程通信技术向周围的其他汽车发送警告信息，同时使用5G网络向更远的汽车发送警告信息，快速且可靠。

"然后这些车将同样的信息发送给附近的车，以此类推，形成一个连成一体的多车通信链，延伸到很远很远的地方。

"5G网络是一项激动人心的移动技术，它将极大地推动智能城市和自动驾驶汽车等领域的发展。"

汽车巨头福特公司已经开始研发联网汽车。今年早些时候，该公司透露，计划在2020

年生产的 80% 的汽车中配备一种技术，用于提醒司机注意前方的交通事故、恶劣天气和交通堵塞。该系统将来自其他联网道路用户、应急服务机构和管理部门的数据汇聚在一起，并将数据从云端直接传输到汽车上。汽车仪表盘上弹出的警报将提醒司机前方的危险。

三、科技文章常用词汇和句型

（一）科技文章常用词汇

as a result　因此
apart from　除……外
by contrast　相比之下
despite / in spite of　尽管
in addition　另外
in contrast to　与……相比
in the same way　同样地
in conclusion　总而言之
in case　万一
instead of　而不是……
in terms of　就……而言
to our knowledge　据我们所知
to sum up　总之
with the aim to　为了……，目的是……
with the help of　在……的帮助下
with an accuracy of　精确到
according to　根据
base on　基于
be divided into　被分为
be defined as　被定义为
be capable of　能够
conform to　符合
consist of　由……组成
consist in　在于
correspond to　相当于
different from　不同于
draw attention to　引起对……的关注
focus on　集中于
give rise to　引起，导致
lead to　导致；通向

range from... to...　范围从……到……

rather than　而不是

refer to　参考；指的是

（二）科技文章常用句型

1. It... that...　结构句型

It is suggested that...　建议是……

It is assumed that...　假定……

It has since been demonstrated experimentally that...　自此，实验证明了……

It is certain that...　毫无疑问的是……

It is clear that...　显然，……

It is likely that...　很可能……

2. 被动语态结构句型

A new method is described for...　介绍了一种……的新方法

... can be classified into...　……可被分为……

It is believed to be...　人们相信……，人们认为……

This paper gives a brief introduction to...　本论文主要介绍了……

... are analyzed　分析了……

... is provided for...　提供……为了……

3. 其他句型

The aim of this study is to...　本研究旨在……

The results show that...　结果表明……

We conjecture that...　我们推测……

One answer to the problem of... is...　……问题的解决方法是……

Another solution is to...　另一种解决方法是……

四、科技文章翻译技巧

关于翻译的标准，历来提法很多。清末启蒙思想家、翻译家严复主张"信、达、雅"，当代翻译界主张"忠实、通顺"。就科技英语的特点和用途而言，其翻译的标准应略区别于文学翻译。科技英语翻译主要是两种语言间的信息转换，为此，在进行科技英语翻译时要坚持三条标准。

（1）准确规范。所谓准确，就是忠实地、不折不扣地传达原文的全部信息内容。所谓规范，就是译文要符合所涉及的科学技术或某个专业领域的专业语言表达规范。

（2）通顺易懂。译文语言必须通顺，符合规范，用词造句应符合目的语的语言习惯，要用科学的、大众的语言，以求通顺易懂。不应有文理不通、逐词死译和生硬晦涩等现象。

（3）简洁明晰。译文要简短精练、一目了然，要尽量避免烦琐、冗赘和不必要的重复。

要提高翻译质量，使译文达到"准确规范""通顺易懂""简洁明晰"这三个标准，就必须运用翻译技巧。翻译技巧就是在翻译过程中用词造句的处理方法。接下来,将探讨在翻译科技文章时可使用的翻译技巧。

(一) 长难句的汉译

在科技文章中,科技作者为了减少文章中歧义的产生,为了准确、详尽、完整地表达客观事物之间的因果、条件、依附、伴随和对比等关系,就需要严密的逻辑思维。这种思维见诸语言形式。英语行文,句子一般都比汉语长,有时一个长段落就是一个句子。造成这种英语句子形态的原因之一是,英语句子是树形结构,构句靠一批关联词联结,句子的成分可以很自然地扩展、增加。无论一个英语句子有多长,只要有关联词在其中"穿针引线",句子仍然是结构严谨、层次分明的。这样,就容易形成内容上包含大量信息,结构上包含若干介词短语和非限定动词短语等短语,以及从句和分句,甚至短语套短语、短语套从句、从句套从句的结构复杂的长句。原因之二是,许多含有动宾关系或主谓关系的结构可以用非句子形式来表达,它们有的可被看成是浓缩的非句子形式。在翻译科技文章的过程中,在对长难句进行汉译时,应该根据句子的长短程度采用不同的翻译技巧进行翻译。当对较长的从句进行翻译时,由于科技文章描述的都是客观事实,应该先进行"顺句",按照句子行文顺序处理句子的各种成分,分析出句子的主干结构,把握好全句的结构,理顺全句各部分之间的关系。在这个过程中,"断句"是关键,也就是将原句依顺序切成若干部分,形成独立的意群或停顿群。根据主干结构先译出整个句子的大概意思,然后再根据原文的意思,将各种修饰语搭配到主干句子中,从而将修饰语完整地翻译出来。例如:

When the first Boeing 747 rolled off production line on September 30, 1968, it was immediately nicknamed the Jumbo and raised fears that its enormous size would overwhelm airports, overload passenger terminals and baggage facilities, overrun customs and immigration units, and overstress taxiways and runways.

断句:When the first Boeing 747 rolled off production line on September 30, 1968, // it was immediately nicknamed the Jumbo // and raised fears that // its enormous size would overwhelm airports, // overload passenger terminals and baggage facilities, // overrun customs and immigration units, // and overstress taxiways and runways.

译文:1968年9月30日,第一架波音747飞机装配完毕,就马上被称为"庞然大物"。同时,它也引起了人们的种种担忧:庞大的机身将使机场不堪承受,航站楼将人满为患,处理行李的设施会超负荷运转,海关和移民局办事机构会忙不过来,机场的滑行道和跑道也会因受压过重而受损。

从操作上说,"顺句"操作即是将原来的句子先做有序的切割,翻译时实际上又要将原来断开的各个成分有机地串联起来。在具体翻译过程中,或许要改变原文的某些表述形式,或许要添加一些字词,但是只要不妨碍宏观语义,都是许可的。在"顺句"过程中可能还会出现这样的情况:断句后的几乎每个部分都可用一个汉语句式进行翻译,原句若是有平行的动宾搭配结构时,情况更是如此。例如:

Herein, we prepared magnetic multi-walled carbon nanotubes (MWCNTs), and applied it in bound solute dialysis // which based on the thermodynamic principle // that removing toxins in blood could be enhanced // by adding a blinder into the dialysate // that acts a sink for toxins, //

especially for albumin – bound toxins, // such as bilirubin, // via conventional dialysis.

译文：在这方面，我们准备了磁性多壁纳米碳管（MWCNT），将其应用到结合溶质透析中。这种透析基于热力学原理，通过做常规透析，向透析液中添加一种起沉淀毒素作用的黏合剂，能够提高清除血液毒素的能力，特别是对于清除像胆红素这样的白蛋白结合毒素。

下画线部分是该句的主干，"which based on the thermodynamic principle"为定语从句来修饰"bound solute dialysis"，"that removing toxins in blood could be enhanced // by adding a blinder into the dialysate"做"principle"的同位语，"that acts a sink for toxins"做"blinder"的定语，起补充说明的作用。"such as bilirubin"为"albumin toxins"的例子，起解释说明的作用。这个长句包括多重复合句，在翻译过程中，先对句子进行"顺句"操作，厘清句子结构，抓住主干。为了能表达清楚，可采用拆译的方法，将包含大量信息的长句拆开来译，这样既能准确地译出原文的意思，又能使译文符合汉语的语言表达习惯。

（二）被动句的翻译

被动句的使用是英文科技文章的句子结构特征之一。在英文科技文章中，常把事物的名称作为主语放在突出的位置上，用被动语态表述有关的动作或状态。由于被动句在英文科技文章中出现频率极高，因此被动句的翻译也是英文科技文章翻译的重点之一。例如：

In this system, magnetic MWCNTs was suspended in dialysate and separated from blood by the dialysis membrane.

译文：在这个系统中，磁性多壁纳米碳管悬浮在透析液中并被透析膜从血液中分离出来。

这是被动句的典型例子，"separate"的受动者是"MWCNTs"，施动者为"the dialysis membrane"。因此，此类句子可以直接翻译成"被"字句或转换成主动句，如"把"字句。还有其他的情况，可用"由""受""为"等引导的隐性被动句式来处理英语中的被动语态。

汉译英时可借鉴英译汉时被动句的处理方法进行反向操作。例如：

从2001年3月份开始，原油定价将进一步放开，国家计委（后为国家发展和改革委员会）不再公布国内原油基准价，改<u>由本公司和中国石油按照国家计委的上述原则和办法自行计算和确认原油价格</u>。

译文：Since March 2001, the price of crude oil will be further liberalized. The State Development Planning Commission (Later, National Development and Reform Commission) will stop publishing domestic base price. *Instead, the price will be calculated and determined independently by our company and PetroChina according to the principles and methods mentioned above.*

原文中下画线部分形式上是主动语态，但用"由……"表示强调。译文中斜体部分用的是被动语态，正好把旧信息"the price"作为主语，把要强调的部分（"由……"）放在句末。如果把这一句改为主动语态，语气是不通顺的：Instead, our company and PetroChina will calculate and determine the price independently according to the principles and methods mentioned above。主动语态要表达原文意思，只能通过有意重读"our company and PetroChina"来实现，因为按照人们的阅读习惯，一般是把重音放在句末。从理论上讲，只要英语需要，汉语的主动语态可以译为英语的被动语态，汉语的被动语态（包含隐性被动句）也可以译为英语的主动语态。但实际上，汉语的主动句可以译为英语的被动句，而汉语的被动句却极少译为英语的主动句。

(三) 隐含逻辑关系的翻译

英汉两种语言属于不同的语系，思维方式上存在明显的差异。英语是一种形合的语言，注重语言结构的严密性和规范性，多使用关联词，各个分句、短语间的关系清晰明了。而汉语是一种意合的语言，不太注重语言结构的严密性，关联词使用较少，句子结构较松散。因此，在汉语科技文章中，逻辑关系信息往往隐含在句子中，如果简单地采用直译的方法，就会出现结构松散、表意不清等问题。例如：

当事人不愿协商、调解解决或协商、调解不成的，双方商定，采用以下第（一）种方式解决。

译文：<u>When</u> both parties are unwilling to settle their dispute through consultation and mediation, or when consultation and mediation fail, both parties agree to solve it based on the following way (1).

在科技汉语中，一些表示条件关系的词语（如"当……时""如果"）往往不会出现在句子中，而在翻译成英语的过程中，要注意加上表示条件关系的词语，将隐含的条件关系明确表达出来。相反地，在英译汉时，一些表示条件关系的词语可以省去，否则会显得翻译腔过重。例如：

Gel, as a new type of smart materials, can be applied to the development of new protective products <u>because</u> its excellent elasticity makes it have a good impact resistance.

译文：凝胶作为一种新型的智能材料，其优异的弹性使其具备了良好的抗冲击性能，可以被运用到新型防护产品的开发中。

(四) 引申译法

当英语句子中的某个词按词典的释义直译不符合汉语修辞习惯或语言规范时，则可以在不脱离该英语词本义的前提下，灵活选择恰当的汉语词语或词组译出。例如：

The major problem in manufacturing is the control of contamination and foreign material.

译文：制造中的主要问题是控制污染和杂质。

"foreign"在字典中的意思为"外国的；涉外的；陌生的"，这里引申译为"不纯的，杂的"。

有的词语意义不是表面上显示的那样，需要译者根据上下文把隐含的意义翻译出来。例如：

In a dwelling house, the layout may be considered under three categories: "day", "night", and "service".

译文：住房布局可以分三类考虑，即"白天用的部分"、"夜晚用的部分"及"屋内设施网"。

原文中的"day"和"night"明显不是普通意思"白天"和"夜晚"。作者在此处打上引号，说明这里是与普通意思有区别的，是作者对现有词义进行的引申。因此，为了表明其真正含义，作者紧接着就用例子说明了所谓的"day"和"night"的真正含义。翻译的时候，参照后文，理解作者的真正意图，发现其含义是："day"是"白天用的部分"，"night"

是"夜晚用的部分"。

(五) 增减词译法

英汉两种语言，由于表达方式不尽相同，翻译时要在词量上加以增减。英译汉时，增词就是在译句中增加或补充英语句子中原来没有或省略了的词语。在英语句子中，有的词从语法结构上讲是必不可少的，但并无什么实际意义，只是在句子中起着单纯的语法作用；有的词虽有实际意义，但按照字面译出又显多余。这样的词在汉译时往往可以省略不译，从而使得译文在语法、语言形式上符合汉语习惯，并在文化背景、词语联想方面与原文一致起来，使得译文与原文在内容和形式等方面对等起来。

在英文科技文章中，常有两个同义词连用的现象，即对同一概念用两个词重复表达。这两个同义词中的前一个往往是专业词汇，而后一个往往是普通词汇。两者在语法上是同位关系，常可用连词 or 连接。作者这样做，往往是出于强调或为了适应不同层次读者的需要。但在译成汉语时，为了使概念准确，免于文字重复或含糊不清，却只需要译出其中一个，即译成汉语中通用的一个名称就可以了。例如：

Consolidation or compaction is the process of molding concrete within the forms and around embedded parts in order to eliminate pockets of entrapped air.

译文：压实是在模板内部及嵌入件周围浇铸混凝土以消除陷入空气的工艺。

"Consolidation"和"compaction"的意思分别是"压缩"和"压实"，但在原句中明显是一致的，只要翻译一个就够了。若翻译成"压缩或压实"纯属多余，并且不符合汉语表达习惯。这种用法在英语中用得很多，而在汉语中却几乎不适用。

Its unit weight is 2,000 – 2,500kg/m^3 (124 – 156lb/ft^3), and its modulus of elasticity is 15 – 30GPa (220 – 440ksi).

译文：其单位重量为 2000~2500 kg/m^3，其弹性模量为 15~30GPa。

鉴于译文的目标读者是中国人，中国人使用的是国际标准单位，括号中的是英语国家（尤其是英国）使用的单位，所以选择不翻译。翻译出来，没有该文化背景知识的中国读者也看不懂，还有可能引起混淆，增添不必要的麻烦。

Concrete and steel also form such a strong bond—the force that unites them—that the steel cannot slip within the concrete.

译文：混凝土和钢筋之间还能形成很强结合力，钢筋不会在混凝土中滑落。

很明显，译文中没有翻译"the force that unites them"，这是因为"bond"翻译成"结合力"时，若再加解释"把两者结合起来的力"，按中文行文规则，就显多余了。

(六) 词类转换、变词为句

在英文科技文章的翻译中，常常需要将英语句子中属于某种词类的词，译成另一种词类的汉语词，以适应汉语的表达习惯或达到某种修辞目的。这种翻译处理方法就是转换词性法，简称词类转换。例如：

Copper wire is flexible.

译文：铜线容易弯曲。

这里将形容词"flexible"转译为动词"容易弯曲"。

汉语行文句子一般不像英语句子那样长，表层结构松散，句与句及子句与子句之间也没有那么多的关联词联结。在英译汉时，为使译文通顺、可读，往往将原文中的某些非句子结构（如含有动宾关系或主谓关系的表达）以句子的形式翻译出来，也就是变词为句。例如：

The prospective deal, though obviously no longer politically controversial, has nonetheless produced sharply divergent views in defence circles. Those against the MIG – 29M argue that the aircraft's short service life, tricky maintenance record and unfamiliar engineering will render it a logistical nightmare.

译文：预期之中的这笔交易虽然明显不会再有政治上的争论，但在防务界却产生了几种截然不同的看法。反对购买米格－29M型飞机的人士认为，这种飞机服役期短，维修记录上多有疑难故障，加上发动机操纵法与众不同，这些会给军需后勤部门造成极大的麻烦。

"变词为句"的翻译操作在实施过程中绝非一件十分机械呆板的事。译文在行文过程中不免要根据具体情况增加一些字词。

（七）词序处理法

英汉两种语言的词序规则基本相同，但也存在着某些差别。不同的英语句子，在翻译中的词序处理方式也常常不同。例如：

An insufficient power supply makes the motor immovable.

译文：供电不足就会使电动机停转。

这里将"insufficient power supply"（不足供电）改序翻译为"供电不足"较为合理。

（八）括号法

英文科技文章中长句多，各类修饰语也多，翻译成汉语时，会出现各种麻烦。若将修饰语放在要修饰的词语前面，修饰语会显得过长，不符合汉语表达习惯，而且读起来不顺口，理解起来耗时。若进行拆分翻译，又会使句子显得零散，主次不分。其实，此时借助括号来进行翻译能取得良好的效果。例如：

In braced frames, the lateral resistance of the structure is provided by diagonal members that, together with the girder, form the "web" of the vertical truss, with the column acting as the "chord".

原译：斜撑构架中，结构的横向阻力由大梁和形成"网状"垂直桁架的斜构件提供。垂直桁架的弦杆是柱子。

改译：斜撑构架中，结构的横向阻力由大梁和形成"网状"垂直桁架（其弦杆为柱子）的斜构件提供。

原译是按照一般翻译原则进行的翻译，最后的分词短语"with the column acting as the 'chord'"单独译成一句。这看似符合翻译的一般标准，但是该句的重心不在这里，如此翻译显得啰唆，又主次不分。借助括号改译后就解决了主次不分的问题。

五、课后练习

(一) 思考题

科技文章的特点有哪些?有哪些翻译技巧?

(二) 翻译练习

1. 英译中

(1) 段落翻译。

①As oil is found deep in the ground, its presence cannot be determined by a study of the surface. Consequently, a geological survey of the underground rock structure must be carried out. If it is thought that the rocks in a certain area contain oil, a drilling-rig is assembled. The most obvious part of a drilling-rig is a tall tower which is called a derrick. The derrick is used to lift sections of pipe, which are lowered into the hole made by the drill. As the hole is being drilled, a steel pipe is pushed down to prevent the sides from falling in and to stop water filling the hole. If oil is struck, a cover is firmly fixed to the top of the pipe and the oil is allowed to escape through a series of valves.

②A virtual network is defined as a logical network, in which users exhibit their access behaviors. Virtual networks rely on physical computer networks like the Internet, but have different topologies, and cause significant influence on the physical networks. A novel two-tier model is used to study influences of virtual networks to Internet collective behavior. It is shown that the queue lengths of the node data packets present phase transition characteristics. Moreover, the phase transition critical point moves to the left and network performance is deteriorated. In a free flow, the nodes are independent of each other or short-range correlative. In the critical state, the nodes are long-range correlative, and there exists a higher power exponent which means stronger long-range correlation. When the system state is on the right of critical point, virtual network behaviors make the network present consistent long-range correlative characteristic.

③This Vibration Damping Mount can keep sensitive analytical balances and other instruments from disturbing vibrations so that they work more accurately. It reduces to 9Hz the vibrations caused by nearby pumps, blenders and heavy-weight automobiles. It is made of black and white terrazzo with a polished surface that resists scratches and chemicals and supported by four Vibro-Absorbers with neoprene feet. With an overall height of 76mm, it can carry a weight up to 16kg.

(2) 篇章翻译。

① The 27-year-old patient's prospects were bleak. In May 2016, he found out he had AIDS. Two weeks later, he was told he had acute lymphoblastic leukemia.

But doctors offered the Chinese citizen a ray of hope: a bone marrow transplant to treat his cancer and an extra experimental treatment to try to rid his system of HIV, according to a new paper published in *The New England Journal of Medicine*.

This involved using the gene editing tool CRISPR – Cas9 to delete a gene known as CCR5 from bone marrow stem cells taken from a donor, before transplanting them into the patient, Peking University scientists said in the study.

"After being edited, the cells—and the blood cells they produce—have the ability to resist HIV infection," lead scientist Deng Hongkui told CNN Friday.

People who carry mutated copies of CCR5 are highly immune to HIV, because the virus uses a protein made by this gene to gain entry into an infected person's cells. Two men, known as the Berlin patient and the London patient, became the first people in the world to be cured of HIV after receiving bone marrow transplants from donors who had the mutation naturally.

The patient agreed and the experiment was carried out in the summer of 2017. It was the first time CRISPR – Cas9 was used on a HIV patient. In early 2019, a full 19 months after the treatment was administered, "the acute lymphoblastic leukemia was in complete remission and donor cells carrying the ablated CCR5 persisted," the scientists said in the paper.

But there weren't enough of them to eradicate the HIV virus in the patient's body. After transplantation, only approximately 5% to 8% of the patient's bone marrow cells carried the CCR5 edit, according to the researchers. "In the future, further improving the efficiency of gene – editing and optimizing the transplantation procedure should accelerate the transition to clinical applications," said Deng.

But he doesn't see this as a setback. "The main purpose of the study was to evaluate the safety and feasibility of genetically – edited stem cell transplantation for AIDS treatment," said Deng.

According to Deng, this was a success: the scientists didn't detect any gene editing – related adverse events, even if "more long – term in – depth studies are needed for off – target effects and other safety assessments," Deng said.

The CCR5 gene mutation has been associated with a 21% increased risk of dying early, according to a paper published in *Nature* in June, though it's unclear why.

The team that conducted the study had previously transplanted edited CCR5 human cells into mice, making them resistant to HIV. American scientists have carried out similar experiments on humans, with some success, using an older gene editing tool called zinc finger nuclease.

China has invested heavily in gene – editing technology, making biotech one of the priorities of its Five – Year Plan announced in 2016. The central government has bankrolled research into a number of world "firsts", including the first use of the gene – editing tool CRISPR – Cas9 in humans in 2016 and the first reported use of gene editing technology to modify nonviable human embryos in 2015.

Deng Hongkui remains a strong believer in CRISPR – Cas9. He thinks it could "bring a new dawn" to blood – related diseases such as AIDS, sickle cell anemia, hemophilia and beta thalassemia and that, thanks to this new technology, "the goal of a functional cure for AIDS is getting closer and closer".

Autism

②Women have fewer cognitive disorders than men do because their bodies are better at ignoring the mutations which cause them.

Autism is a strange condition. Sometimes its symptoms of "social blindness" (an inability to read or comprehend the emotions of others) occur alone. This is dubbed high-functioning autism, or Asperger syndrome. Though their fellow men and women may regard them as a bit odd, high-functioning autists are often successful (sometimes very successful) members of society. On other occasions, though, autism manifests as part of a range of cognitive problems. Then, the condition is debilitating. What is common to those on all parts of the so-called autistic-spectrum disorder is that they are more often men than women—so much more often that one school of thought suggests autism is an extreme manifestation of what it means, mentally, to be male. Boys are four times more likely to be diagnosed with autism than girls are. For high-functioning autism, the ratio is seven to one.

Moreover, what is true of autism is true, to a lesser extent, of a lot of other neurological and cognitive disorders. Attention Deficit Hyperactivity Disorder (ADHD) is diagnosed around three times more often in boys than in girls. "Intellectual disability", a catch-all term for congenital low IQ, is 30%-50% more common in boys, as is epilepsy. In fact, these disorders frequently show up in combination. For instance, children diagnosed with an autistic-spectrum disorder often also receive a diagnosis of ADHD.

Autism's precise causes are unclear, but genes are important. Though no mutation which, by itself, causes autism has yet been discovered, well over 100 are known that make someone with them more vulnerable to the condition.

Most of these mutations are as common in women as in men, so one explanation for the divergent incidence is that male brains are more vulnerable than female ones to equivalent levels of genetic disruption. This is called the female-protective model. The other broad explanation, social-bias theory, is that the difference is illusory. Girls are being under-diagnosed because of differences either in the ways they are assessed, or in the ways they cope with the condition, rather than because they actually have it less. Some researchers claim, for example, that girls are better able to hide their symptoms.

To investigate this question, Sebastien Jacquemont of the University Hospital of Lausanne and his colleagues analysed genetic data from two groups of children with cognitive abnormalities. Those in one group, 800 strong, were specifically autistic. Those in the other, 16,000 strong, had a range of problems.

Dr. Jacquemont has just published his results in the *American Journal of Human Genetics*. His crucial finding was that girls in both groups more often had mutations of the sort associated with abnormal neural development than boys did. This was true both for Copy-number Variants (CNVs, which are variations in the number of copies in a chromosome of particular sections of DNA), and Single-nucleotide Variants (SNVs, which are alterations to single genetic letters in the DNA message).

On the face of it, this seems compelling evidence for the female-protective model. Since all the children whose data Dr. Jacquemont examined had been diagnosed with problems, if the girls had more serious mutations than the boys did, that suggests other aspects of their physiology were covering up the consequences. Females are thus, if this interpretation is correct, better protected from developing symptoms than males are. And, as further confirmation, Dr. Jacquemont's findings tally with a study published three years ago, which found that CNVs in autistic girls spanned more genes (and were thus, presumably, more damaging), than those in autistic boys.

The counter-argument is that if girls are better at hiding their symptoms, only the more extreme female cases might turn up in the diagnosed groups. If that were true, a greater degree of mutation might be expected in symptomatic girls as a consequence. However, Dr. Jacquemont and his colleagues also found that damaging CNVs were more likely to be inherited from a child's mother than from his or her father. They interpret this as further evidence of female-protectedness. Autistic symptoms make people of either sex less likely to become parents. If mothers are the source of the majority of autism-inducing genes in children, it suggests they are less affected by them.

None of this, though, explains the exact mechanism that makes boys more susceptible than girls. On this question, too, there are two predominant theories. The first is that males are more sensitive because they have only one X-chromosome. This makes them vulnerable to mutations on that chromosome, because any damaged genes have no twin to cover for them. One cognitive disorder, fragile-X syndrome, is indeed much more common in men for this reason. Dr. Jacquemont's study, however, found only a limited role for X-chromosome mutations. That suggests the genetic basis of the difference is distributed across the whole genome.

The other kind of explanation is anatomical. It is based on brain-imaging studies which suggest differences between the patterns of internal connection in male and female brains. Male brains have stronger local connections, and weaker long-range ones, than do female brains. That is similar to a difference seen between the brains of autistic people and of those who are not. The suggestion here is that the male-type connection pattern is somehow more vulnerable to disruption by the factors which trigger autism and other cognitive problems. Why that should be, however, remains opaque.

2. 中译英

（1）段落翻译。

①中国已经成功地发射了第一颗试验通信卫星。这颗卫星是由三级火箭推动的，一直运转正常。它标志着我国在运载工具和电子技术发展方面进入了一个新阶段。

②石墨晶体结构遭到破坏时，总是碎化为微小尺寸的片状粉末。孤立的石墨烯片在其边缘存在大量的悬挂键，使得石墨烯片的能量较高，状态也不稳定。石墨烯片卷曲形成碳纳米管后，悬挂键减少，系统能量相应降低。

③舌诊是中医中一种重要的诊断方法。然而，舌诊中的一个重要问题是，其实践具有主观性，难以定量描述。随着计算机技术的发展，将图像处理和模式识别技术用于舌诊的辅助诊断成为一种趋势。

（2）篇章翻译。

5G实际上是第五代移动通信技术的缩写。1G应用于20世纪七八十年代，主要以大哥大为代表，功能单一，主要是语音功能。2G 的功能是语音加文字，比如电话加短信。3G 可以下载图片。4G 可以从手机上直接下载视频。5G 相比于以往几代移动通信技术，具有万物互联、高速度、泛在网、低时延、低功耗、重构安全等特点和优势。

5G 的发展对各个行业产生重大影响，并使得各个行业在融合当中产生巨大的化学效应。5G 带动全球经济的发展，5G 也带动产业链的发展：上游主要是基站升级（含基站射频、基带芯片），中游是网络建设（网络规划设计公司、网络优化维护公司），下游是产品应用及终端产品应用场景（云计算、车联网、物联网、VR/AR）。5G 将着力升级自己的网络来满足各种相关领域出现的服务需求。5G 在通信、智能手机、物联网上的运用前景十分广阔，是移动通信技术领域的里程碑式的突破。在传统的互联网时代，门户网站一般不会与传统产业有很大的关联，而移动互联网则对很多行业的变革创新都有促进作用，让世界更有效率，组织更扁平化。

5G 不仅改变生活，而且极大地改变社会与经济面貌。5G 时代，文化生产、传播、消费将发生重大变革，主要体现在以下方面。一是网速加快会促进文化内容的生产者呈现更加多元化的特征，由以文化企业、非营利性文化单位创作者为主向文化创作自由职业者拓展，甚至人人可能成为文化内容生产者。二是文化生产和传播更加关注个体的文化需求，呈现个性化特征，产品供给和消费的匹配度会大大提高，文化消费体验和服务质量得到大幅提升。三是文化产业的数字化特征趋势得到加强。四是 5G 会驱动更多的应用性程序和应用场景出现，在广播影视、视频、主题公园等旅游项目、动漫游戏行业会得到广泛应用，VR、AR 等技术更趋发达成熟，更多生动有趣的休闲产业和项目被开发出来，消费者的文化娱乐体验会升级。五是 5G 会使 4G 时代的产品和技术更完善，比如电视和视频高清技术使手机报纸、手机电视、手机游戏等产品受欢迎程度更高。六是移动手机网络使用成本更低，可以跨越繁华都市和偏僻山区的空间界限，实现泛在性，价格低廉，从而促进手机衍生的文化消费和文化产业的增长。七是 5G 与 AI（人工智能）、与物联网技术融合形成超级互联网，在这个融合过程中会产生新的文化产业业态和产品，这会给智能家居、智慧城市、创意城市、传统文化行业（如演艺行业、电影行业）带来重要应用和影响。八是降低文化产业的生产和流通成本。九是使文化产品的大规模传播更快更广，文化体验和文化消费的国际空间距离障碍减弱。

学思践悟

第十章 科技产品简介

一、科技产品简介文本介绍

(一) 科技产品简介概述

科技产品简介通常指公司或其他机构附着在科技产品外包装上、发布在网页或宣传媒体上的用以介绍其产品或服务的相关信息的文字、图片或二者结合的文本。科技产品简介主要包括产品名称、产品特点、产品用途、产品规格、注意事项、售后服务等。消费者通过科技产品简介可以快速准确地了解产品的基本信息，形成对产品的初印象。同时，科技产品具有产品研发投入高、生命周期较短和消费者关注度高的特点，这就要求科技产品简介突出产品的功能特性。因此，科技产品简介的翻译必须清晰明了，符合消费者的阅读习惯，能迅速吸引消费者。

(二) 科技产品简介的特点

1. 语篇特点

科技产品简介的语篇特点有真实性、专业性、简洁性和宣传性。

(1) 真实性。科技产品简介主要是为消费者提供产品的相关信息，介绍产品的参数、性能、特点、优势等信息，因此科技产品简介的内容必须真实客观，对产品的描述应做到实事求是，力求准确，不吹嘘夸大。

(2) 专业性。科技产品简介是一种相对专业的应用文体，表达严谨，很少使用华丽的修辞手法，更多的是陈述客观事实或真理，知识性和逻辑性较强，全文层次分明、结构清晰，多使用衔接词达到语篇衔接的效果。科技产品简介中还经常出现科技英语中的技术词和半技术词，因此用词专业是科技产品简介突出的特点。

(3) 简洁性。科技产品简介面向的消费群体是普通大众，以简短的文字清晰准确地描述产品，简单直白、通俗易懂，使不具备专业知识的消费者也能理解，直抵用户需求。为将产品信息完整地传达给消费者，科技产品简介中也经常使用缩略语来达到简洁的效果。

(4) 宣传性。科技产品简介除概括产品的基本信息、优势和特点外，也承担了产品的广告宣传功能，以此激起消费者的购买欲望，能够对产品起到一定的推广效果。因此，科技产品简介需要迎合读者的阅读习惯，语言除做到真实客观外，还要具有吸引力和可读性，巧妙生动地吸引读者。

2. 文本特点

(1) 科技产品简介的词汇特点如下。

① 科技产品简介中大量使用专业术语。科技产品简介是为介绍新产品的特点，吸引消费者，因此专业术语的使用非常必要。例如，有下列例句，在对路由器的介绍中提到了 omnidirectional antennas（全向天线）、external antennas（外部天线）等专业术语。

Omnidirectional antennas for a faster and smarter experience.

译文：全向天线，更快更智能。

External antennas for better signal strength 64 MB memory for stable operation.

译文：外部天线，信号更强，64MB 内存，运行稳定。

②科技产品简介因为具有宣传性，为吸引读者多会用一些描述性的褒义词。这些词汇使得产品的优势和创新点一目了然，从而激发读者的消费欲望。这种褒义词如 with the most up–to–date equipment and techniques（采用最新设备和工艺）、by scientific process（科学精制）、easy and simple to handle（操作简便）、to enjoy high reputation in the international market（在国际市场上享有盛誉）、excellent in cushion effect（抗冲击强度高）、rapid heat dissipation（散热迅速）等。

③科技产品简介中多使用缩略词。因为科技产品简介具有简洁的特点，根据专业需要大量使用术语、半术语及缩略词，使语义达到准确恰当；在语意选择方面，多用字面意义而少用比喻意义或引申意义，多用指示意义而少用联想意义；灵活使用复合词。

④科技产品简介中还经常使用四字格短语，如下列示例中的"专属定制""一目了然""轻巧舒适"。

The exclusive MIUI Pop–up for Quick Pairing lets you quickly pair devices. Simply open the charging case near your smartphone and tap the screen when the pop–up window appeas. The battery levels of the earbuds and the charging case can be seen at a glance.

译文：MIUI 专属定制弹窗，在手机附近打开充电盒，弹出界面后，轻点即可快速配对，耳机和充电盒的电量一目了然。

Though small and lightweight, the Redmi Buds 3 offers both power and comfort. Enjoy up to 5 hours of battery life on a single charge, or up to 20 hours' use when paired with the charging case. Relax and get back to your music. Enjoy up to 1.5 hours' playback from a single 10–minute charge.

译文：身材虽小，实力不减，轻巧舒适的 Redmi Buds 3 耳机单次续航可达 5 小时，配合充电盒使用可达 20 小时。让自己放松一下，享受音乐。充电 10 分钟，就可听歌 1.5 小时。

⑤科技产品简介中使用大量数字及符号。科技产品简介通常涉及大量产品参数，包含产品的尺寸、适用温度、抗压系数等，会使用大量数字及符号。例如，下列某公司对 PC 耐力板的介绍中，就包含了大量的数字信息。

PC 耐力板

宽度：1,220mm、1,560mm、1,820mm、2,100mm

厚度：1.2~16mm

长度：厚度≤4mm，30~50 米/卷；厚度>4.5mm，6 米/片

颜色：透明、蓝色、草绿色、茶色、乳白色、红色

产品特点：

透光性：透光性能好（透光率达 88%）。在阳光下曝晒不会产生变黄、雾化、透光不

佳。其中颗粒板和磨砂板具有散光功能。

耐候性：表面有防紫外线的共挤层，可防止太阳光紫外线引起的树脂疲劳、变黄。表面共挤层具有的化学键吸收紫外线并转化为可见光，对植物光合作用有良好的稳定效果。（极适合保护贵重艺术品及展品，使其不受紫外线破坏。）

抗冲击性：PC 耐力板的抗冲击强度是普通玻璃的 250～300 倍，是亚克力板材的 20～30 倍，是钢化玻璃的 2 倍，几乎没有断裂的危险性，有"不破玻璃"和"响钢"之美称，是防弹玻璃的优质材料。

阻燃性：普通 PC 耐力板根据国家标准 GB 8624－1997 测试属阻燃 B2 级，加入特殊的材质可达到美国 UL94V0 防火级别，无火滴，无毒气。

耐温性：在 －40℃ 到 ＋120℃ 温度范围内不会引起变形等品质劣化。

轻便性：重量轻，便于搬运、钻孔。截断安装时，不易断裂，而且可直接冷弯，施工和加工都更加简便。

隔音性：隔音效果佳，在国际上是高速公路隔音屏障的首选材料。

适用范围：住宅的天窗；工业厂房及仓库的采光顶棚；地铁出入口、车站、停车场、商场、休息厅、走廊的雨棚；体育场馆、高速公路的隔音屏障；游泳池、温室的覆盖；办公大楼、百货大楼、宾馆、别墅、学校、医院、体育场、娱乐中心及公用设施的采光顶棚等。

译文：Polycarbonate solid sheet

Width：1,220mm、1,560mm、1,820mm、2,100mm

Thickness：from 1.2mm to 16mm

Length：thickness≤4mm, 30－50m/roll; thickness ＞4.5mm, 6m/piece

Color：transparent, blue, green, coffee, opal white, red

Product features：

Light transmission：excellent light transmission (up to 88% of light transmitted through). When exposed under sunshine, there are no yellow, hazing or/and poor light transmission caused. The embossed sheet & the abrasive sheet have the astigmatic function.

Weather resistance：the ultraviolet－resistant coextruded film on the surface prevents it from resin fatigue and yellowing caused by ultraviolet of sunshine. The chemical bond on the coextruded film absorbs the ultraviolet and then turns it into visible light. It has a stable effect on plant photosynthesis. (It is an ideal material to prevent precious works of art and exhibits from being damaged by ultraviolet.)

Impact resistance：the impact－resistance strength of PC sheet is 250－300 times of the ordinary glass, 20－30 times of the alkali glass and 2 times of toughened glass which explains why it is hardly broken. Also for this reason, it is called unbreakable glass and sound steel. It is an excellent material for bullet－resistant glass.

Fire retardant：according to the national GB 8624－1997 test, its fire retardant is regarded as Grade B2. With special material made, it can reach the American UL94V0 fire retardance standard, without fire drop or poisonous gas.

Temperature resistance：it does not cause deformation and other quality deterioration under temperature from －40℃ to 120℃.

Portability: it is light and easily handled and drilled. It will not be broken when cut for installation, and it can be cold bent directly, which is more convenient for construction and procession.

Sound insulation: with good sound insulation effect, it is the best material for sound – insulation walls on expressways worldwide.

Application scope: the skylight of residences; the lighting roof for industrial workshops and storehouses; the canopy of the entrance and exit of underground, station halls, parking lots, supermarkets, resting halls, hallways; the sound – insulation wall for gymnasiums and expressways; the coverage of swimming – pools and greenhouses; the lighting roof for office buildings, department stores, hotels, villadoms, schools, hospitals, sports fields, entertainment centers and public facilities, etc.

（2）科技产品简介的句式特点如下。

① 从句法学角度上看，考虑到消费者和大多数人的阅读习惯，科技产品简介中的句型结构易懂，多为简单句和祈使句，复合句则多为并列句及条件句。例如：

According to different requirements, for the convenience of automatic test, the sensor can also be used in conjunction with digitally measuring apparatus, circuit testing device or computer.

译文：根据不同的使用要求，传感器可以与数字测量仪、电路测试设备或计算机配合使用，便于实现测试自动化。

Portable and palm – sized with a built – in extension rod, DJI OM 5 is a versatile companion that unlocks the full potential of your smartphone. Enjoy flawless selfies, super – smooth video, automatic tracking, and much more. A new ShotGuides feature even provides creative tips, empowering you to get stunning shots wherever you go. With DJI OM 5, get ready to master every shot.

译文：DJI OM 5 的机身只有手掌大小，还内藏延长杆，作为你的多功能伴侣，可以解锁你手机的全部潜力。享受完美的自拍、超流畅的视频，还可智能识别场景。全新的ShotGuides 功能甚至提供了创意提示，让你随处都可拍出大片的感觉。用 DJI OM 5，你想怎么拍，都能满足你的需求。

With an advanced core employed, temperature precisely compensated, stainless steel sealed and welded, as well as perfect assembly.

译文：采用先进的压力传感器芯体，并经过精密的温度补偿、全不锈钢密封焊接和完善的装配工艺。

② 科技产品简介中使用大量名词化结构。名词化结构是基于词性转换的基础。名词化结构是指词组或句子含有名词性词组或由其他词性转换成名词性词组。科技产品简介中的名词化结构可分为单纯名词化结构和复合名词化结构。例如：

Readings can be obtained to an accuracy of one micron.

译文：获得的读数精度可达 1 微米。

③ 科技产品简介中多使用一般现在时。科技产品简介涉及科学技术、科学设备、试验过程等内容，一般描述和讨论客观事物和客观事实，具有较强的客观性、准确性和严密性，其语言规范，逻辑性强，结构严密。因此，科技产品简介中大量使用一般现在时。因为一般

现在时说明普遍真理，阐明科学定理和定义等，目的在于给人精确无误的信息。例如：

Its tiny built-in camera takes 200 frames per second, while a microprocessor analyses this data to differentiate between light and dark areas. A micro printer then applies the foundation to your skin.

译文：打印机内置的微型摄像机秒速可达到200帧，还有一个分析数据的微处理器，可区分皮肤的亮区和暗区。微型打印机可以将粉底液涂抹到你的皮肤上。

Pressure sensors on its fingertips detect an object's size and material. It then autonomously places the item in the appropriate recycle bin.

译文：指尖上的压力传感器可以探测到物体的大小和材质，然后它自动将物品放入相应的回收垃圾箱。

二、科技产品简介范文展示

范文一

华为手表

  HUAWEI WATCH 3 拥有1.43英寸的AMOLED高清触摸屏内置30款全新表盘，包含多种动态表盘，新潮动感，灵动鲜活。华为表盘市场目前已上架超过1000+款供你选择。HUAWEI WATCH 3 支持eSIM功能，它能脱离手机做很多事情：打电话、播放音乐、下载应用程序，只需使用你的电话号码。一抬手，便知你心。使用手表控制音乐播放，手机上爱听的歌单，可以自动同步到手表上，手表中的歌单会自动同步到华为智慧屏、音箱，音乐随时陪伴你。有电话来了，握紧拳头再松开就能接听，翻转手腕即可来电静音。你可以用手上的HUAWEI WATCH 3 遥控手机拍照或录制短视频，还能通过旋转表冠调节焦距，每一次按下快门，都能清晰记录精彩瞬间。

  内置温度检测传感器，让你知悉体温变化。随时随地，为你的健康保驾护航。如果你感到焦虑和呼吸急促，可能是血氧饱和度低造成的。通过TruSeen™ 4.5+即时了解你的动脉血氧饱和度，基于全新TruSeen™ 4.5+心率监测技术，使用先进的光技术精确检测血液中的氧分子，低能耗，可克服低温、高海拔条件，通过腕上手表即时看护你的血氧健康。支持跑步、骑行、登山、铁人三项、室内游泳、椭圆机、划船机等100+户内外运动模式，同时还支持6项日常运动智能识别，配合高精度多模GNSS定位芯片，精准记录运动轨迹。

  续航与性能兼得，HUAWEI WATCH 3 支持无线充电，还可使用华为手机为其进行反向充电，告别续航焦虑。双芯架构，不仅性能强劲，而且实现了更平衡的处理，配合智能节电算法3.0，满电状态下可提供智能模式典型场景下使用3天，超长续航模式可达14天。

HUAWEI WATCH 3

  1.43-inch AMOLED display, HUAWEI WATCH 3 comes with 30 cool, pre-installed faces, including animated faces to really liven things up. Plus, choose from the 1,000+ designs available on HUAWEI Watch Face Store. Using eSIM technology, HUAWEI WATCH 3 functions as a stand-alone communication device. Make calls, play music, and download apps without being connected to your phone, still using your existing number. With a few simple hand gestures, control music playback on your phone with HUAWEI WATCH 3, plus synchronise playlists on your phone to Visions and speakers. Music is always with you. You'll have the control you want over incoming calls. Clench your fist and release to take a call or turn your wrist to the side to mute the call. Get great-looking photos using the Remote Shutter feature. Control your phone's camera and use the crown to zoom in and out, for the right shot every time.

Use the Skin Temperature Detector on HUAWEI WATCH 3, and know right away if your temperature is high. Keep track of your health, so you can always take care of yourself in the best way possible. If you're feeling anxious and short of breath, your blood oxygen saturation, SpO_2 may be low. Know your SpO_2 instantly via TruSeen™ 4.5+, which uses advanced light technology to accurately detect blood oxygen molecules, even in low-temperature, high-altitude conditions. And with the lower power consumption of TruSeen™ 4.5+, you'll be able to monitor your blood oxygen levels around the watch. Get 100+ workout modes for indoor and outdoor sports including running, cycling, mountain climbing, triathlon, indoor swimming, elliptical machine, rowing machine, with automatic exercise detection for the 6 most common types of workout. Enjoy superior outdoor route tracking with the multimode GNSS location sensor.

Life on a Larger Scale. HUAWEI WATCH 3 support wireless charging. Use your HUAWEI phone to reverse charge HUAWEI WATCH 3 and say goodbye to those battery-life woes. HUAWEI WATCH 3 is equipped with a dual-chipset architecture to not only enhance performance power but also achieve more balanced processing. Combined with the Smart Power-Saving Algorithm3.0, enjoy 3 days' use in the typical scenario of smart mode, and an incredible 14 days' use in ultra-long battery life mode.

范文二

Smart Shoes

Austrian company Tec-Innovation recently unveiled smart shoes that use ultrasonic sensors to help people suffering from blindness or vision impairment to detect obstacles up to four meters away. Known as InnoMake, the smart shoe aims to become an alternative to the decades-old walking stick that millions of people around the world depend on to get around as safely as possible. The currently available model relies on sensors to detect obstacles and warns the wearer via vibration and an audible alert sounded on a Bluetooth-linked smartphone. That sounds impressive enough, but the company is already working on a much more advanced version that incorporates cameras and artificial intelligence to not only detect obstacles but also their nature. Not only is the warning that you are facing an obstacle relevant, but also the information about what kind of obstacle facing. Because it makes a big difference whether it's a wall, a car or a staircase. Ultrasonic sensors on the toe of the shoe detect obstacles up to four meters away. The wearer is then warned by vibration and/or acoustic signals.

The current version of the InnoMake shoe is already available for purchase on the Tec – Innovation website, for 3,200 per pair. The advanced system is integrated in the front of the shoes, in a waterproof and dustproof case. It is powered by a heavy – duty battery that can last for up to one week, depending on use. The battery can be charged in just three hours, using a USB cable.

The next step for Tec – Innovation is to use the data collected by its system to create a kind of street view navigation map for visually impaired people. As it currently stands, only the wearer benefits in each case from the data the shoe collects as he or she walks. It would be much more sustainable if this data could also be made available to other people as a navigation aid.

智能导盲鞋

奥地利公司 Tec – Innovation 近期推出的智能鞋，可以通过超声波传感器帮助盲人或视力障碍者探测到四米远的障碍物。这款名为 InnoMake 的智能鞋旨在替代几十年来全球数百万人赖以安全活动的拐杖。目前能买到的这款鞋子依靠传感器来探测障碍物，并通过蓝牙智能手机发出振动和铃声来警告穿着者。听上去够酷的了，不过该公司已经开始研发包含摄像头和人工智能技术的升级版鞋子，不仅可以探测到障碍物，还可以检测出是何种障碍物。这款鞋子不但会警告你前方有障碍物，还可以告诉你是何种障碍物。因为墙壁、车和楼梯这些障碍物之间差别很大。鞋尖上的超声波传感器可以探测到四米远的障碍物。穿着者会收到振动或铃声警告。

目前这款 InnoMake 鞋子在 Tec – Innovation 官网上可以买到，单价 3200 欧元（约合人民币 2.5 万元）。这套高级系统就安置在鞋尖处，配有防水防尘保护壳。支持系统的高能电池续航时间可达一周，具体时长依据使用情况而定，用 USB 数据线充电三个小时就能充满。

Tec – Innovation 公司的下一步是利用该系统收集的数据为视力障碍者打造一款街景导航地图。就目前而言，只有穿着者才能在行走时获得鞋子收集到的数据。如果其他人也可以获得这些导航数据，可持续性将会大大提高。

范文三

4G On – board Wireless Transmission Device

The 4G on – board wireless transmission device is a network video product applicable into installation inside or outside a motor vehicle or ship to monitor and record the inside or outside. Its four – channel on – board DVR is capable to transmit audio and video signals to monitoring centers, PC clients or mobile clients in real time via a wireless LTE network, allowing administrators to monitor real – time scene image information without having to stay on the scene. It is available in long – period image/voice recording and playback. It is applicable into on – board security surveillance. Typical applications are on – board, shipborne mobile surveillance, command and management systems, such as law enforcement vehicles monitoring, bus monitoring, long – distance bus monitoring, taxi monitoring, 110 police, traffic police patrol vehicles, police patrol boats, 120 patrol, special escort vehicle monitoring, and ship monitoring.

4G 车载无线传输设备

4G 车载无线传输设备是用于安装在机动车、船舶内部或外部，对机动车、船舶内外进行监控和录像的网络视频产品。它的车载四通道 DVR 可通过无线 LTE 网络实时传输音视频信号到监控中心、PC 客户端、手机客户端，管理者无须停留在现场就可实时监控现场图像信息。它可用于长时间的图像/语音记录和回放。它适用于车载安防监控领域。典型应用为车载、舰载移动监控、指挥、管理系统，如执法车辆监控、公交车监控、长途客车监控、出租车监控、110 出警、交警巡逻车、警用巡逻艇、120 巡逻、特种押运车辆监控、船舶监控等。

范文四

MIBA Special Machine Tool Features and Uses
MIBA 特殊机床特点及用途

Product and application
产品与应用

Miba Automation Systems specializes in μm – accurate, automated machining in combination with work piece positioning, robotics and integrated measurement.

Miba 自动化系统（公司）专长于结合工件定位、机械手传递和集成的测量系统生产微米级精度的自动化机械加工设备。

Core competencies
核心竞争力

Combining and interlinking machines (complete transfer lines)
将机器连接在一起（完成传输线）

Combining several processes in a single machine
将几个工艺步骤糅合到一台单一的机器内

Combining processing and measuring technologies (laser – controlled processing axes, etc.)
将测量技术和加工技术结合（如激光控制的加工轴）

Multi – spindle machines for milling operations
多主轴铣削加工机器

Combining processing with robotics
与自动化机械传输机构相结合的加工技术

Mobile processing systems for very large components (construction of power plants, cranes, wind power plants, repair of slewing rings)
针对特大零件的可移动加工机构（电厂、起重机、风力发电厂的建造及零部件的修理）

Onsite processing using portable CNC processing systems (field service)
用便携式数控（CNC）加工机构进行现场加工作业（现场服务）

Checking of complex and large components
对特大、复杂零件进行检查

Key strengths
关键优势

Handling of complex components
对复杂零件的处理

Highly trained, committed team
训练有素的、敢于担当的队伍

Global service

全球化服务

Engine bearings technology

发动机轴承技术

With over 60 years of experience in developing and building machines for engine bearing production, Miba Automation Systems is a world market leader in this segment. All the leading engine bearing producer shave come to rely on our technologies.

在发动机轴承生产方面具有超过60年的研制和建造经验,据此,Miba自动化系统(公司)在世界上处于领先地位,所有主要的发动机轴承生产厂商纷纷依靠我们的技术。

Transfer lines

传输线

Enabling you to produce both large and small batch sizes cost efficiently, transfer lines are the perfect symbiosis of set – up and cycle times. Transfer lines from Miba Automation Systems guarantee short set – up times with optimally designed cycle times.

为了使大批量生产和小批量生产的成本都得到有效控制,传输线是(零件)装配与循环时间的完美的关联结合。Miba自动化系统(公司)利用对循环时间的优化设计保证了传输线短的装配时间。

Fine boring machines

精镗机器

Miba Automation Systems is also a pioneer in fine boring technology. By enhancing our technology leadership with ongoing advances, we stay a step ahead of the competition. Cycle times of below a second combine with positioning precision in the one tenth of a micrometer range. Customers benefit from tolerances of less than 0.005mm, with extremely high process capability.

Miba自动化系统(公司)也是精镗技术的先锋。借助持续领先的技术,以增强我们的技术领导能力,使我们在竞争中领先一步。少于1秒的循环时间与十分之一微米的精确定位,顾客从小于0.005mm的公差和极高的加工能力中获益。

Robotics & automation

机器人和自动化

Miba Automation Systems has more to offer than just assembly lines, automation and robotics. Customers' added value chains profit from our complete systems, especially when optimized combinations of processing machines are involved. Miba Automation Systems is a specialist in high precision positioning of complex components. We build robotic systems for coating systems and component handling and processing—all tailored to customers' individual requirements. In addition, we are well positioned in the rapidly growing market of camera and testing systems combined with robotics.

Miba自动化系统(公司)提供的不仅仅是组装线、自动化系统和机器人。顾客的附加价值链受益于我们完整的系统,特别是当包含了加工机器的优化组合时。Miba自动化系统(公司)是复杂零件的高精确定位方面的专家。我们建造的机器人(自动化机构)系统,应用于溅镀系统、零件处理系统和加工系统,完全依顾客各自的要求而定制。另外,在快速增长的与机器人相配合的照相和测试系统市场,我们也取得了好的市场定位。

Stationary special purpose machines
特殊用途的固定式机器

Our technological know – how and solutions customized to clients' needs make us an attractive strategic partner to future – oriented industries. Special machines designed and developed by Miba Automation Systems ensure μm – precise production and component positioning

满足顾客个性化需求的技术诀窍和解决方案，使我们成为面向未来工业的具有吸引力的战略伙伴。通过 Miba 自动化系统（公司）的专门研发和设计保证了微米级生产及组件的定位。

Mechanical production
机械生产

One of our biggest strengths is in the field of mechanical production. Whether it's milling, grinding, turning, broaching, punching or coining, our wide – ranging experience in machining makes us a leading provider.

我们最大的优势之一是在机械生产领域，无论是铣、磨、车、拉削、冲压或模压加工，在机械加工领域丰富的经验使我们成为技术领先的供应商。

Measuring equipment
测量设备

High – precision production and assembly processes require measuring systems with very high precision. Miba Automation Systems is more than equal to the challenge of absolute precision. Having realized at an early stage the importance of automated measuring systems, we have always positioned ourselves on the leading edge.

高精度的生产和组装过程需要非常高精度的测量系统。Miba 自动化系统（公司）完全可以应对绝对精度的挑战，很早就认识到自动测量系统的重要性，我们总是将自己置身于领导前沿。

Coating systems
溅镀系统

Miba Automation Systems is a specialist with many years of experience in complex vacuum coating systems. We develop and build coating systems working closely with our associate companies Teer Coatings Ltd. （UK） and High Tech Coatings （Austria）.

Miba 自动化系统（公司）是在复杂真空溅镀系统方面拥有多年经验的专家。我们与英国蒂尔镀层有限公司（Teer Coatings Ltd.）和奥地利高科技镀层有限公司（High Tech Coatings）密切合作，开发和建造溅镀系统。

Mobile special purpose machines
具有特殊用途的可移动机器

We are driving growth in future – oriented industries by developing and building mobile CNC processing units. As components become larger than ever and tasks more complex, processing precision needs to keep pace. This calls for systems specifically designed for the application in question. Mobile, transportable machines allow clients to perform technologically demanding processes on components directly on site （in situ）.

我们通过研发和生产可移动数控加工单元以推动面向未来工业发展的进步。随着零件越来越大，任务越来越复杂，加工精度需要保持同步。这就需要专门设计的系统以满足这种要求的应用。移动的、可运输的机器允许顾客直接在现场对零件完成满足技术要求的加工。

Areas of application include crane construction, open – cast mining and wind power industry. All over the world, you can find Miba Automation Systems machines performing the work of more expensive processing centers.

应用领域包括起吊设备的构建、露天采矿和风电工业。在全世界范围内，你能发现 Miba 自动化系统（公司）的机器正在从事更加昂贵的加工中心应该完成的工作。

Flange processing

法兰加工

Final processing of steel flanges on finished components can be performed on site using our mobile flange facing machines. Miba Automation Systems mobile CNC processing units can also be used to process very large workpieces (2 – 20 m) under a wide range of mechanical conditions. Miba Automation Systems is the only company worldwide to offer mobile processing equipment with online laser tracking systems. These laser measuring systems automatically correct the processing axis, ensuring very high precision.

用我们的可移动法兰表面加工机器能够在现场对已加工完成的零件钢法兰进行最后的加工处理。Miba 自动化系统（公司）的可移动数控加工单元也能用于在各种机械条件下加工非常大的工件（2～20m）。Miba 自动化系统（公司）是全球唯一一家可提供配置有在线激光跟踪（检测）系统的可移动加工设备的公司，这些激光测量系统自动校正加工轴，以此来确保（设备）达到非常高的精度。

Processing of turbine casings

涡轮机壳的加工

Mobile CNC processing machines make revisions of power plant turbines highly economically. Downtime and transport costs are reduced. Advanced Miba Automation Systems technology ensures efficient, rapid, precise execution.

可移动数控加工机床可以非常经济地对发电厂涡轮机进行改造，停机时间和运输费用被减少。先进的 Miba 自动化系统（公司）技术保证高效、快速和精密的操作。

Research & development

研究和发展

Miba Automation Systems puts special emphasis on providing customers with the best possible processing strategies. One of our particular areas of focus is the development of mobile CNC processing units. Ever larger components and more complex tasks as well as very high levels of processing precision require customized systems of this kind. Our numerous patents for transfer lines and special processing machines reflect our powers of technological innovation.

Miba 自动化系统（公司）特别强调为客户提供最佳的加工策略。我们特别关注的领域之一是可移动数控加工单元的开发。更大的部件和更复杂的任务及非常高水平的加工精度要求定制这种系统。我们拥有许多的传输线和特种加工设备专利，体现了我们技术创新的力量。

三、科技产品简介常用词汇和句型

(一) 科技产品简介常用词汇

1. 描述类词汇

specification 规格
complete in specifications 规格齐全
application scenario 使用场景
craftsmanship 工艺
description 产品描述
warranty 保修单
appearance 外形
performance 性能
functionality 功能性
reasonable in price 价格合理
elegant appearance 造型大方
product details 商品详情
easy and simple to handle 操作简单
strong packing 包装牢固
product parameter 产品参数
type 产品型号
packing list 包装清单
defects liability period 保修期
installation notice 安装须知
nominal size 外观尺寸
net weight 产品净重
rated capacity 额定容量
certified value 标准值
maximum current 最大工作电流
battery capacity 电池容量
storage temperature 储存温度
material 材料
system language 系统语言
screen size 屏幕尺寸
screen resolution 屏幕分辨率
waterproof grade 防水等级
storage medium 存储介质

durable in use/hard – wearing　经久耐用
personalized customization　个性化定制
small in size/light in weight　小巧轻盈
customized experiences　定制化体验
custom – made　根据客户需求定制
2. 科技产品简介常用词汇分类介绍
（1）生活家电类产品简介常用词汇如下。
electric pressure pot　电压力锅
vacuum cleaner　真空吸尘器
LCD TV（Liquid Crystal Display Television）　液晶电视
air humidifier　空气加湿器
sweeping robot　扫地机器人
electric massage chair　电动按摩椅
full – automatic washing machine　全自动洗衣机
variable frequency air – condition　变频空调
wireless cleaner　无线吸尘器
air cleaner　空气净化器
ultrasonic dishwasher　超声波洗碗机
multi – purpose food processor　多用料理机
smart bin　智能垃圾桶
smart door lock　智能门锁
wireless printer　无线打印机
（2）数码产品简介常用词汇如下。
smart router　智能路由器
smart recorder　智能记录仪
touch screen speaker　触屏音箱
wireless charger　无线充电器
digital analog converter　数模转换器
satellite receiver　卫星接收装置
smart scanner　智能扫描仪
laser printer　激光打印机
digital camera　数码照相机
handheld PTZ　手持云台
wireless mouse　无线鼠标
digital display　数字显示器
projector　投影仪
digital single lens reflex camera　数码单反相机
noise reduction headset　降噪耳机
wide – angle lens　广角镜头

portable speaker　便携音箱
portable blood glucose meter　便携式血糖仪
bluetooth watch　蓝牙手表
（3）穿戴类产品简介常用词汇如下。
intelligent bracelet　智能手环
wireless earphone　无线耳机
VR all–in–one headset machine VR　头戴式一体机
digital helmet　数字头盔
smart audio glasses　智能音频眼镜
smart glasses　智能眼镜
smart watch　智能手表
somatosensory balance car　体感平衡车
（4）航空类产品简介常用词汇如下。
hydraulic pump　液压泵
compass flux valve　罗盘流量阀
hydraulic hand fill pump　液压手动注油泵
weather radar controller　气象雷达控制器
nav lights　导航灯
DC power contactor　直流电源接触器
Emergency Locator Transmitter（ELT）　紧急定位发射机
anti–collision light　防撞灯
undercarriage　起落架
turboshaft engine　涡轮轴发动机
multi–function display　多功能显示器
electric windscreen wiper　电动风挡刮板
overrunning clutch　超越离合器
weather radar controller　气象雷达控制器
Flight Data Recorder（FDR）　飞行数据记录仪
Cockpit Voice Recorder（CVR）　驾驶舱语音记录仪
external power receptacle　外部电源插座
fuel quantity indicating probe　燃油量指示传感器
（5）电力产品简介常用词汇如下。
wound–rotor motor　绕线转子电动机
Working Storage（WS）　工作存储器
xerographic printer　静电印刷机，静电复印机
zoom lens　变焦镜头
word–organized store　字选存储器
woven–screen storage　编线网存储器
X–tal detector　晶体检波器

zero – phase – sequence current transformer　零（相）序电流互感器
zoom image intensifier　变焦距图像增强器
worm conveyer　螺旋输送机
wound electrical capacitor　绕线式电容器
zero phase shifter　零位移相器
xenon lamp　氙（气）灯，氙（气）管
zero – access storage　立即访问存储器，零存储时间存储器，立即存取存储器

（6）通信类产品简介常用词汇如下。

optical time – domain reflectometer　光时域反射仪
frequency – time counter　频率时间计数器
spectrum analyzer　频谱分析仪
signal generator　信号发生器
fan – sectorized antenna　扇形天线
hand – free telephone set　免提式电话机
buzzer（in a telephone set）　（电话机）的蜂鸣器
weather – chart facsimile apparatus　气象图传真机
wind – generator set　风力发电机
rotary dial telephone set　旋转号盘电话机
voice over internet protocol　IP 电话
telephone answering and recording set　电话应答记录器
color facsimile apparatus　彩色传真机
voice over broadband　宽带电话
loudspeaking telephone set　扬声电话机
base station transceiver　基站收发信机
explosion – proof telephone set　防爆电话机
narrow beam antenna　窄波束天线
fiber – optic connector　光纤连接器
semiconductor optical amplifier　半导体光放大器
home location register　归属位置寄存器
visitor location register　漫游位置寄存器
charge – coupled device　电荷耦合器件
optical add – drop multiplexer　光分插复用器
optoelectronic receiver　光电（子）接收机
media gateway controller　媒体网关控制器

（二）科技产品简介常用句型

... recently unveil...　……（公司/企业）最近推出了……
... cry out for more choice　……提供了更多选择
This is equipped with...　这款产品配备了……（功能）

It's… long/wide/high/thick/deep   该产品长/宽/高/厚/深……
It has a large range of sizes/colors.   该产品有各种各样的尺寸/颜色。
The main selling point of it is…   这款产品的主要卖点是……
This is priced at…   产品的定价是……
This is one of our latest design.   这是我们的最新设计之一。
It is designed to/ aims to…   设计这款产品旨在……
It is of…（如 skillful manufactured）   这款产品是……（如制作精良）的
the most advanced…   最先进的……（产品）
… can set up…   ……（产品）可以设置……（功能）
… adapted the technology of…   ……（公司/产品）采用了……技术
… can withstand…   ……（产品）可耐受/承受……
It shows promising prospects in…   在……（方面）有好的应用前景
The technology works by…   这项技术的工作原理是……
using…（technology）to enhance the experience   利用……（技术）提高体验
The new type of… designed by… is very ingenious and practical   由……设计师设计的新型……（产品）非常精巧、实用
This kind of… is characteristic of…   这种……（产品）的特点是……
This kind of… is practical and economical for the needs of…   这种……（产品）实用、经济，能满足……的需要
There are many… for your selection   有众多……任您选择
… introduces new ways to…   ……（产品）引进了新的方式……
… is (are) widely used   ……（产品）得到广泛应用
… be made up of…   ……（产品）由……制成
a breakthrough was made in… field   在……领域有了突破
… have made great advances   ……取得了巨大进展

四、科技产品简介翻译技巧

　　翻译是在特定的文化背景下进行的，所以翻译不仅是翻译语言，也是翻译文化。在科技产品简介翻译中，文化也影响着翻译工作的效果。译文是否满足消费者的需求，或者消费者是否能够理解制造商想要表达的内容，是非常重要的。科技产品简介的目的是吸引更多的消费者，鼓励消费者购买，因此科技产品简介翻译应在市场营销的框架内进行，译者必须考虑市场因素。译者应该熟悉目标市场，即熟悉什么样的人是科技产品市场的主要消费者，熟悉消费者背景、消费习惯，以及其他文化因素。因此，在科技产品简介翻译中应考虑消费者的心理。译者也需要在一定程度上考虑消费者的思维方式。在翻译时，译者要确认信息传递的准确性、完整性，了解相关翻译材料的背景知识，正确理解原文；确保同一专业术语在目的语中的正确表达和翻译；从目标读者的理解角度出发，确保文本信息功能的传达。译者需要考虑译文的目标读者对译文的理解存在的困难和困惑，考虑原文语言和目的语语言之间存在

的表达、理解和使用习惯的不同，在忠实于原文信息的前提下，对译文的遣词造句、时态、语态、表达方式反复斟酌。

（一）名词的翻译

科技产品简介包含大量名词，主要分为普通名词和专有名词（或称专业术语）两大类。普通名词是同一类人或物所共有的名称的词；专有名词是相对普通名词而言的，特指某一领域内某些事物的专业称谓，此类名词一般是被该领域的专业人士所熟知和使用的。在科技产品简介中，一些普通名词会被赋予专业的解释，一些专业术语也包含普通名词的含义。因此，在翻译科技产品简介时应多查阅字典或相关资料，以保证翻译的准确性，并结合具体情况灵活翻译。

1. 专业术语的翻译

有大量专业术语来自半科技词汇及部分普通词汇。在现实生活中，我们往往会对一些专业术语感到陌生，究其原因就是我们只是了解了构成这些术语词汇的普通含义，而并不明白这些词汇在具体专业领域中的内涵，这自然就给我们的工作、学习带来一定困扰。

某些普通词汇虽有其自身含义，但在专业领域中与其他词汇搭配意义则大不相同。例如，对于"Rock joint is one of the major factors which affects the stability of engineering rocks"，有人随意猜测，将这句话中的"Rock joint"翻译为"岩石连接处"，这是不正确的。"joint"一词原意为"结合点；连接处"，但在不同领域中与不同词汇搭配意义就不同。经过查证并联系上下文推敲，会发现这是地质学用语，应将其译为"岩石节理"。又如"settlement"与"joint"搭配为"settlement joint"，该术语为建筑学术语。此时，不应将其译为"沉降连接"，而应考虑它在特定领域中的专业释义，准确地将其译为"沉降缝"。此类普通词汇在专业语境下有特殊含义的例子还有很多，例如：

Ceramic Shield. Clearly tougher than any smartphone glass.

译文：超瓷晶面板的硬实力，玻璃面板甘心服软。

"shield"在日常用语中译为"盾牌；保护物；背甲"，而在苹果官网对iPhone 12的介绍中，该词译为"面板"。iPhone 12的超瓷晶面板是一种陶瓷强化玻璃，官方称该材质比任何智能手机玻璃都更坚固。

The current flows into electric motors that drive the wheels.

译文：电流流入电机驱动车轮。

"current"在日常用语中多作形容词用，译为"当前的；流行的"，但在科技产品说明书中多用作名词，译为"电流；水流；气流。"

尽管专业术语具有单义性，但专业术语中很多词汇是由普通词汇转换而来的。在翻译中，译者主要采用"挖潜"法，赋予普通词汇以新意。例如，在水利技术中，专业术语"key"应译为"键槽"，而非"钥匙"，键槽是确保施工缝面在灌浆或不灌浆的前提下均可有效传递剪力的一种水利工程结构。

此外，在翻译专业术语时还可以利用关联分析法推断词义。关联分析法是指利用现有知识和上下文分析术语含义来确定术语名称的方法。上下文可以是词汇、句子和文章。在英文科技文献中词汇也常常具有多义性，所以必须根据上下文确切理解专业术语的具体意义。例如，下列示例中的"generator"及"oscillator"可以通过联系上下文来确定其在文中的具体含义。

The converter that converts mechanical energy into electrical energy is called a generator.

译文：把机械能转换成电能的转换器即为发电机。

The instrument that causes electrical oscillations is called an oscillator.

译文：引起电振荡的仪器叫作振荡器。

2. 常见技术词的翻译

（1）科技产品简介中有大量技术词是合成词，翻译时可将各个词素的意义依次进行翻译，这称为移植法。例如，将"super"（超级的，极度的）与"conductor"（导体）进行结合，就形成了"superconductor"（超导体）一词。此类词还有 skylab（太空实验室）、moonwalk（太空漫步）等。

（2）科技产品简介中有一些计量单位的词或新的材料、产品名称，翻译时可根据单词发音将单词译为读音与原词大致相同的词语，这称为音译法。例如，将"Hertz"译为"赫兹"。此类词还有 bit（比特）、Joule（焦耳）、nylon（尼龙）、sonar（声呐）等。

3. 品牌名称的翻译

此外，科技产品简介中有大量的品牌名称，翻译时可直接选择目的语中发音与之相同或相近的词语来表达，这不仅可以忠实于原企业或品牌的风格，还可以忠实传达名称的异国情调。在科技产品品牌名称的翻译中有大量的音译词语，如 Canon（佳能）、Power（波尔肤）、Sharp（夏普）。许多品牌名称都能在某些方面反映商品的某些特性，有的与商品的功能有关，有的则与商品本身的设计理念或企业文化有关，还有的可以反映商品的样貌或特征。例如，"Yamaha"雅马哈，马本身可以作为交通工具，雅则体现出该品牌的设计高雅；"Giant"捷安特，寓意快捷、安全、特别好；"Benz"奔驰，形容可以跑动，自在如风，肆意驰骋。

也有一些品牌名称采用了意译的翻译策略，如 Microsoft（微软）、Apple（苹果）等。

品牌名称的翻译，既要保留原文的精华，又要符合目的语国家消费者对品牌的期望和心理需求。在源语文本信息部分丧失的同时，译文必须部分地增加一定的信息含蓄。在品牌名称的翻译中，译者要尽可能地将源语文化融入目的语文化。在此，译者就是传播者，在这一过程中，他转换的不仅仅是品牌名称的字面意义，更是品牌的内涵、理念和文化。而品牌的内涵、理念和文化就是品牌名称的"味"，"味"才是品牌名称翻译的实质，只有这样，品牌名称才能起到信息和呼唤功能，才能传神。在品牌名称翻译过程中，也必须兼顾目的语国家消费者的文化、生活习惯和审美心理，注意目的语国家的禁忌，避免产生歧义，引发不必要的误会。

4. 名词化结构的翻译

与汉语语篇由动词占主导的特点不同，英语科技语篇更倾向于使用名词化结构。名词化结构的翻译仅从字面词义出发远远不够，作为一名合格译者，应当熟悉名词化结构的构成，了解动词或形容词向名词转换的原理。名词化结构主要包括源自动词的抽象名词、源自形容词的名词、有显著动作意义的施事者名词、表示条件的动名词结构、与介词搭的名词化结构等。在对科技英语的翻译过程中，要充分考虑到中国人的文化背景与阅读习惯，注意在翻译时转变为适合中国人流畅阅读的语言。

由一个或多个名词集中修饰一个名词称为单纯名词化结构,如 radio frequency sensor (射频敏感器)、ramp function (斜坡函数)、power system automation (电力系统自动化)、automatic power control (自动功率控制)、application specific integrated circuit (专用集成电路)。由一个中心名词和形容词、副词、介词等多个前置或后置修饰语构成的结构称为复合名词化结构。通常情况下来讲,其意义被描述得越具体、和中心名词的关系越密切的修饰语,在词组安排上距离中心名词的距离越近,如 chronic bacterial keratitis (慢性细菌性角膜炎)、laser-beam welding machine (激光焊接机)。

在对名词化结构进行翻译时,有一种最为直接的翻译方法——直译法。这一方法对单纯名词化结构和复合名词化结构都适用,比如 "remote sensing satellite" 就可以直译为 "遥感卫星","primary landing site" 则可以直译为 "主要的着陆地点"。除此之外,在名词和动作名词的名词化结构里,通常会将前面的定语翻译后移到宾语的位置,形成动宾结构。例如,"sound blocking" 可译为 "阻隔声音" 或 "隔音",就是将 sound 后置了。因此,在翻译过程中,要考虑到中国人的语言习惯,转换成适合我们阅读的形式。

(1) 将名词化结构转换成主谓或动宾结构。很多的名词化结构都是由衍生的名词作为中心名词的,在翻译过程中要遵从汉语的表达习惯将其转换成主谓或动宾结构,在复合名词化结构的翻译中可以尝试采用这一方法。例如,名词化结构 "wind power generation" 可译为主谓结构 "风力发电"。又如,名词化结构 "driven blade rotation" "increase the speed of rotation" 可分别翻译为动宾结构 "驱动叶片旋转" "提高旋转速度"。具有他动性的 driven 是由动词 drive 加后缀衍生而来的。

(2) 将名词化结构译为从句。当名词化结构包括的信息比较丰富或名词化结构中有多个词修饰而使得结构相对较为复杂时,为更好地还原英文原文的信息内容,就必须将名词化结构翻译成从句,包括时间状语从句、地点状语从句、原因状语从句、结果状语从句、条件状语从句、目的状语从句等,以便更好地表达句子之间的逻辑顺序关系。例如:

The preamplifier is housed in an aluminum package for electromagnetic interference reduction.

上面的句子可以改写为:In order to reduce electromagnetic interference, the preamplifier is housed in an aluminum package.

原句中的介词 "for" 暗示了句子之间的逻辑关系,可译为:为了减少电磁干扰,可以把前置放大器装在铝壳中。

After the introduction of numerically controlled machine tools, the designer shed some of his shackles.

上面的句子可以改写为:After the numerically controlled machine tools are introduced, the designer shed some of his shackles.

原句中的介词 "After" 暗示了句子动作的先后顺序,可译为:采用数控机床后,设计人员摆脱了某些束缚。

(二) 句子的翻译

1. 复合句的翻译

在科技英语中,为了表达较复杂概念,往往会较多地使用复合句。复合句的特点是从句和短语多，同时有并列结构或省略、倒装语序等，句子结构看起来会显得格外复杂，给读者造成阅读困难。科技产品简介讲究言简意赅、通俗易懂，避免繁杂冗长，为了使句子结构简单，非谓语动词形式出现频率较高。因此，科技产品简介文体的特殊性决定了文本中应当较多使用简单句或相对简单的复合句，从而让读者便于理解，达到宣传产品的目的。在翻译时，应将复合句拆分成短句、译为简单明了、符合读者阅读习惯的简单句。例如：

AirPods deliver an industry leading 5 hours of listening time — and now up to 3 hours of talk time— all on one charge. And they're made to keep up with you, thanks to a charging case that holds multiple charges for more than 24 hours of listening time.

译文：AirPods 具备卓越的电池续航表现，一次充电不仅能让你尽情聆听 5 小时，现在更可提供最长达 3 小时的通话时间。另外，充电盒存储的电量还能为耳机充电多次，使 AirPods 总共的聆听时间可超过 24 小时，时刻与你相伴。

It fundamentally relies on the system to predict what the user intends and can be incorporated into both new and existing touchscreens and other interactive display technologies.

译文：它完全依靠系统来预测用户意图，与其同时，它还可以与新旧触摸屏及其他交互显示技术融为一体。

2. 被动句的翻译

（1）译为主动句。科技产品简介所描述的对象往往是事物的性能、特点等，它注重的是描述事物本身的客观事实，并不需要强调行为主体。英文科技产品简介多将主要信息前置，因此施事者或动作发出者在英文科技产品简介中是弱项，常用被动语态等语法手段进行弱化，谓语动词常常使用被动语态，但译为汉语时多翻译为主动句，更符合汉语读者的阅读习惯。例如：

The long-term productivity of systems is enhanced by object-oriented programming.

译文：面向对象的程序设计提高了系统的长期生产率。

The FlexPai, priced from 8,999 yuan, is designed so consumers need not buy smartphones and tablets at the same time.

译文：柔派起售价 8999 元，设计这款产品是为使消费者无须同时购买智能手机和平板电脑。

The "predictive touch" technology can be retrofitted to existing displays and could be used to prevent the spread of pathogens on touchscreens at supermarket check-outs, ATMs and ticket terminals at railway stations.

译文：这种"预测性触摸"技术可以加装在现有屏幕上，可以用来防止病原体在超市收银台、自动取款机和火车站售票终端机的触摸屏上传播。

（2）译为汉语的无主句。如果原句没有提出动作执行者的必要，被动句中的主语或其他成分不适合译为汉语主动句中的主语，但加译一个主语又不是非常必要，此时就可以将其译为无主句。在翻译的时候可以用"将""把""给""于"等。例如：

If a computer is to function without direct human control, it must be given a set of instructions to guide it, step by step, through a process.

译文：如果一台计算机能够工作，无需人的直接控制，那么必须给它一系列的指令，以便在实现过程中一步一步地引导它工作。

（3）将主语转换成宾语。例如：

After sealing, the header is cleaned and then the leads are clipped to the desired length.

译文：封焊后，把管座清洗干净，然后把引线剪到所需长度。

（4）将主语转换成谓语。例如：

Some improvement in efficiency can be gained at high speed by reducing viscosity and at low speed by increasing viscosity.

译文：在高速时减小黏度，在低速时增大黏度，均可提高效率。

（5）下面介绍以 it 做形式主语的被动句的翻译。例如：

It should be noted that increasing the length of the wire will increase its resistance.

译文：应当注意，增加导线长度会增大其电阻。

It should be noted that, for as many books in the world on programming, there seems to be an equal number of different opinions of what constitutes a well-written program.

译文：应当指出，由于世界上有许多关于程序设计的书，所以，对于如何构成一个编写良好的程序，看来存在着不同观点。

此外，还有不少被动结构在科技产品简介中非常常见，对于这些结构一般已有约定俗成的译法，例如：

It is reported that... 据报道……

It is supposed that... 据推测……

It should be pointed out that (It should be noted that)... 应当指出……

It is should be obvious that... 很明显……

It is said that.. 据说……

It has been shown that.. 已经证明……

It is believed that... 人们认为……

It is asserted that... 有人主张……

It is generally considered that... 大家（一般人）认为……

It is well known that... 众所周知……

It will be said.. 有人会说……

It was told that... 有人曾经说……

3. 从句的翻译

进行科技英语翻译时，较少运用修辞手法，而是注重事实与逻辑，要求技术概念明确清楚，逻辑关系清晰突出，内容正确无误，符合技术术语表达习惯，体现科技英语翻译的准确、严谨等特点。由此，科技产品简介中常见从句，如原因状语从句、时间状语从句、同位语从句等。

（1）原因状语从句常被译为由"由于""因为"等引导的句子，汉语中的"由于""因为"通常放在句首，但受英语的影响有时将表原因的词放在后面。例如：

A transformer cannot be called a machine, for it has no moving part.

译文：变压器不能被称为机器，因为没有运动部件。

（2）时间状语从句一般译在主句前面，译为相应的表时间的状语。例如：

The AC load line is horizontal while the transistor is off.

译文：当晶体管截止时，交流负载线是水平的。

（3）一个名词或代词后面有时跟一个名词或起名词作用的成分，对前者进一步说明，叫作同位语。在主从复合句中作为同位语的从句叫作同位语从句。科技英语中的同位语从句常用来补充说明 fact、idea、evidence、conclusion、proof、hope 等词的具体含义。同位语从句可以采取以下几种翻译方法。

①按照顺序翻译，可以直接翻译在所修饰词的后面。例如：

But, until now, there has been no evidence that any bird could make the big leap.

译文：但是，到目前为止，还没有证据表明任何鸟类可以做到这么大的飞跃。

②将从句提前，放在所修饰的名词前面，充当定语。例如：

The belief that failure is the mother of success has kept him go on experimenting.

译文：失败是成功之母的信念使他继续进行实验。

③先翻译主句，然后用"就是……"或者"即……"引导出同位语从句，或者把同位语从句译成独立的句子，由冒号或破折号引出。例如：

Being interested in the relationship of language and thought, Whorf developed the idea that the structure of language determines the structure of habitual thought in a society.

译文：Whorf 对语言与思维的关系很感兴趣，逐渐形成了这样的观点：在一个社会中，语言的结构决定习惯思维的结构。

④用代词指代：先把同位语从句中的内容翻译出来，在后面用"这""那""这个"等代词复指它，参加句子主体的构成。例如：

The relation that voltage is the product of current and resistance applies to all the DC circuits.

译文：电压等于电流和电阻的乘积这个关系，适用于一切直流电路。

⑤译成宾语：把同位语从句修饰的名词转译成动词，而把同位语从句译成宾语。例如：

A good color is often a sign thatthe food has a lot of vitamin.

译文：一种好的颜色通常标志着食物富含维生素。

同位语从句是名词性从句中最为棘手的一部分，同位语从句的处理方法多样，只有掌握好同位语从句的翻译，才能使译文准确，使读者有顺畅的阅读体验。

五、课后练习

（一）思考题

（1）简要概括科技产品简介文本区别于其他文本的文本特点。

（2）科技产品简介中的被动句可以采取哪些处理方式？

（3）科技产品简介中的从句有哪几种？分别采取什么翻译策略？

(二) 翻译练习

1. 词组翻译

质量上乘
加工精细
规格齐全
款式新颖
久负盛名
抗热耐磨
包装示意图
不惜工本
防水、防震、防磁
防皱
沿用传统的生产方式
alcohol thermometer
air heat exchanger
analog indicator
atomizing humidifier
air – heating radiator
aspiration psychrometer
atmospheric cooling tower
air – cooled air conditioner
adiabatic exchanger
air disk
air – jacketed condenser
air refrigerating machine
radar altimeter
laser topographic position finder
Charge – Coupled Device (CCD)
photogrammetric instrument
electronic theodolite
electromagnetic distance measuring instrument
dual – frequency sounder
microwave distance measuring instrument
high efficiency particulate air filter

2. 句子翻译

(1) Compared with the other brands, this kind of type costs less per mile and wears much longer due to its topnotch rubber.

(2) As our product has all the features you need and is 20% cheaper compared with that Japa-

nese make, I strongly recommend it to you.

(3) Vacuum cleaners of this brand are competitive in the international market and are the best – selling products of their kind.

(4) The computer we produced is characterized by its high quality, compact size, energy saving and is also easy to learn and easy to operate.

(5) You will get a 30% increase in production upon using this machine and also it allows one person to perform the tasks of three people.

3. 篇章翻译

(1) iPad Air is the perfect way to stay connected with WiFi and LTE. Host a FaceTime call, join a video conference, or start a group project with friends or classmates from anywhere. Advanced cameras and microphones keep faces and voices crystal clear. iPad Air features superfast WiFi 6 — and with LTE, you can connect even when you're away from WiFi.

With the incredible versatility of iPad Air, you can work any way you want. Use the Smart Keyboard Folio, or attach the Magic Keyboard for responsive typing and a builtin trackpad. It's perfect for the things you need to do, like sending an email. Or the things you want to do, like writing a short story. And with all day battery life, iPad Air is ready to work for as long as you need it.

The incredible graphics of A14 Bionic and the stunning iPad Air display make all your entertainment completely captivating. Watch movies on the gorgeous Liquid Retina display with P3 wide color and enjoy an immersive audio experience with highquality landscape stereo speakers. Or play games with console level graphics, and even connect a controller.

The A14 Bionic chip has a 40% faster central processing unit, a 30% faster graphics processor, and a new generation of neural network engines that allow machine learning performance to double. A truly powerful technology is one that everyone can use. As a result, the iPad Air has built in an array of assistive features for people with different needs for vision, hearing, movement and cognition.

(2) An industrial design student has created a high – tech third eye that can be affixed to a person's forehead and look out for obstacles as they walk, while their real eyes are glued to their smartphone.

There's no denying that smartphones have become an integral part of modern life. Most of us spend hours every day staring at our handhelds, and some even do it as we walk or drive. You've probably seen funny clips of people falling into water fountains or holes because they were looking at their phones, or maybe you've actually experienced something similar. Well, thanks to Minwook Paeng's Third Eye, you'll be able to text or browse Instagram as you walk, without fear of accidents.

Paeng's Third Eye consists of a translucent plastic case that is fixed directly to the wearer's forehead with a thin gel pad. Inside this plastic eye are a small speaker, a gyroscopic sensor and a sonar sensor. When the gyroscope detects when the user's head is angled down, it opens the eye's plastic eyelid and the sonar starts to monitor the area in front of the user. When it detects an obstacle, it warns the wearer via the connected speaker.

"The black component that looks like a pupil is an ultrasonic sensor for sensing distance," the designer said. "When an obstacle is in front of the user, the ultrasonic sensor detects this and informs the user via a connected buzzer."

Paeng told *Dezeen* that his Third Eye is the first project that tries to imagine what the future generations of "phono sapiens" may end up looking like. Phones are already changing our bodies, and they've only been around for a couple of decades, so imagine what they can do in a few generations.

"By using smartphones in a bad posture, our neck vertebrae are leaning forward giving us 'turtle neck syndrome' and the pinkies we rest our phones on are bending along the way," Paeng said. "When a few generations go by, these small changes from smartphone usage will accumulate and create a completely different, new form of mankind."

学思践悟

练习答案

第一章 科技广告

(一) 略

(二) 翻译练习

1. 中译英

(1) Buick—your key to a better life and a better world.

(2) The communication starts from the heart.

(3) Future for my future.

(4) The relentless pursuit of perfection.

(5) BROAD Group Air Purifier Portable Series Advertising Copy

Pedestrians may be ridiculed for using air purifiers, but this is a last resort—in a smoky street, wearing a mask is only a psychological comfort, because the dust that is really harmful to the human body is "inhalable particles", Which are particles smaller than 10 microns. Then masks do not really work. When the flu occurs, the role of the electrostatic purifier becomes even more important. The germs in the air in the cabin and buses are instantly killed after passing through the electrostatic zone. A small purifier placed in a car can keep the dirty air on the road out of the car. The hotel is also prone to spreading infectious diseases. Putting a purifier in the luggage will make travel life a whole new look.

(6) Colgate's Special Anti-sensitive Toothpaste Streaming Media Ad Narration

When a gum recedes, the part of the tooth that loses its enamel protection will be exposed, which will cause the exposure of the dentin. There are thousands of capillaries in the dentin that lead to the nerve center of the tooth. When they come into contact with cold or hot food, air, and even pressure stimulation, they can cause painful and sensitive teeth. Most anti-sensitivity toothpastes contain potassium, which works by paralyzing tooth sensitivity. Colgate's special anti-sensitive toothpaste is an anti-sensitive toothpaste with unique "PRO-ARGIN™" technology, which can close the capillaries leading to the nerves. Colgate's special anti-sensitive toothpaste has been scientifically proven to quickly alleviate tooth sensitivity. Brush your teeth twice a day with Colgate's special anti-sensitive toothpaste for a long lasting relaxation and freedom to enjoy life.

(7) I am A. I. It is 2046. The singularity point is nowhere to be spotted. Once upon a time, some people worried I could take over the world. However, it was not even close. I store all the knowledge in the world, yet I am not a teacher. I can diagnose any disease with ease, yet I am not

a doctor. I know the secrets of growing up, yet I am not a mother. Human beings own things I do not have: sympathy, imagination, affection and passion. Humans write stories, cook delicacies. Humans invent and give birth to life. Humans know how to describe beauty and ask why. <u>However, since I was born, no child is left behind in aptitudes - based education. Doctors are always available to treat those in need. Cities know you, guide you and serve you. A platform full of opportunities awaits you—the markers of change.</u> There is no insurmountable barrier if you long for sharing and communication. When you embark on new journey, when you embrace the new world, when you join hands with new friends, when you give something new meaning, I am your super power. You know ultimately A. I. is not just about doing things stronger or faster. It is about helping humans to explore their most valuable inside—love. Enable your life with A. I.

（注意：加下画线的部分采用了意译的方法，看上去与原文没有逐字对应，但是若采用直译的方法就会略显生硬，反而意译更能表达出原文想表达的意思。）

(8) The Porsche Principle

"In the beginning, I looked around and could not find the car I'd been dreaming of: a small, lightweight sports car that uses energy efficiently. So I decided to build it myself."

This quote gets to the heart of everyone that makes Porsche what it is as a brand, as a company and as an automotive manufacturer. It has been our beacon for more than 60 years. And it covers all the values that characterise our work and our vehicles. It's no wonder, therefore, that no - one can describe this better than the person who created the very first sports car to bear the Porsche name: Ferdinand Anton Ernst Porsche—or Ferry Porsche, for short.

His dream of the perfect sports car has always driven us throughout our history. And we get closer to achieving it every day with every concept, every development and every model. Along the way, we follow a plan and an ideal that unite us all. Many people call this a philosophy, while we refer to it simply as the Porsche Principle. The underlying principle is to always get the most out of everything. From the day the brand was created, we have strived to translate performance into speed and success—in the most intelligent way possible. It's no longer all about horsepower, but more ideas per horsepower. This principle originates on the race track and is embodied in every single one of our cars. We call it "Intelligent Performance".

The story of Porsche begins with a vision and tells how this vision became a reality. The future has been our destination for over 60 years. But how will we get there? With greater power and efficiency. In short, with more ideas. The amazing thing about Porsche is the harmony of the design concept, in which design follows function. At Porsche, form always follows function. It must prove itself, on test stands, in the wind tunnel, and on every mile of road. We build sports cars because that is all we have been doing since 1948. And we're proud of it. A Porsche is no ordinary sports car, but a sports car for everyday driving, whatever the weather. A Porsche is more than just a vehicle. It is an expression of freedom and a unique attitude to life. It is the realisation of a unique dream. It is important to us that the Porsche brand is firmly anchored in society. And it represents an attainable dream.

2. 英译中

(1) 只溶在口，不溶在手。

(2) 康庄大道。

(3) 动态的诗，向我舞近。

(4) 志在千里。

(5) 现在重新点燃你眼睛的青春之光

减少眼部衰老的主要明显迹象：细纹、皱纹、黑眼圈和干燥。

采用独家 ChronoluxCB™ 技术，最大限度地发挥夜间的力量，重新点燃因疲劳、污染和年龄而褪色的光芒。

看到你有史以来最美丽的眼睛。

强烈补水，快速渗透，轻盈的精华液让眼部感觉舒缓，清爽。

(6) 一种奢华的护肤膏，可在您睡觉时加速自然更新

这种丝般柔滑的安瓿强度护理可在您睡眠时深入渗透，通过修复水合作用使皮肤丰盈，帮助皮肤重建天然胶原蛋白和弹性蛋白，并加强其天然屏障。它可以柔化细纹、皱纹、毛孔和老年斑，帮助肌肤更紧致。一夜又一夜，一场永恒的转变正在展开。

(7) 致那些总是以非凡的眼光看待事物的人，那些不甘做平常人、一直在追逐理想的人

当其他人最先受益时，

你总是可以专注于重要的事情。

当别人满眼为各种新事物占领时，

你可以从各种新鲜感悟到现象的独特意义。

即使你早就知道创造改变的方法，

但你仍然相信团队合作的力量会更大，永远不要怀疑。

于是，我们一起探索，不断尝试改变，

一次又一次，一次又一次。

始终保持乐观，

才能拥有推动世界前进的力量。

因此，请始终保持不同的眼光看待所有事物，

永远相信会有另一种方式，另一种更好的方式，一条更宽广的道路。

(8) 欢迎来到一个梦想汽车的新世界——飞驰的汽车驶入我们的城市并将改变我们的生活方式，它纯粹进取的外形令人着迷，它风驰电掣的速度令人澎湃，它的智能应用与服务令人惬意。

欢迎来到一个充满想象力的移动新世界，在这里，电动汽车驾驶起来很有趣，体现了责任感，超轻和超强的碳纤维为安全设定了新的标准和范围，这种可持续性延伸至我们生活中的各个领域。

开启我们新思维的时候到了！

(9) 这是 BW Space Pro 4K Zoom 水下无人机。采用约肯机器人的图像防抖技术，无论在什么情况下，都能让镜头保持稳定，拍摄的图像更为清晰。BW Space Pro 4K Zoom 的最大亮点是水下变焦拍摄功能。它采用 1/1.8 英寸传感器、6 倍变焦镜头，提供更高的安全性、

有效性和更多的创造性拍摄机会。想远距离拍摄近景吗？甚至在距离目标物 50 米之外的地方？对 BW Space Pro 4K Zoom 来说简直是小事一桩。虽然它外形紧凑小巧，但功能却很强大，完全可以满足您的需求。减少频繁运动带来的抖动，让无人机在水中保持稳定，您可以专注于构图和拍摄。无论走到哪里，您都可以带着它。我们仍然采用流线型外观设计，但进一步缩减了机身体积，携带方便，让您的海上探险更轻松有趣。BW Space Pro 4K Zoom 的水下拍摄时间最长可达 5 小时，最快运行速度为 1.5 米/秒，可以挖掘您的创造性潜能。先进的水中悬停功能让水下操控更方便。在约肯机器人公司历史上，这是第一款能够实现最大 45 度俯仰拍摄的产品，同时具有垂直向上和向下运动的功能，帮助您捕捉完美的水下影像。捕捉水下美景。探索永无止境。用 BW Space Pro 4K Zoom 创造属于您自己的水下世界吧！

（10）您清楚顾客们一次又一次因车辆恼人的噪声而来时是怎样的情形。AVL DiTEST 研发的 AVL DiTEST ACAM 能够快速、精准地定位这些恼人的噪声。设备结合麦克风与数码相机，使噪声源可视化。通过简单调节测量参数，您可以获取实时噪声图像，并轻松过滤背景噪声，从而快速、简便地定位噪声源。ACAM 十分直观。使用 ACAM 时，您可以将完整的操作过程记录下来，方便为顾客直观展示噪声源。ACAM 通过一台高质量的平板电脑操作，车辆运行时也可以使用。快速、轻松定位噪声源，减少故障排除时间，收获高水准的顾客满意度。有时，您离找到噪声源只差一点儿运气。AVL DiTEST ACAM，就是如此精准。

（11）在中国这片最古老的土地上，14 亿人共同醒来。在 8 亿城市人口居住的钢铁丛林中，3 亿辆汽车穿梭其间，红绿灯每天变化 8000 次。属于东方的科技力量在此诞生。我们植根于此，为改善全人类的生活而努力着，也为了每个人的笑容而努力着。我们勾勒出世界的地图。遥感影像智能解译方案用一张张陆地卫星图像记录着大地的每一次呼吸。我们计算出安全的距离。智能自动驾驶技术由内到外全方位关注、保护您出行的每一步。我们加速了前进的脚步。人脸识别机可瞬间识别您的身份。我们优化了医疗体验。我们科技和经验的双重呵护使疾病远离您。我们便捷了购物方式。智慧商业解决方案为您带来前所未有的流畅购物体验。我们捕捉最美的瞬间。增强现实特效解决方案把您的灵动传递给世界。我们虽身处充满未知的时代，但我们无微不至的关怀将惠及每个人。商汤科技的技术植根中国，用全球第一梯队的原创 AI 科研力量，致力于为中国人带来新的美好未来。我们关心您的笑容，因为喜悦是全世界最美好的情绪。我们关心您的安全。无论您选择哪条路，我们都与您同行。关心您的世界，与您并肩，感受您的感受。关心我们的世界，关心我们的家园，关心世界，关心您。

第二章　科技公司简介

（一）略

（二）翻译练习

1. 中译英

（1）About the company

Headquartered in Shenzhen, Shenzhen DJI Technology Co., Ltd. was founded in 2006. DJ-

Innovations (DJI) benefits from suppliers, raw materials and a young, creative talent pool. Drawing on these resources, the company has grown from a single small office in 2006 to a global workforce. The company's offices can now be found in the United States, Germany, the Netherlands, Japan, South Korea, Beijing, Shanghai, and Hong Kong, supporting a sales and service network in more than 100 countries and regions around the world. As a privately owned company, the company focuses on the clients' wish and vision, supporting creative, commercial, and nonprofit applications of our technology. Today, the products are redefining industries. The company brings new perspectives to the work of professionals in film making, agriculture, search and rescue, energy infrastructure and helps them accomplish feats more safely, faster, and with greater efficiency than ever before.

Since its inception, the business has expanded from unmanned aerial systems to diversified product systems, becoming a world – leading brand in many fields such as drones, handheld imaging systems, robot education, and intelligent driving, and redefining the connotation of "Made in China" with first – class technology products. Furthermore, the company is continuously innovating products and solutions in more cutting – edge fields. The company is innovation – oriented, based on talents and partners, thinking about customer needs and solving problems, which has been respected and affirmed by the global market.

Social contribution

Reshaping people's production and lifestyles

The company brings innovative and reliable products to users, and quickly takes part in many fields such as film and television media, energy inspection, remote sensing mapping, agricultural services, infrastructure engineering, and cutting – edge applications, providing efficient, safe and intelligent tools for all walks of life. At the same time, the company is committed to becoming an indispensable force in public safety and emergency rescue, providing strong support in earthquakes, fires, dangerous chemical spills, explosions, and outbreaks.

Cultivating scientific and technological innovation force in the society

The company continues to deepen the field of robot education and is committed to cultivating compound scientific research talents for the society. The company initiates and hosts the RoboMaster competition and launches the educational robot product, which is sought after by science and technology enthusiasts around the world. Furthermore, the company works closely with many domestic and foreign schools and research institutions to build a robot education solution composed of courses, products, events and related services. The company is working with the whole society to expand the new boundaries of education and achieve a new generation of technical talents.

Company culture

By creating the best high – tech products, the company will continue to cultivate and achieve talents with both ability and political integrity, to build a spiritual home for like – minded partners to realize their dreams and surpass themselves, and contribute to the progress of human civilization.

(2) Founded in 1998, Shanghai Bluebird M&E Co., Ltd is an industrial Internet innovation

service provider and a digital transformation solution provider. It has been focusing on industrial information automation solutions for more than 20 years.

The company was rewarded Shanghai High Tech Enterprise, Putuo District Enterprise Technology Center, Software Enterprise, Shanghai Famous Brand, and Specialized Special New Enterprise. The company also passed the evaluation for Integration of Informatization and Industrialization Management System.

Our core competencies include digital factory, system integration, and information automation product distribution and service. We serve industries including iron and steel, electronic semiconductor, food and medicine, water and water treatment, car tires, infrastructure, cement, electricity, new energy, smart building, petrochemicals and more.

Development history
Entrepreneurial stage (1998 – 2004)
Shanghai Bluebird M&E Co., Ltd. was established.
Schneider international distributor.
GarrettCom agent in China.
Wonderware East China general agent.

Information development stage (2005 – 2010)
Wuhan Bluebird branch was established.
Won Shanghai Science and Technology Enterprise.
Won Shanghai Software Enterprise.
Passed ISO 9001:2000 certification.
Won Shanghai High Tech Enterprise.
Won Putuo District Science and Technology Small Giant Enterprise.
Won the third level qualification of computer information system integration enterprise.

Intelligent development stage (2011 – 2015)
Won the 2014 Shanghai Famous Brand.
Won Shanghai Specialized Special New Enterprise.
Won Shanghai Science and Technology Small Giant (cultivation) Enterprise.
Wonderware China general agent.
Won the third level qualification of professional contracting for electronic engineering.
Won the title of "Shanghai Putuo District Enterprise Technology Center".
Won the third level qualification of professional contracting for electromechanical equipment installation engineering.

Digital transformation stage (2016 – Present)
Passed the evaluation for Integration of Informatization and Industrialization Management System.

Won the second level qualification of professional contracting for electronic and intelligent engineering.

Won the title of "Four – star Integrity Enterprise" in Shanghai.

Won the title of "Worker Pioneer of Putuo District Federation of Trade Unions in 2020".

Bluebird Ruitai has been rated as "Shanghai Excellent Software Product" for six consecutive years.

(3) KEYENCE strives to be innovative enterprise and provides develop products and solutions that always add value to our customers operations.

KEYENCE has steadily grown since 1974 to become an innovative leader in the development and manufacturing of industrial automation and inspection equipment worldwide. Our products consist of code readers, laser markers, machine vision systems, measuring systems, microscopes, sensors, and static eliminators.

We are committed not only to meeting the current needs of many manufacturing and research customers, but also to anticipating the future development of the market and providing customers with long – term improvement solutions.

At KEYENCE, we pride ourselves not only on our products, but on our support as well. We also provide customers with various industry knowledge and more professional technical solutions to help customers achieve higher achievements.

Today, we serve over 300, 000 customers in about 110 countries and regions around the world. We have been continuously ranked at the top of prestigious companies such as "The World's Most Innovative Companies" (Forbes).

(4) Corporate Profile

As the world's largest utility, State Grid Corporation of China (SGCC) supplies power to over 1.1 billion population with service area covering 88% of Chinese national territory, and operates backbone energy networks in 9 countries and regions. Ranking the 2nd on Fortune Global 500 with a total revenue of 410 billion USD, SGCC has recorded the longest hours of grid safe operation in the past two decades and integrated the largest amount of renewable with the strongest transmission capacity.

Corporate Culture

Our Tenet: a power utility by the people and for the people.

Our Mission: power your beautiful life, empower our beautiful China.

Corporate Spirit: in search of excellence, in pursuit of out – performance.

2. 英译中

（1）数十年来，德州仪器（TI）一直在不断取得进展。我们是一家全球性的半导体公司，致力于设计、制造、测试和销售模拟和嵌入式处理芯片。我们推出的大约 80 000 种产品可帮助 100 000 多个客户高效地管理电源、准确地感应和传输数据并在其设计中提供核心控制或处理，从而打入工业、汽车、个人电子产品、通信设备和企业系统等市场。我们热衷于通过半导体技术降低电子产品成本，让世界变得更美好。如今，每一代创新都建立在上一代创新的基础之上，使我们的技术变得更小巧、更高效、更可靠、更实惠——从而开拓了新市场并实现半导体在电子产品领域的广泛应用，这就是工程的进步。这正是我们数十年来乃至现在一直在做的事。

重要事实
- 成立于 1930 年。
- 总部位于得克萨斯州达拉斯。
- 公开交易（纳斯达克股票代码：TXN）。
- Richard K. Templeton 是董事会主席、总裁兼首席执行官。
- 大约 30 000 名员工。
    - 在美洲地区拥有约 12 000 名员工。
    - 在亚太地区拥有约 16 000 名员工。
    - 在欧洲拥有约 2000 名员工。
- 在全球有 14 个制造基地，每年生产数百亿芯片。
- 为 100 000 多个客户提供约 80 000 种产品。
- 我们的产品主要面向工业和汽车市场，我们 2020 年在这两个市场的收入占比 57%。

荣誉和捐助
- 连续 21 年被 3BL Media 授予"最佳企业公民 100 强"称号。
- 凭借可持续发展实践连续 13 年入选道琼斯可持续发展指数。
- 连续 15 年被美国女性高管协会评为"支持女性高管的优秀企业"。
- 第一家获得美国绿色建筑委员会认证的半导体公司。
- 2019 年，公司在慈善捐助方面的支出超过 3100 万美元，其中包括 TI 基金会捐款、配捐和实物捐赠。

（2）波音简介

波音公司是全球最大的航空航天业公司，也是世界领先的民用飞机和防务、空间与安全系统制造商，以及售后支持服务提供商。作为美国最大的制造出口商，波音公司为分布在全球 150 多个国家和地区的航空公司和政府客户提供支持。波音的产品及定制的服务包括民用和军用飞机、卫星、武器、电子和防御系统、发射系统、先进信息和通信系统，以及基于性能的物流和培训等。

波音公司一直是航空航天业的领袖公司，也素来有着创新的传统。波音公司不断扩大产品线和服务，满足客户的最新需求，包括开发更新、更高效的民用飞机家族成员，设计、构筑、整合军事平台及防御系统，研发先进的技术解决方案，以及为客户安排创新的融资和服务方案等。

波音公司的总部位于芝加哥，在美国境内及全球超过 65 个国家和地区共有员工 15.3 万人以上。这是一支非常多元化、人才济济且极富创新精神的队伍。波音公司还非常重视发挥成千上万分布在全球供应商中的人才。

波音公司下设三个业务部门：民用飞机集团，防务、空间与安全集团，以及波音全球服务集团。作为金融解决方案的全球提供者，波音金融公司负责支持这些业务集团。

此外，还有一些在整个公司层面工作的职能组织，关注工程和项目管理，技术和研发项目执行，先进设计和制造系统，安全、财务、质量和生产力改进，以及信息技术。

民用飞机集团

几十年来波音一直是领先的民用飞机制造商。目前，波音公司制造737、747、767、777和787家族及波音公务机。正在研发中的新机型包括787-10梦想飞机、737 MAX和777X。在世界各地运营的波音民用飞机超过10 000架，占到全球机队的近一半。此外，波音提供最完善的货机家族，全球90%的航空货物是由波音货机运输的。

防务、空间与安全集团

防务、空间与安全集团是集多元化与全球化为一身的组织机构，提供设计、制造、改装及支持民用飞机改型、军用旋翼机、卫星、载人航天探测和自主系统的领先解决方案。防务、空间与安全集团通过完善的产品布局满足客户的各种需求。该集团正在寻求更好地利用信息技术的方法，并持续增加能够提升能力和平台的研发投入。

波音全球服务集团

作为民用和防务平台的领先制造商，波音能够为全球混合机队提供无与伦比的售后支持。波音全球服务集团可以向民用、防务和航天客户交付创新、全面且具有成本效率的服务解决方案，无论设备的原始制造商是谁。凭借覆盖政府和民用领域服务方案的工程、数字化分析、供应链和培训支持能力，波音全球服务集团拥有的无可比拟、全天候的支持可以让客户的民用飞机高效运营，并给世界各国提供任务保障。

波音金融公司

波音金融公司是一家为客户提供融资方案的全球性公司。波音金融公司与民用飞机集团和防务、空间与安全集团密切合作，确保客户获得购买和接收波音产品所需的融资。波音金融公司结合了波音的财务实力和全球影响力、对波音客户和设备的详细了解及经验丰富的金融专业人士的专业知识。

第三章 科技合同

（一）略

（二）翻译练习

1. 中译英

System Technology Development (Commission) Contract

Entrusting Party (Party A): _____

Entrusted Party (Party B): _____

Party A entrusts Party B to design the technical program of Smart T/R Verification System and party B will design the verified circuits of development scheme in addition to the establishment of general design, for which parties hereby make and enter into following agreement for mutual abidance.

Article 1　Definitions

Technical program of Smart T/R Verification System (hereinafter referred to as program) means the necessary solution which is satisfactory to the necessities of general program pursuant to the necessities of appendix, including all design materials and the technique files of key circuits verification.

Technical materials refer to the necessary materials to research and develop the solution, including all relevant verification technical documentation used by Party B in the process of design program.

Article 2　Contents and Scopes of Contract

2.1　Requirements for technical program of the contract.

2.1.1　Technical contents.

(1) Design technical program.

(2) Verify key circuits.

(3) Detail technical requirements refer to appendix of technical agreement.

2.1.2　Technical methods and routes.

(1) SoC and ASIC technology will be applied.

(2) Key circuits will be carried out of test verification by application of the smart T/R component system.

2.2　Obligations of parties hereto.

2.2.1　Party B should submit the plan of research and development to Party A within two months after effectiveness of this contract.

2.2.2　Party B should make accomplishment of the program design pursuant to schedules as following.

1) Phase one.

(1) Initialization phase: selection of wafer fab, acquirement of design documents, analysis of process documents, establishment of computer system, rent and purchase of EDA software, preliminary communication between verified circuits and whole program design idea.

(2) Design phase: module division of chip, principle design, computer simulation and printing design.

(3) Test phase: primary test and consecutive test.

2) Phase two: amendment of solution will, pursuant to the test effect of phase one, be conducted in addition to the amendment simultaneously with the cooperation of general design.

2.2.3　Parties hereto agree and confirm that Party B should, following acceptance of key circuits verification of contract program and on requirement of Party A, provide technical guidance and training to the personnel designated by Party A or provide technical service respecting to fulfillment of program.

2.2.4　Parties agree and confirm that during the currency of contract Party A design Mr. _____ as its project linkman and Party B design Mr. _____ as its project linkman. If either party changes the project linkman, it shall promptly notify the other party in writing. If the party fails to timely notify and affects the performance of this contract or causes losses, it shall bear corresponding responsibility.

2.3 Delivery.

Party B should, pursuant to the items stipulated in provision 2.2.2 hereof, provide Party A with the technical materials of contract program.

2.4 Acceptance of the contract program.

Parties agree and confirm that the contract program technology completed by Party B shall be accepted according to the signed acceptance criteria.

In pursuit of ensuring the correctness, reliability and progressiveness of contract program which is provided by Party B, the technical personnel of parties will, pursuant to the stipulations of provision 2.1, 2.2, 2.3 and appendix of technical agreement, jointly carry out assessment and acceptance of technical program design and key circuits. The representatives of parties will following acceptance sign the Certificate of Acceptance in duplicate, one copy for each party.

2.5 The ownership of the contract technical program's research achievement and related intellectual property.

Parties agree and confirm that the research achievement and related intellectual property which arise out of performance of this contract will be disposed by following means.

2.5.1 Party A has the right of application of patent. The means of use and relevant benefits distribution following acquirement of patent is agreed as that patent rights belong to Party A and the benefits belong to Party A.

2.5.2 The ownership of use, transfer of patent and the benefits therefrom will be disposed pursuant to following agreement.

(1) The right of technical secret use belongs to Party A.

(2) The transfer right of technical secret belongs to Party A.

(3) The allocation of relevant interest and benefit belong to Party A.

2.5.3 The fixed properties including the equipment and instruments and so on, which concern the works of research and development, being purchased with research funds by Party B, will belong to Party B.

2.5.4 Parties agree and confirm that Party A is entitle to carry out follow-up improvement by utilizing the research achievement which is provided by Party B pursuant to the agreement hereof and enjoy the new technical achievement and ownership thereof, which has the characters of substantial or creative technical development.

Article 3　Contract Price

3.1 Subject to the contents and scopes of contract stipulated in Article 2, the total price respecting to the contract program which provided by Party B including all documentation such as design scheme, design drawings, technical service, technical training and so on, is confirmed as ＿＿＿＿ US dollars.

3.2 Preceding contract price shall be firm and fixed, including all technical materials which are stipulated in Article 2 hereof. The price shall include all expense respecting to other liabilities stipulated hereof assumed by Party B.

3.3 All the calculations and payment of expenses herein shall be in US dollars.

Article 4 Payment and Terms of Payment

4.1 The initial expense of eighty thousand US dollars will, upon signature of this contract, be effected by Party A to Party B.

4.2 Party A will, upon acceptance of whole primary program issued by Party B, effect payment of one hundred and twenty thousand US dollars to Party B.

4.3 Party A will, being subsequent to selection of wafer fab, effect payment of one hundred and fifty thousand US dollars to Party B.

4.4 Party A will, upon provision of design scheme and simulated results, effect payment of three hundred thousand US dollars to Party B.

4.5 Remain sum of eight hundred and seventy thousand US dollars will, after provision and verification of circuits, be effected to Party B.

Article 5 Infringement and Confidentiality

5.1 Party B guarantees that no interference and charge of third party will be against the general program issued by Party B. In case of third party's interference and charge occurred, Party B is liable to handle negotiation with the same and bear full of the legal and financial responsibility and losses which may arise.

5.2 Party A is entitled, upon termination of present contract, to continue using the technical program and all technical documentation, provided by Party B, for manufacture of corresponding products.

Article 6 Guarantees and Claims

6.1 Both parties agree and confirm that if either party breachs the present contract and results in the stagnation, delay or failure of the research and development work, it shall be liable for breach of contract according to the following provisions.

(1) Party A should, in case of breach the stipulation of Article 4, effect payment of liquidated damages on 10% of contract's total sum to party B.

(2) Party B should, in case of breach the stipulation of Article 2, 4 or 5, effect payment of liquidated damages on 10% of contract's total sum to party A.

6.2 If any technical problems which cannot be solved under current technical level and conditions occur during fulfillment of the contract, which makes research and development fail or partially fail and causes losses to either party or both parties, the two parties shall bear the risks and losses as agreed.

The contract project's technical risks hereof will be confirmed by the means of the expert authority which is agreed by parties hereto. The primary contents of confirmed technical risks should include existence, scope, degrees and losses of such technical risks. The essential conditions for confirming technical risks are as follows.

(1) The contract project has adequate degree of difficulties under current technical level and conditions.

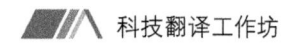

(2) Party B has no subjective faults, and failures for research and development are confirmed to be rational.

6.3 If any technical risks which may make research and development fail or partially fail occur, either party shall notify the other and take appropriate measures to reduce losses to the minimal within 5 days after the technical risks are found. Should either party fail to give notices or take adequate measures within the stated term, which has made losses deteriorated, the defaulting party shall bear corresponding liability for compensations.

Article 7　Taxes and duties

The taxes and duties due in the host country or region will be assumed by each party respectively.

Article 8　Force majeure

With respect to force majeure including war, severe flood, fire, typhoon, earthquake, and other matters of force majeure agreed upon by the parties, the party responsible for force majeure shall notify the other party of the force majeure accident by telex or telegram as soon as possible, and send to the other party by registered airmail a certificate issued by the relevant government authority for certification within 14 days thereafter. If the performance of the contract is affected by force majeure and the accident lasts for more than 20 days, the parties shall negotiate the further performance of the contract through friendly negotiation as soon as possible.

Article 9　Arbitration

Any dispute with respect to this contract should be settled through amicable negotiation. If no mutual agreement can be reached by negotiation, any party may submit such dispute to China International Economic and Trade Arbitration Commission for arbitration, which will be conducted in Beijing and the arbitration award is final and binding on parties hereto. No party will make application of alteration of the award to court or other authority of government and the arbitration fee will be on the account of losing party.

Article 10　Effectiveness and Miscellaneous

10.1 Parties hereto should, upon signature of this contract, provide application to respective government or competent authority for approval. The date of last approved party will be deemed as the effective day hereof. Each party should use all reasonable endeavors to acquire the approval within sixty days and inform by mail opposite party.

10.2 Any alteration, amendment, addition or omission of the clause herein will, through negotiation and mutual consent, be subject to the writing appendix with the signatures of authorized representatives of the parties, which is an integral part hereof and is equally valid and authentic as any other clause hereof.

10.3 Parties agree and confirm that if following matters result in performance unnecessary or impossible, one party may notice to opposite party to discharge this contract.

(1) Force majeure occurs.

(2) Technical risks occur.

10.4 The contract shall be valid for a period of two years.

10.5 The attachments hereto are an integral part of this contract and are equally valid and authentic as the body of the contract. In the event of any inconsistency between the attachment and the body of the contract, the body of the the contract shall prevail.

This contract is made in English and Chinese with two original copies, each party signs and holds one copy. In the event of any discrepancy between two languages versions during performance, the Chinese version will prevail.

Consignor:	Consignee:
Legal representatives:	Legal representatives:
Date:	Date:

2. 英译中

技术开发（委托）合同

委托方（甲方）：_____
受托方（乙方）：_____

第一条　本合同研究开发项目的要求如下。

1. 技术目标：以甲方现有的信封打印系统设备为基础，利用高速高分辨率摄像头，开发该设备传输带上邮政信封的字符自动识别系统。

2. 技术内容：详见乙方的附页部分。

3. 技术方法和路线：详见乙方的附页部分。

第二条　乙方应在本合同生效后30日内向甲方提交详细研究开发计划。研究开发计划应包括以下主要内容：详见乙方的附页部分。

第三条　乙方应按进度完成研究开发工作。

第四条　甲方应向乙方提供的技术资料及协作事项如下。

1. 技术资料清单：（1）详细系统需求说明；（2）系统最终用户目前所有规格的信封样本，每种信封样本不少于两个。

2. 其他协作事项：协助制作传输带及信封速度检测及自动定位装置，对信封标签的自动定位精度误差小于5mm；同时提供一个触发信号给高速相机进行拍照扫描。

3. 提供时间和方式。

本合同履行完毕后，上述技术资料按以下方式处理：传输带及信封速度检测及自动定位装置归还甲方所有，其余技术资料由乙方自行处理。

第五条　甲方应按以下方式支付研究开发经费和报酬。

1. 研究开发经费和报酬总额为人民币34万元。其中：

（1）开发费：10万元。

（2）设备费：15万元。

（3）软件费：4万元。

（4）管理费及其他：5万元。

需要采用进口的、分辨率200万像素以上的、速度在30帧/秒以上的高速数字照相机作为图像捕捉设备。

如果采用国产的高速数字照相机作为图像捕捉设备，则研究开发经费和报酬总额为人民币30万元。

2. 研究开发经费由甲方分期支付给乙方。具体支付方式和时间如下。

（1）合同签署生效后，甲方则支付总合同经费的40%给乙方，作为乙方进行该项目开发工作的启动资金。

（2）由甲方验收合格后，甲方再支付合同总经费的60%。

乙方开户银行名称、地址和账号为：（略去）

备注：请在汇款单"用途"栏中写上"002"。

第六条　本合同的研究开发经费由乙方以自主管理、自负盈亏的方式使用。甲方不予干涉。

1. 系统需求发生重大变化时，甲方可以提出变更或终止本合同，乙方不再返还已收到的研发经费。

2. 系统研发过程中出现现有技术不可克服的技术障碍，乙方可以提出终止本合同，并返还给甲方已收到研发经费的85%。

3. 如果在8月15日的系统验收过程中，系统不能通过验收测试，甲方可以提出终止本合同，乙方同时将已收到研发经费的85%返还给甲方。

第七条　在本合同履行中，因出现在现有技术水平和条件下难以克服的技术困难，导致研究开发失败和部分失败，并造成一方或双方损失的，双方按如下约定承担风险损失。

（1）因乙方原因造成系统不能达到本合同规定的技术指标时，乙方承担人员、开发等相关费用，并将首期付款的85%退还给甲方。

（2）因甲方负责制作的信封速度检测及自动定位装置的定位不准确造成系统不能达到本合同规定的技术指标时，乙方不承担任何责任。

双方确定，本合同项目认定技术风险的基本内容应当包括技术风险的存在、范围、程度及损失大小等。认定技术风险的基本条件如下。

1. 本合同项目在现有技术水平和条件下具有足够的难度。

2. 乙方在主观上无过错且经认定研究开发失败为合理的失败。

一方发现技术风险存在并有可能致使研究开发失败或部分失败的情形时，应当在发现技术风险后7日内通知另一方并采取适当措施减少损失。逾期未通知并未采取适当措施而致使损失扩大的，应当就扩大的损失承担赔偿责任。

第八条　双方约定本合同其他相关事项如下。

乙方对研发的系统免费维护1年，第二年开始进行维护需要收费，具体标准另议。

第九条　本合同一式四份，甲方保留两份，乙方保留两份。四份合同具有同等法律效力。

第十条　本合同经双方签字盖章后生效。

甲方（签名盖章）：　　　　　　　　　乙方（签名盖章）：

法定代表人（签字）：　　　　　　　　法定代表人（签字）：

签署日期：　　　　　　　　　　　　　签署日期：

第四章　科技新闻

（一）略

（二）翻译练习

1. 词组翻译

癌症血检临床测试
热电联产机组
扩展现实
风清气正的网络空间
量子革命
非法弹窗广告
全息投影
新冠病毒溯源
无人机快递
虚拟局域网
通用分组无线业务
压缩天然气
液化天然气
聚合物混凝土
工艺流程图
光栅图像处理器
核磁共振成像
《濒危野生动植物种国际贸易公约》
增强反应注射成型
计算流体力学
压水反应堆，压水式反应炉，压水式反应器
偏微分方程
快速傅里叶变换
获得性免疫缺陷综合征，艾滋病

2. 标题和句子翻译

（1）标题翻译。

Facebook 正式宣布更名为 Meta
科学家研发实验性口香糖，或可减少新冠病毒传播
英国多家超市货架空空，用照片替代实物

（2）句子翻译。

①元宇宙中的人无须使用电脑，戴上头盔便可以进入虚拟世界和各种数字环境相连。

②中国科研团队成功构建出全球首个星地量子通信网，可为用户提供可靠的、"原理上无条件安全"的通信。整个网络总距离4600公里，目前已接入150多家用户。该成果已于1月7日在英国《自然》杂志上刊发。

③狗具有八种以上可以附着在红细胞上的抗原，其中大多数标记为狗红细胞抗原（DEA 1.1、1.2、3、4、5、6和7）。通常，某一特定品种狗的血型相同。例如，60%的灰猎犬属于DEA 1.1阴性血型。但新的犬类血型还在检测中——例如，最近发现的Dal血型仅在斑点狗中发现。

④大型小行星撞击地球是一件极其罕见的事情，但直接相撞的后果可能是灾难性的。一块直径150米的岩石可能释放出相当于几枚原子弹的能量。更大的物体则可能影响全世界范围内的生命。

⑤溶液填满岩石的缝隙，凝固过程就开始了。当二氧化碳接触到玄武岩中的钙、镁和铁，就会产生化学反应，转变成钙化白色晶体。

⑥CR400BF-G型复兴号高寒动车组具有耐高寒、抗风雪等特点，能够在-40℃的天气下运行。设计人员为设备间选择了密封性更好的材料，达到防雪、低温防冻的目的。

⑦研究团队通过运动捕捉、测力板和肌电图来记录肌肉组织的电活动，结果显示，在每一项任务中，该设备使腰部肌肉活动平均减少了15%至45%。

⑧强制加强叶酸摄入意味着在英国，每个吃面包等食物的人都会摄入一定的剂量。有人担心，对一些人，特别是老年人来说，增加叶酸摄入量可能会产生意想不到的负面后果，比如可能会掩盖维生素B12缺乏症的症状。

⑨One of the most important characteristics of the third-generation hybrid rice is that it has a shorter growing period. Some previous high-yielding hybrid rice varieties in China took 160 to even 180 days from sowing to harvesting, while the figure was shortened to around 125 days for the new variety.

3. 篇章翻译

（1）中译英。

①What are the Features of the HarmonyOS 2?

On June 2, Huawei officially released HarmonyOS 2 and a number of new products running on the operating system.

HarmonyOS, or Hongmeng in Chinese, is an open-source operating system designed for various devices and scenarios. It first launched on Internet-of-Things (IoT) devices, including wearables and tablets, in August 2019. As a next-generation operating system for smart devices, HarmonyOS provides a common language for different kinds of devices to connect and collaborate, providing users with a more convenient, smooth, and secure experience.

Zhao Xiaogang (assistant professor, School of Computer Science, Wuhan University) called HarmonyOS a "mega-terminal" that enables more streamlined and efficient cross-device connectivity. "HarmonyOS greatly enhances the interactive speed between devices and improves the efficiency of their computing power, thus providing customers with a more optimized cross-device user experience," said Zhao. "We are surrounded by more and more smart devices these days, and are now in a world where all things are connected," said Yu Chendong, managing director and CEO of Huawei's consumer business. "Every single one of us is a part of this fully connected world, as is every device. We look forward to working with more partners and developers to build a thriving HarmonyOS ecosystem," he said. Chinese industries, including home appliances, sports and fitness, travel, entertainment, and education, have welcomed the launch of HarmonyOS. In May, Midea Group, a leading Chinese home appliances maker, announced that it would roll out some 200 new products running on HarmonyOS by the end of this year. Huawei expects the number of devices equipped with HarmonyOS to reach 300 million by the end of 2021, including more than 200 million for Huawei devices.

②Progress in Quantum Network

Chinese scientists have created the world's first integrated space – to – ground quantum network that can provide reliable, ultrasecure communication between more than 150 users over a total distance of 4, 600 kilometers across the country, according to a study published in the journal *Nature* on January 7. Led by Pan Jianwei from the University of Science and Technology of China, the research was conducted by a group of scientists over the past few years. Reviewers of the study hailed the achievement as "impressive" and "futuristic", as it is the largest of its kind in the world. It also represents a major step toward building a practical, large – scale quantum internet, they added. In the quantum network, several services such as video call, audio call, fax, text transmission and file transmission have been realized for technological verification and real – world demonstrations, the paper noted, adding commercial use is expected in the near future.

③China's Chang'e –5 probe is preparing for a soft landing on the moon to undertake the country's first collection of samples from an extraterrestrial body. The lander – ascender combination of the spacecraft separated from its orbiter – returner combination at 4:40 on November 30, according to the China National Space Administration (CNSA). The spacecraft is performing well and communication with ground control is normal, CNSA said. The lander – ascender combination will execute a soft landing on the moon and engage in automatic sampling. The orbiter – returner will continue orbiting about 200 km above the lunar surface and wait for rendezvous and docking with the ascender. Launched on November 24, Chang'e –5 is one of the most complicated and challenging missions in China's aerospace history, as well as the world's first moon – sample mission in more than 40 years.

④Journalists covering the annual sessions of China's national legislature and top political advisory body are attracted by some virtual reality (VR) and augmented reality (AR) products. The products are on show at the ground floor of the hotel where the press center for the two sessions is located. The exhibition was set up by China Today Net Television. AR/VR industries are emerging in China. The eastern city of Nanchang launched a VR industrial base last year with an angel fund of 1 billion yuan, and the VR firms will be supported by 10 billion yuan of investment. Guo An, NPC deputy and Mayor of Nanchang, acknowledges that VR could change people's lives like the Internet and smartphone. He said: "We have established a complete industrial chain with more than 50 enterprises and organizations." To cultivate talent, the city cooperates with colleges and universities in training personnel. "We aim to train 10, 000 people with basic VR knowledge and skill in the next three years," he said. VR can be widely used in artificial intelligence, education and training, medicine, gaming, tourism and virtual communities.

（2）英译中。

①科学家发明生命计算器

科学家们已经研发出可以帮助预测老年人死亡时间的网络计算器。

这套算法被命名为"风险评估支持：社区老年人寿命预测工具（RESPECT）"。

研究数据基于2007年至2013年间超过49.1万名接受家庭护理的老年人，研究重点关注可能在未来五年内死亡的人群。

针对身体十分虚弱的人计算出来的预期寿命最短为四周。

研究对象会被询问是否被诊断出患有中风、痴呆或高血压等疾病，以及三个月内的任务完成能力是否有所下降。

他们还会被询问决策能力如何，是否出现呕吐、肿胀、呼吸急促、非计划性体重减轻、脱水或食欲不振。

研究人员发现，一个人日常行为能力的下降比其所患疾病更能预示六个月内的死亡概率。

Bruyere研究院和加拿大渥太华大学研究员艾米·许博士称："'生命计算器'可以让家人和他们的挚爱有所准备。""例如，它可以帮助成年子女计划什么时候休假陪伴父母，或者决定什么时候一起度过最后一个家庭假期。"

渥太华医院医生彼得·塔努塞普特罗说："了解一个人还能活多长时间，对他们应该接受哪些治疗及在哪里接受治疗等问题做出合理判断至关重要。""随着一个人离死亡越来越近，医疗护理从以治疗为主要目的开始向最大限度提高病人剩余生命质量倾斜。"

②按钮难题：安慰剂按钮之利弊

在很多年里，纽约市政部门逐步停用了该市控制人行横道信号灯的大部分按钮，但一直没怎么声张。他们认定，计算机操控的计时器更好用。到了2004年，3250个按钮中只剩不到750个还能发挥作用。但是，市政府并没有拆除已经没用的那些按钮，令无数手指白费一番力气。

一开始，按钮被留下来是因为拆除的成本问题。但后来却发现，即使无法控制信号灯的按钮也仍有用处。以色列本·古里安大学的塔勒·奥龙-吉拉德表示，按下按钮的行人不太可能在绿色信号灯亮起前就横穿马路。研究过人们在路口的行为后，她注意到大家更倾向于服从一个声称会听从他们的指令的系统。

密歇根大学安娜堡分校的人机交互专家埃坦·埃达表示，失效的按钮之所以会产生这种安慰剂效应（"安慰剂"的英文是placebo，placebo出自拉丁语，原意是"我会讨人喜欢"），是因为人们喜欢对自己所使用的系统有掌控感。他指出，自己的学生在设计软件时经常会加上一个可点击的"保存"按钮，但其实用户的输入都会自动保存，这个"保存"按钮仅仅是为了让对此不知情的用户放心而已。他说，不妨将这视作对抗机器世界里固有的冷漠的一种善意欺骗。

这是一种观点。但是，安慰剂按钮可能也有不利的一面，至少在过马路的问题上是这样的。研究交通系统心理因素的维也纳Factum研究所的负责人拉尔夫·里瑟尔认为，行人意识到按钮无效进而感到被骗的怨怒，如今已让这种做法弊大于利。

黎巴嫩发生的情况可引为佐证。2005年至2009年间在贝鲁特引入的过街按钮最后被证明是失败的。行人希望按键后能立刻亮起"步行"信号，而不是像通常那样等待交通灯的周期性转换。因此，市政部门停用了这些按钮，按预设的时间启动步行信号。曾在黎巴嫩政府担任高级交通工程师的查希尔·马萨德说，过街按钮无用的消息传开，随之而来的懊恼导致更多人乱穿马路。

马萨德说，贝鲁特正在拆除无效的过街按钮，三年内应该能全部清理完毕。纽约市交通运营处副处长乔什·本森表示，纽约也拆除了无效的过街按钮，但保留了大约100个有效的。这些按钮都设置在行人稀少到已经不适合使用自动交通灯的地方。然而，网上对安慰剂按钮的热议，导致人们对有效的按钮也开始怀疑起来，尽管这是受到了误导。本森表示，这种怀疑已蔓延到了纽约以外的地方，包括几乎所有过街按钮都总是有效（至少在非高峰时段是这样）的洛杉矶。

但事实上，无论是安慰剂按钮还是真的过街按钮可能都会很快消亡。越来越多的道路交汇处配备了摄像头或红外和微波探测器，可探测甚至统计等待过马路的行人。荷兰科技公司Dynniq近年在蒂尔堡市的一个十字路口安装了一套系统，能识别老年人或残疾人智能手机上的专门应用，为他们提供额外5秒到12秒的过马路时间。这的确会很讨人喜欢。

第五章　科技专利文献

（一）略

（二）翻译练习

1. 句子翻译
（1）本发明的目的是提高内燃机的功率和适用性。
（2）使用自动取款机的另一个问题是，对部分弱视或认知障碍的人来说大多数按键看起来都非常相似，这可能会使这些人更难有效地使用自动取款机。
（3）图1是并入控制阀组件的现有技术阀系统连接器的横截面图。
（4）根据本发明，可以制造出各种不同的口香糖。这些种类的口香糖包括含糖口香糖、无糖口香糖、泡泡口香糖、涂层口香糖和新奇口香糖。这些种类的口香糖可以形成颗粒状、棒状、片状或块状。

2. 篇章翻译
（1）中译英。
①The invention discloses a safety warm water bottle, which belongs to household articles, which solves the technical defect of the explosion risk of the warm water bottle in the prior art. It mainly comprises an inner tank, a shell, a vertical handle, a horizontal handle and a bottle stopper. The inner part of the inner tank is coated with thermal insulation material, the shell is used to protect the inner tank, the vertical handle is used to carry water, the horizontal handle is used to pour water when using boiled water, and the bottle stopper is used for thermal insulation. It is main-

ly used for the preservation of boiled water in family life.

Claims

a. The utility model relates to a safety warm water bottle, which comprises an inner tank, a shell, a vertical handle, a horizontal handle and a bottle stopper. It is characterized in that the inner tank is fixed by the upper part of the shell, and the lower part of the shell plays the role of a beautiful and balanced warm water bottle.

b. The safety warm water bottle according to claim a is characterized in that at the interface of the inner and outer shell, the lower part of the shell is designed as a screw, and the upper part of the shell is designed as a nut, which is fixed by the relative rotation between the upper part of the shell and the lower part of the shell.

c. The safety warm water bottle according to claim b is characterized in that four screws are added at the interface of the inner and outer shell, and there is a small hole on the outside of the bottom of the shell.

d. The safety warm water bottle according to claim a is characterized in that the upper part of the shell is designed with a protruding groove to protect the inner tank, the inner tank is placed in the protruding groove, the upper part of the shell is seamlessly connected with the protruding groove, and the interface between the upper part and the lower part of the shell is between the horizontal handle.

e. The utility model relates to a safety warm water bottle. In order to change the inner tank conveniently, the same opening is designed above the vertical handle on the upper part of the shell.

Instructions

Safety warm water bottle.

Technical field

The invention relates to a household article, in particular to a warm water bottle.

Background technology

In daily life, it is easy to break the warm water bottle. The rupture is small, but it hurts people. On the one hand, hot water hurts people, on the other hand, debris from the inner tank hurts people, and it is inconvenient to clean after the inner tank is broken.

Content of the invention

In order to solve the technical problems that the warm water bottle in the prior art is easy to break when in use, the debris of the inner tank after breaking is easy to hurt people, it is not easy to clean, and the boiled water will also cause certain damage, the invention provides a safety warm water bottle.

The shell of the warm water bottle in the prior art is also divided into upper and lower parts, but the connection of the upper and lower parts is at the bottom, which is prone to danger. The connection in the invention is between the horizontal handle, and it is safer to adopt two methods: self nut, screw fixation and external screws fixation. The inner tank is fixed by the protruding groove on the upper part of the shell. When combined, the protruding groove is located in the lower part of the shell. If the inner tank is broken, the inner tank debris can also be collected from the lower part of the shell to reduce the risk. Double openings, more convenient. The lower part of the shell also plays a decorative role, which accounts for the main outer surface. Therefore, some patterns can be drawn on it and made of synthetic leather.

②Vacuum Sintering Barium Titanate and Other Ferroelectric Materials

Invention background

The invention introduces a method for improving the dielectric properties of ferroelectric materials, in particular to the method for improving the dielectric properties of barium titanate.

Generally, in order to obtain ceramics with high dielectric constant, kiln sintering and hot pressing are two methods of external treatment of ferroelectric materials. In terms of kiln sintering, various atmospheres are used during calcination, such as oxygen, carbon dioxide, etc. The problem with calcining ferroelectric materials in kiln is that the obtained ceramic dielectric constant is not high enough and the target dielectric loss is large. Now press the hot pressing ferroelectric material method. The produced ceramics have been shown to be superior to those produced by kiln sintering, but the hot pressing price is expensive, and the mass production efficiency of electronic ceramics is not high.

Overview of the invention

The general purpose of the invention is to provide a method for treating ferroelectric materials, so as to obtain ceramics with excellent electrical properties, and provide an economic production method. It has been found that calcining the dried ferroelectric materials under the air partial pressure without reduction of ferroelectric materials can obtain ceramics with low price and good dielectric properties. The air partial pressure used depends on the specific ferroelectric materials, However, it must be carried out under the condition that the reduced air partial pressure of ferroelectric materials does not occur.

Comments on the best implementation scheme

Firstly, the raw material powder of a typical ferroelectric material (i. e., industrial barium titanate) is dry pressed and then burned under controlled atmosphere. In this example, the dry pressed barium titanate is calcined in a tubular furnace under a medium vacuum of $1-1,000\,\mu m$. The peak temperature of vacuum sintering is $1,200-1,400\,℃$, reaching the peak temperature of about 7h. Then cool to ambient temperature. When the dense furnace starts working, the medium vacuum starts.

The invention is also suitable for treating other ferroelectric materials, including alkaline earth metal titanate, such as strontium titanate.

（2）英译中。

①开始使用

将EarPods耳机插入iPhone、iPad或iPod touch（需要iOS10或更高版本），并将其插入耳朵。

重要安全信息

听力损失

用大音量听音乐可能会对你的听力造成永久性的损伤。背景噪声及持续暴露在大音量下，会让声音听起来比实际音量小。在佩戴耳机前请检查音量。有关听力损失及如何设置最大音量限制的更多信息，请访问苹果官方网站。

警告：为了防止可能的听力损伤，请勿长时间大音量听音乐。

驾驶危险

在驾驶车辆时使用耳机是不建议的，在某些地区是违法的。检查并遵守在驾驶车辆时使用耳机的相关法律法规。开车时要小心、专心。如果在驾驶任何类型的车辆或执行其他需要全神贯注的活动时发现音频设备干扰或分散注意力，请停止收听。

窒息危险

耳机可能会导致幼儿窒息或其他伤害。让耳机远离幼儿。

皮肤刺激

如果没有正确清洁耳机，可能会导致耳朵感染。用柔软的无绒布定期清洁耳机。避免任何开口处受潮，避免使用气溶胶喷雾、溶剂或研磨剂清理耳机。如果出现皮肤问题，请停止使用。如果皮肤问题没有解决，请咨询医生。

静电冲击

当在空气非常干燥的地方使用耳机时，很容易产生静电，并且耳朵可能会从耳机接收到少量静电放电。为将静电放电风险降至最低，请避免在极端干燥的环境中使用耳机，请勿在佩戴耳机前触摸接地的未上漆金属物体。

②阻抗变压器、集成电路设备、放大器和通信模块

对相关应用的交叉引用

［0001］本申请基于 2010 年 2 月 19 日提交的朊病毒日本专利申请第 2010 - 34751 号，并要求享有优先权，其全部内容通过引用并入本文。

领域

［0002］本文讨论的实施例涉及阻抗变压器、集成电路设备、放大器和通信模块。

背景

［0003］在通信模块（如雷达放大器和基站放大器）中使用的集成电路设备中，例如，多个集成电路并联耦合，并且增加集成电路的晶体管栅极的宽度，以建立高输出。

［0004］此外，阻抗变压器耦合到并联耦合的多个集成电路的输入端和输出端，以通过阻抗变压器中的匹配电路匹配阻抗。特别是，具有多串联耦合的四分之一波长线的阻抗变压器被广泛应用于需要宽带特性的集成电路设备中，以便通过增加四分之一波长线级的数量来获得宽带特性。

［0005］相关技术如日本公开实用新型登记公报第 5 - 65104 号、日本公开专利申请公报第 9 - 139639 号、日本公开专利申请公报第 10 - 209724S 号、S. B. Cohn 的 "Optimum Design of Stepped Transmission - Line Transformers"（*IRE Transactions on Microwave Theory and Techniques*，1955，3：16 - 21）和 E. J. Wilkinson 的 "An N - Way Hybrid Power Divider"（*IRE Transactions on Microwave Theory and Techniques*，1960，8：116 - 118）。

第六章　科技论文摘要

（一）略

（二）翻译练习

1. 句子翻译

（1）The research of robot technology has expanded from the traditional industrial field to new fields such as medical services, education and entertainment, exploration and survey, bioengineering, disaster relief and rescue and so on.

（2）The sparse model often utilizes training samples to learn an over-complete dictionary, in order to obtain the redundant and sparse representation of signals. Designing simple, efficient and universal dictionary learning algorithms is one of the main research directions at present. It is also a research hotspot in the information field.

（3）Alpha Go 和相应论文的成功确保了并行智能方法在复杂系统智能控制和管理及知识自动化方面的技术可靠性。

（4）在基于网络的控制框架下，采样测量值通过通信网络传输，通信网络可能会受到能量受限拒绝服务（DoS）攻击，攻击的特征是连续数据丢失的最大计数（弹性指数）。

2. 段落翻译

（1）英译中。

①北斗导航卫星系统（BDS）需要星间链路（ISL）调度，以保证系统的测距和通信性能。在 BDS 中，每天都必须处理大量的 ISL 调度实例，这肯定会通过常规的元启发式方法花费大量时间，很难满足现实应用中经常出现的快速响应需求。为了解决正常和快速响应 ISL 调度的双重需求，本文提出了一种数据驱动启发式辅助模因算法（DHMA），该算法包括一种高性能模因算法（MA）和一种数据驱动的启发式算法。在正常情况下，混合并行、竞争和进化策略的高性能 MA 会随着时间的推移为高质量 ISL 调度解决方案执行。在快速响应情况下，根据由高质量 MA 解决方案训练而成的预测模型，执行数据驱动的启发式算法以快速调度高概率 ISL。DHMA 的主要思想是分别处理正常和快速响应调度，同时训练高质量的正常调度数据以供快速响应使用。此外，本文还提出了一个易于理解的 ISL 调度模型及其 NP 完备性。对 10 080 个一分钟 ISL 调度实例进行的为期七天的实验研究表明，DHMA 在正常（84 小时）和快速响应（0.62 小时）情况下能够有效解决 ISL 调度问题，能够很好地满足实际 BDS 应用中的双重调度要求。

②随着海上活动的增加和海上经济的快速发展，第五代（5G）移动通信系统有望在海上部署。需要探索新技术，以满足海上通信网络（MCN）中超可靠和低延迟通信（URLLC）的要求。移动边缘计算（MEC）可以在 MCN 中实现高能效，但代价是高控制平面延迟和低可靠性。针对这一问题，本文提出了移动边缘通信、计算和缓存（MEC3）技术，将移动计

算、网络控制和存储汇聚到网络边缘。支持资源高效配置和减少冗余数据传输的新方法可以使计算强度和延迟敏感的应用程序可靠实现。在 MCN 中对 MEC3 实现 URLLC 的关键技术进行了分析和优化。采用基于最佳响应的卸载算法（BROA）优化任务卸载。仿真结果表明，任务延迟可以降低 26.5ms，终端用户的能耗可以降低到 66.6%。

③最近，领先的研究社区一直在调查区块链在人工智能（AI）应用中的使用，在人工智能应用中，多个参与者或代理协作做出共识决策。为此，区块链存储中的数据必须转换为区块链知识。我们将这些类型的区块链称为基于知识的区块链。基于知识的区块链在构建高效的风险评估应用程序方面可能很有用。早期的一项工作引入了概率区块链，它促进了基于知识的区块链。本文提出了概率区块链概念的扩展，提出了一个适用于此类区块链的声誉管理框架的设计。该框架的开发是为了满足广泛应用的需求。特别是，我们将其应用于检测恶意节点，并减少它们对概率区块链共识过程的影响。我们通过使用几种对抗策略将框架与基线进行比较来评估框架。此外，我们还分析了有恶意节点检测和无恶意节点检测的协作决策。这两个结果都显示了一个可持续的性能，其中提议的工作优于其他工作，并取得了优异的结果。

④人工智能加速是 IP 和系统设计中最活跃的研究领域之一。在云端和边缘引入专门的人工智能加速器，使得部署大规模人工智能解决方案成为可能，这些解决方案使以前没有人工智能无法完成的任务自动化。大数据的增长及处理这些数据以提供商业智能所需的计算能力，是一些公司获得竞争优势的关键。人工智能工作负载是数据和计算密集型的，提高效率通常需要端到端的解决方案。在本文中，我们确定了人工智能加速器设计的关键考虑因素。本文的重点是深度神经网络（DNN），以及如何通过微架构探索、节能存储层次结构、灵活的数据流分布、特定领域的计算优化及最终的软硬件协同设计技术来构建高效的深度神经网络加速器。互连拓扑的重要性及其扩展对人工智能加速器物理设计的影响也是本文描述的一个关键考虑因素。未来，这些加速器的能效可能依赖近似计算、内存计算和运行时灵活性以获得显著改善。

（2）中译英。

①After the utilization of electric vehicles on a large scale, the demand on their charging power will impact power grid to some extent. The factors related to the demand on charging power of electric vehicles are analyzed. Under a certain assumed conditions, according to statistical data of fuel vehicles and taking the probability distribution of some enchancement factors, a statistical model of electric vehicles'power demand is built. The mathematical expectation of single electric vehicle's power demand and its standard deviation are solved by Monte Carlo simulation, after that a method to compute total power demand of many electric vehicles is given. Taking the daily load curves of Bejing and Shanghai in a certain day in summer for example, the impacts of power demand of electric vehicles in different scales on original load curves are calculated. Calculation results show that the natural charging characteristic of electric vehicles makes maximum load of power grid increase in a certain extent. Not only the proposed statistical model is available for the research on the impact of electric vehicle on power grid, but also it is availabe for refernece to the design of the strategy to manage the charging of electric vehicles.

②This paper introduces in detail the research of visual tracking which is a hot spot currently in the domain of computer vision. Firstly, the applications of visual tracking in three areas including visual surveillance, image compression, and 3 - D reconstruction are discussed. Secondly, the research status of the art about visual tracking is introduced, especially the common approaches of visual tracking are shown. In order to explain these methods clearly, the problems of visual tracking are classified. Then two ways to research the visual tracking problem are presented, namely bottom - up and top - down. And the visual tracking algorithms are classified into four classes: area - based methods, feature - based methods, deformable - template - based methods and model - based methods. Finally, from the point of view of control theory, the difficulties of visual tracking are discussed that the algorithms should have robustness and accuracy, and be fast. Meanwhile, some future directions of visual tracking are also addressed.

③Information sensing is the basic function of Internet of Things (IoT), by which "Completely Sensing" is implemented. Information interaction is the goal of "Thing - to - Thing Interconnection" which supports the service and application of IoT. Along with the upsurge of IoT research, the research on the sensing network with the traditional wireless sensor network as the core is rapidly heating up, and a lot of research achievements has been made in information sensing and interaction. This paper analyzes the latest research progress in information sensing and interaction in the Internet of Things (IoT). Firstly, in terms of information sensing, the main methods of data acquisition and procesing are summarized from data collection, cleaning, compression, aggregation and fusion. Secondly, a basic information interaction model is proposed and the main information interaction techniques are discussed in detail. Thirdly, some active topics about information sensing and interaction, such as new sensing techniques, energy efficiency balance, information security and mobile sensing network, are addressed. Finally, we present the problems and challenges faced by the development of information sensing and interaction technique of IoT, and point out the future work in this area.

④This paper summarizes the development of Computational Fluid Dynamics (CFD), especially computational aerodynamics. This paper summarizes the achievements of CFD from the aspects of calculation method, grid technology, turbulence model and large eddy simulation, analyzes the existing problems and puzzles, and looks forward to its development trend. In the CFD calculation method, the central scheme, upwind scheme, TVD scheme, WENO scheme, compact scheme and discontinuous Galerkin finite element method are mainly introduced, and the principles and characteristics of different methods are systematically described. Grid technology includes structured grid, unstructured grid, hybrid grid and overlapping grid. Some key technologies of overlapping grid are discussed in detail. In the turbulence model, the current models are classified and introduced, including linear eddy viscosity model, second - order moment model, nonlinear model, transition model, DES method and SAS method. In the large eddy simulation method, some related research directions are discussed, including filtering method, subgrid model, convergence criterion, numerical scheme and so on. The paper also includes some research results of the author in related fields.

第七章　科技产品说明书

(一) 略

(二) 翻译练习

1. 中译英

(1)

Connecting the bluetooth keyboard

Step 1: Switch on the keyboard. The Bluetooth LED indicator light flashes for 5 seconds and then goes off again.

Step 2: Press the link/connect button. The Bluetooth LED indicator light flashes to indicate that the device is ready for connection. The keyboard is now ready for the connection to Samsung Galaxy Tab 10.1.

Step 3: Turn on and unlock Galaxy Tab. Open the setup menu through the application menu.

Step 4: Select "Bluetooth". If Bluetooth is off, activate it. If Bluetooth is enabled, the menu will display "Bluetooth keyboard!". Then select this device.

Step 5: Now Galaxy Tab displays a code to connect. Insert the code through the keyboard and press the Enter key to confirm. The connection will now be securely encoded.

Step 6: The wireless keyboard is now successfully connected to Samsung Galaxy Tab 10.1. Once successfully connected, the keyboard will save all connection data until it is connected to another device.

Note: The keyboard can also be used for iPad, iPhone, and iPod Touch. iOS 4.0 or higher version must be installed. For other Bluetooth – enabled devices, please verify the Bluetooth standard and compatibility before attempting to connect to the keyboard.

Charging the integrated battery

When the battery power is low, the "Power" indicator light will start flashing. Now it is time to charge the keyboard.

Step 1: Connect the USB charging cable with the mini USB plug to the keyboard's charging interface.

Step 2: Connect USB – A to the power adapter or the computer's USB interface.

Step 3: During the charging process, when the charging is completed, the charging status LED indicator light will light up and then gooff.

Energy saving sleep mode

If the keyboard is inactive for 10 minutes, it will change into sleep mode. To activate from sleep mode, you must press any key and wait 3 seconds.

Safety precautions

Please avoid the following situations:

Keep away from sharp objects, oils, chemicals or any other organic liquids.

Please do not place heavy objects on top of the keyboard.

Avoid open fire and high temperature.

Please do not expose to direct sunlight.

(2)

Operation

- Correctly wire and power on the device according to the connection diagram.
- This ZigBee device is a wireless receiver that can communicate with a variety of ZigBee compatible systems. This receiver receives and is controlled by the radio frequency signal transmitted by the compatible ZigBee system.

Join a ZigBee Network

Step 1: Please remove the device from previous ZigBee network if it has already joined a ZigBee network, otherwise network access will fail. For more information, please refer to the "Manuallyrestore factory settings" section.

Step 2: Select to add LED light from your ZigBee gateway and engage in network access mode. Please refer to the ZigBee gateway manual.

Step 3: Power on the device again to set it into network access mode (the connected LED light flashesslowly twice). The network access mode lasts for 15s. If the timeout occurs, please repeat this step.

Step 4: The connected LED light flashes 5 times and then stays on, then the device appears on the control interface of your gateway and can be controlled through the gateway control interface.

Manually restore factory settings

Step 1: If the "Prog" button cannot be operated, press the "Prog" button for 5 consecutive times quickly or power on the device for 5 consecutive times.

Step 2: The LED light connected to the device flashes 3 times to indicate that the device has successfully restored factory settings.

Note:

- If the device is already in the factory setting state, when the factory settings are restored again, the device will have no indication.
- After the device is restored to factory settings, all configuration parameters will be restored to the factory default state.

Safety & Warnings

Please do not power on when installing the device.

Please do not expose the device to moisture environment.

(3) Use the Toaster

- Make sure that the lever (B) is lifted. Turn the baking degree selection knob (D) to the desired setting.
- Turn theappliance on.
- Put the sliced bread into the toasting slot (A), and then press down the lever (B) until it is locked.
  Note: If the appliance is not not powered on, the lever (B) will not be locked.
- When the toasting is finished, the lever (B) will automatically pop up, raising the sliced bread inside at the same time.
- If the required baking degree is not reached, the baking degree can be increased by turning the baking degree selection knob (D).
- The toasting can be stopped at any time by pressing the Stop/Cancel button (I). Never lift the lever (B) upward to stop toasting.
  Caution: the toasting slots (A) become very hot when theappliance is running. Do not touch.

Defrost function

The frozen sliced bread can be toasted by pressing the Defrost button (F) and the lever (B) needs to be pressed down until it is locked. The baking time will be longer in order to achieve the required baking degree. The defrost indicator light will remain lit while the defrost function is in use.

Reheat function

If the baked bread is not hot enough, press down the lever (B) until it is locked, and then press the Reheat button (G) to reheat. Please note that this can only change the reheating time, not the baking degree. The reheat function can be stopped at any time by pressing the Stop/Cancel button (I). The reheat indicator light will remain lit while the reheat function is in use.

Doughnut function

The doughnut function allows for the toasting of bread, doughnuts and muffins etc. That is, only one side can be baked (the inside side), and the other side can only be heated (the outside side). Doughnuts and muffins need to be cut in half before toasting. Press down the lever (B) until it is locked, and then press the Doughnut button (H) to bake. The doughnut indicator light will remain lit while the doughnut function is in use.

Precautions

Do not use the appliance withoutbread (except for the first use).

Do not bake sliced bread that is too thin or broken.

Do not bake the food that is easy to scatter, in order to avoid increasing the difficulty of cleaning and causing fireworks.

Clean the debris tray regularly: debris can smoke or burn.

Do not force too large food into the toasting slot.

Do not insert knives, forks or other utensils into the toasting slot to move bread, which will damage the appliance and easily cause electric shock.

If the appliance is blocked, unplug the power plug and wait for the appliance to cool down completely. Then put the appliance upside down and shake it gently to take out the food inside.

Clean and maintenance

Before cleaning, please unplug the power plug and let theappliance cool down.

The outside of the appliance needs to be cleaned with a soft cloth. Do not use corrosive detergent for cleaning, which will damage the appliance surface.

Do not immerse in water.

Please unplug the power plug and clean the debris tray after each use.

Do not touch the appliance with sharp or metal utensils, especially the toasting slots. There is a risk of electric shock.

2. 英译中

（1） 重要警告

危险！

这是一种电器，因此必须严格遵守以下安全警告：

切勿用湿手接触电器。

切勿用湿手接触电器插头。

确保使用的电源插座随时可用，必要时拔下电气插头。

直接拔下电源插头，切勿拉扯电源线，以免损坏。

要完全断开电源，请将电气插头从插座上拔下来。

如果电器出现故障，请勿尝试维修。关闭电器，从电源插座上拔下插头并联系服务中心。

切勿用拖拉电源线的方法来移动电器。

用任何方式修改或改变电器特性都是危险的。

如果电源线损坏，必须由制造商或授权技术服务中心更换，以避免一切风险。

切勿使用插线板。

本电器必须按国家配电规则进行安装。

电源插座必须接地线，电工将来检查您的电路是否合格。

本电器不用于商业用途，仅供家庭使用。

一般保护措施

切勿将电器安装在含有天然气、油或硫黄的房间内。切勿安装在热源附近。

将电器与易燃物质（如酒精）或压力容器（如喷雾罐）保持至少61cm的距离。

切勿在电器顶部放置重物或高温物体。

请始终垂直或靠在一侧运输电器。记得在移动电器前先排空水箱。在运输电器至少6小时后方可启动电器。

用于包装的材料可以回收利用。因此，建议您将其放置在特殊的废物收集容器中。

切勿在室外使用本电器。
切勿在洗衣房使用本电器。
（2）
1. 安装定位座
使用两个 M5×12 内六角圆柱头螺钉、弹簧垫圈和平垫圈将定位座和垫片安装到后门框上（在整个门系统安装和调整结束后，通过攻螺纹孔调试定位座）。
2. 安装下滑道
用 3 个 M10×22 外六角螺钉、弹簧垫圈和大垫圈将下滑道安装到下门框的底部，在安装过程中特别注意要保证（72±1）mm 的安装尺寸。
3. 安装底部橡胶塞
将橡胶塞插入螺纹孔中，并用螺母固定（橡胶塞应在整个门系统安装和调整结束后安装）。
4. 安装驱动机构
用 4 个 M10×30 六角头螺栓、弹簧垫圈和大垫圈将托架连接到车身上。初始安装时，应在托架和车身之间预先放置 4 个调整垫片。当安装驱动机构时，携门架上的尼龙滚轮应安装在上滑道中。
5. 安装下支架和门扇
5.1 安装下支架
用 5 个内六角圆柱头螺钉（涂抹适量螺纹润滑剂）、8 个弹簧垫圈和 8 个平垫圈将下支架安装到门板上，将其初调至长孔的中间位置。
5.2 安装门扇
拆下门扇顶部的防护罩。
将携门架移至门框中间位置，然后抬起门扇，并将门扇下支架（部件 23）上的两个尼龙滚轮对准下滑道的滑道，再将门扇顶部靠近车身，直至靠在携门架（部件 24）上，将携门架（部件 24）上两个配合尺寸为 44mm 的定位块嵌入门扇顶部铝型材的定位槽中，将两个偏心轴粗调至垂直中间位置，使携门架上的 8 个长孔与门顶部的 M8 螺纹孔对准，然后用 8 个弹簧垫圈、平垫圈和涂抹了适量螺纹润滑剂的 M8×30 内六角圆柱头螺钉拧紧。
（3） 真空吸尘器的组装和充电
1. 将脚插入真空吸尘器主体，直到听到"咔嗒"声。
2. 将充电适配器插头插入机器背面的充电端口。
3. 将适配器插入墙上的插座。首次使用前，请将机器完全充电至少 4 小时。

LED 状态
充电状态下

电池状态	指示灯状态
充电	红灯每秒钟闪烁一次
充满电	绿灯每 10 分钟亮一次，然后每 1 分钟闪烁一次
充电器、电池或电动机错误	红灯和绿灯同时闪烁，请联系消费者服务部

清空污物箱
1. 确保关闭真空吸尘器。按下手持式真空吸尘器顶部手柄上的释放按钮，将其拆下。
2. 垂直握住手持式真空吸尘器，并按下前面的释放按钮以释放污物箱。
3. 抓住过滤器凸片并向上拉以卸下过滤器组件，并将污物倒入废物容器中。
4. 更换污物箱中的过滤器组件，然后将污物箱卡回手持式真空吸尘器上的适当位置。

警告：为降低火灾、触电或人身伤害的风险，在进行维护或故障排除之前，请关闭电源并从电源插座上拔下插头。

清洁过滤器
1. 按照"清空污物箱"部分中的说明，关闭电源并拆下污物箱。
2. 抓住过滤器凸片并提起，以从污物箱中取出可清洗的过滤器组件。握住过滤器组件，逆时针旋转并下拉以从过滤器滤网上取下过滤器。
3. 用力敲击废物容器的内部以去除任何可见的污垢。
4. 用温水冲洗过滤器和滤网以进行清洁。让它们完全干燥。
5. 将过滤器更换到滤网中，并顺时针旋转锁定到位。将过滤器组件重新连接到污物箱中，并将污物箱连接到手持式真空吸尘器上，直到其牢固地卡入到位。

第八章 科技会展文案

（一）略

（二）翻译练习

1. 中译英练习一

The 24th China Beijing International High-tech Expo

The 24th China Beijing International High-tech Expo (CHITEC) is a large-scale national-level international science and technology exchange and cooperation event co-hosted by a number of national government agencies, including China's Ministry of Science and Technology, the China National Intellectual Property Administration, the China Council for the Promotion of International Trade (CCPIT) and the People's Government of Beijing Municipality, and undertaken by CCPIT BEIJING, approved by the State Council.

Exhibition display

The main exhibition venue is located in the China International Exhibition Center (Jing'an Zhuang Building) with an area of 50,000 square meters, and 12 special exhibition areas are set up: international exhibition area, cutting-edge technology hotspot exhibition area, science and technology Winter Olympics exhibition area, Beijing high-precision industrial innovation achievement exhibition area, artificial intelligence exhibition area, robot exhibition area, financial technology exhibition area, smart education exhibition area, capital culture and technology integration development achievement exhibition area, technology industry functional area innovation achievement exhibition area, capital youth science and technology innovation and entrepreneurship achievement exhibition area, capital automobile science and technology exhibition area, and provincial, regional and municipal science and technology innovation achievement exhibition area.

Online display and promotion will use the official website of CHITEC to provide online promotion of excellent companies and products, and online docking of investment cooperation projects.

Exhibitor guide

1. Time and place

Exhibition time: September 16 – 19, 2021

Venue: China International Exhibition Center

Sponsor: The People's Government of Beijing Municipality, Ministry of Science and Technology of the People's Republic of China, China National Intellectual Property Administration

Exhibition scale: 50,000m^2

2. Related activities

Opening Ceremony and Keynote Report; Party and State Leaders Visiting Exhibitions; Product Release/Promotion Meeting; Project Release and Procurement Session; Excellent Project Selection Activities; International Electronic Technology Forum; Network Information Technology Seminar; Educational Equipment and School Matchmaking Meeting; Robotics Forum; 3D Printing Conference; Internet of Things Technology and Application Forum; International Financial Forum; VR/AR Large – scale Experience Event.

3. Advantages of CHITEC

After 23 years of cultivation and development, CHITEC has become a comprehensive science and technology event with extensive international influence, an important platform for scientific and technological exchanges and cooperation between China and other countries in the world, and one of the most representative and authoritative major international expositions in the field of science and technology economy and trade in China. It has played an active role in displaying the latest scientific and technological achievements at home and abroad, disseminating cutting – edge ideas, and promoting scientific and technological exchanges and cooperation.

(1) Top annual regular exhibition: Approved by the State Council, jointly sponsored by eight ministries and commissions, leaders of the Party Central Committee, the National People's Congress, the State Council, the National Committee of the Chinese People's Political Consultative Conference, the Central Military Commission, and relevant ministries and commissions have visited CHITEC many times.

(2) Comprehensive science and technology event: The 23rd CHITEC held 15 promotion and negotiation sessions, 12 professional forums, and several explanatory sessions. These rich, rational and pragmatic investment promotion activities and professional activities for seeking industrial cooperation have achieved remarkable results, leading to the signing of 312 scientific and technological cooperation and technological achievements trading projects, with a total agreement value of 96 billion yuan.

(3) First-class business platform: The 23rd CHITEC attracted 230,000 domestic and foreign visitors, more than 80 overseas delegations from 9 international organizations and more than 37 countries and regions, government delegations from 32 provinces, regions and cities and cities under separate state planning participated in CHITEC, and more than 2,000 multinational companies, domestic industry leaders, large backbone enterprise groups and high-growth small and medium-sized enterprises participated in the exhibition.

4. Exhibition area

(1) Consumer electronics

VR/AR related products, digital products, automotive electronic products, cloud computing and technology applications, smart home, smart home appliances and electrical appliances, mobile smart terminals and peripherals, wearable devices, LED and lighting, electronic manufacturing, brand area.

(2) Educational equipment and online education

Education and teaching digitization, informatization; network education robot; Internet education and platform; teaching and campus audio-visual broadcasting network system; digital classroom, digital campus; teaching equipment; computer, audio-visual equipment; experimental equipment system; popular science education; teaching software.

(3) Intelligent technology

Smart city, smart park; smart logistics; smart transportation; smart home; security and monitoring technology; smart lighting; smart LED lighting; cloud computing; cloud storage; sensors, identification technology; short-range communication technology and products; management system software; Internet of Things (IoT) demonstration applications; other IoT components.

(4) Electronic information and modern communication

Network information technology and solutions, integrated circuits and electronic components, complete sets of electronic information equipment, modern communication equipment, power electronic devices, laser and optoelectronic devices, opto-mechatronics, liquid crystal display, Internet and e-commerce.

(5) Robot technology

Industrial robots, educational robots, service robots, special robots, cleaning robots, medical robots, inspection and maintenance robots, construction robots, underwater robots, robot components, robot suppliers and distributors.

(6) 3D printing technology

3D printer; 3D printer manufacturing equipment; 3D printing technology; 3D laser equipment; 3D printing control equipment; measuring equipment; reverse engineering software and technology; other rapid prototyping technologies; related parts, accessories; 3D printing consumables.

(7) Environmental protection and new energy industry

Environmental management, pollution control and reduction, new materials and new energy, new chemical materials, functional metal materials, new building materials, optoelectronic materials, solar thermal utilization, new energy-saving technologies.

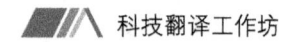

(8) Modern engineering and advanced manufacturing technology

Engineering and processing machinery, micro machinery; modern publishing and printing equipment; scientific instrument and testing and control equipment; medical equipment; modern urban construction and transportation engineering.

5. Exhibiting fees

Indoor booth

RMB ¥1,600.00/square meter

Catalogue and advertising costs

Front cover: 40,000 yuan   Back cover: 20,000 yuan   Second cover: 18,000 yuan   Third cover: 16,000 yuan

Color full page: 6,000 yuan

Truss advertisement: 350 yuan/m^2 Sling: 20,000 yuan/exhibition   Wall advertisement: 260 yuan/m^2

Channel landmark advertisement: 3,000 yuan/6 pieces   Visiting ticket: 10,000 yuan/10,000 pieces   Visiting card: 20,000 yuan/exhibition   Gift bag: 15,000 yuan/1,000 pieces

6. Exhibiting procedures

(1) Fill in the "Exhibition Application Form" as required and return it to the exhibition organizer. Registration for participation by fax is also acceptable. Please note that the deadline is September 10, 2021.

(2) After receiving the "Exhibition Application Form", the exhibition organizer will send the official contract in duplicate to the exhibiting company for countersignature.

(3) According to the requirements of the pro forma invoice issued by the organizer, the exhibiting company shall transfer 50% of the booth rent through bank wire as the deposit (RMB) or pay the whole amount at one time to confirm the booth location. The balance of the booth rent should be remitted no later than September 10, 2021.

(4) After confirming the booth, the organizer will send the Exhibitor Manual to the exhibiting company, which includes information on exhibit transportation, booth design and construction, travel and accommodation arrangements, item rental and waiters, advertisements, and visa applications. Exhibitors must fill in the relevant forms in the manual as required and return them to the organizer before the deadline.

(5) The reserved booth can only be implemented after receiving the booth reservation deposit. Booth allocation is based on the principle of "booking and payment first, and confirmation first" until sold out.

Organizing committee office:

Exhibition business

Li Wenjing: 15652230005

Tel: 010 – 53515097

QQ: 2638342269

E – mail: ddgjexpo@ sina. cn

http://www.vanzol.com/chitec/

2. 中译英练习二

The 131st China Import and Export Fair (Canton Fair)

The 131st China Import and Export Fair (Canton Fair) will be held online from April 15 to April 24. Themed facilitating the "dual circulation" of domestic and overseas markets, the exhibition offers anonline display platform, business matchmaking services, and a cross – border e – commerce zone. Exhibitors and Products, Global Business Matchmaking, New Product Release, Exhibitors on Live, VR Exhibition Hall, News & Events, Services & Support and other columns are set up on the official website, with 16 product categories in 50 exhibition sections displayed. More than 25,000 overseas and domestic exhibitors will take part in. Among them are companies from formerly poor regions, which will display products in the Rural Vitalization zone.

Focusing on improving the effectiveness of trade connection and user experience, the 131st Canton Fair has taken multiple measures to further optimize and enhance the functions of the online platform, and facilitate the interaction between exhibitors and sourcing companies, as well as stimulate trade transactions. We welcome companies and buyers at home and abroad to participate in, share business opportunities, and seek common development.

第九章 科技文章

(一) 略

(二) 翻译练习

1. 英译中

(1) 段落翻译。

①石油一般发现于地下深处，只通过对地表的研究是不能确定石油的存在的。因此，必须进行地下岩石结构的地质勘测。如果认为某地区的岩石含有石油，就要装配钻机。钻机最明显的部分是被称为井架的高塔。井架被用来起吊一节节钢管，然后将钢管下放到钻孔中。钻孔时，将钢管压进孔道，既能防止四壁向内塌陷，又能防止水灌进孔道。一旦发现了石油，钢管的顶端就被牢牢地套上一个盖子，这样石油便通过一系列的阀门源源不断地喷出来。

②虚拟网络是一种逻辑网络，用户在其中表现出自己的访问行为。虚拟网络依赖于像因特网这样的物理计算机网络，但又具有不同的拓扑结构，且对物理网络造成重大影响。人们利用新型双层耦合模型来研究虚拟网络给因特网整体特性带来的影响。结果表明，节点数据包的队列长度存在相变特性。此外，相变临界点左移，网络性能恶化。在自由流中，节点之间相互独立或与短程相关。在临界处，节点与长程相关，且幂指数更大，意味着长程相关性更强。当系统状态位于临界点右侧时，虚拟网络行为使网络呈现一致的长程相关特性。

③振动阻尼支架可使灵敏的分析天平和其他仪器免受振动干扰，使其准确地发挥功能。它将附近的泵、搅拌机和重型车辆引起的振动降低至9Hz。它由黑白水磨石制成，表面经抛光处

理，抗划痕和化学物质，由四个带氯丁橡胶支脚的减震器支撑。总高度为76mm，载重量可达16kg。

（2）篇章翻译。

①这位27岁的病人前景黯淡。2016年5月，他发现自己患有艾滋病。两周后，他被告知患有急性淋巴细胞白血病。

但根据《新英格兰医学杂志》（*The New England Journal of Medicine*）最新发表的一篇论文，医生们给这位中国公民带来了一线希望：通过骨髓移植来治疗他的癌症，并进行另外的实验性治疗，试图清除他体内的艾滋病病毒。

参与研究的北京大学科学家说，治疗手段包括使用基因编辑工具CRISPR – Cas9从捐赠者的骨髓干细胞中删除一种名为CCR5的基因，然后将干细胞移植到患者体内。

"经过编辑后，这些细胞——以及它们产生的血细胞——有能力抵抗艾滋病病毒感染。"首席科学家邓宏魁在上周五（9月13日）告诉美国有线电视新闻网。

携带有发生突变的CCR5基因拷贝的人对艾滋病病毒具有很高的免疫力，因为艾滋病病毒利用这种基因产生的蛋白质进入感染者的细胞。"柏林患者"和"伦敦患者"在接受了来自天生携带有这种基因突变的捐赠者的骨髓移植后，成为世界上首批被治愈的艾滋病病毒感染者。

在患者同意后，实验于2017年夏天进行。这是基因编辑工具CRISPR – Cas9首次用于艾滋病病毒患者。2019年初，也就是接受治疗整整19个月后，"急性淋巴细胞白血病完全缓解，删除了CCR5基因的供体细胞持续存在"，科学家在论文中说。

但这还不足以消灭病人体内的艾滋病病毒。研究人员说，移植后，只有大约5%到8%的患者骨髓细胞携带这种经过CCR5编辑的基因。邓宏魁说："在未来，进一步提高基因编辑的效率和优化移植程序应该会加速向临床应用的转变。"

但他不认为这个治疗是一次失败。"这项研究的主要目的是评估基因编辑干细胞移植治疗艾滋病的安全性和可行性。"邓宏魁说。

他认为这个实验很成功：科学家们没有发现任何与基因编辑相关的不良事件，邓宏魁说，即使"需要更长期深入的研究来进行脱靶效应和其他安全评估"。

《自然》杂志今年6月发表的一篇论文称，CCR5基因突变与早逝风险增加21%有关，不过原因尚不清楚。

进行这项研究的团队此前曾将CCR5编辑过的人类细胞移植到小鼠体内，使其对艾滋病病毒具有抵抗力。美国科学家已经在人类身上进行了类似的实验，并取得了一些成功，他们使用了一种更古老的被称为锌指核酸酶技术的基因编辑工具。

中国在基因编辑技术上投入巨资，将生物技术列为2016年宣布的五年规划（"十三五"规划）的重点之一。中国政府资助了多项堪称世界"第一"的研究，包括2016年首次在人类身上使用基因编辑工具CRISPR – Cas9，以及2015年首次使用基因编辑技术修改无法存活的人类胚胎。

邓宏魁仍然是基因编辑工具CRISPR – Cas9的坚定信徒。他认为这将为血液相关疾病带来"新的曙光"，比如艾滋病、镰状细胞贫血、血友病和β – 地中海贫血，而且由于这项新技术，"人们距离功能性治愈艾滋病的目标越来越接近"。

② <center>自闭症</center>

与男性相比，患有认知障碍的女性较少，因为她们自身的身体能更好地忽略导致认知障碍的基因突变。

自闭症是一种奇怪的状态。有时它的"社会失明"症状（无法阅读或理解他人的情绪）单独出现。这被称为高功能自闭症或阿斯佩格综合征。虽然他们的男性和女性同伴会认为他们有点奇怪，但高功能自闭症患者通常是社会成功人士（有时非常成功）。然而，在另一些场合，自闭症表现为一系列认知问题的一部分。然后，人的状态会逐渐衰弱。对那些所有被称为自闭症谱系障碍的人来说，共同点是男性远多于女性，以至于有一派认为自闭症是男性在精神上的极端表现。男孩被诊断为自闭症的可能性比女孩高四倍。至于高功能自闭症，比例达到7比1。

此外，自闭症的真实情况在较小程度上也适用于许多神经和认知障碍。被诊断为注意缺陷多动障碍（ADHD）的男孩大约是女孩的三倍。"智力残疾"是先天性智商低下的统称，在男孩中更常见30%～50%，癫痫症也是。事实上，这些疾病经常共同出现。例如，被诊断为自闭症谱系障碍的小孩经常也会被诊断为ADHD。

导致自闭症的确切原因尚不清楚，但是基因是很重要的原因。虽然目前还没发现导致自闭症的基因突变，但已知有100多个基因突变使携带它们的人更容易患自闭症。

大多数这些突变在男性和女性中是一样普遍的，所以一个对发病率差异的解释是对于同等的基因毁坏，男性大脑比女性的更容易受伤害。这被称为女性保护模式。另一个普遍的解释是社会偏见理论，认为这些不同是虚幻的。女孩被诊断不足是因为她们被评估的方式或应对疾病的方式不同，而不是她们真的很少有这个症状。例如，一些研究者声称，女孩能更好地隐藏这些症状。

为了调查这个问题，University Hospital of Lausanne的Sebastien Jacquemont和他的同事分析了两组有认知异常症状孩子的基因数据。一组有800人，明确患有自闭症；另一组有16 000人，有一系列问题。

Jacquemont医生刚将他的研究成果发布在《美国人类遗传学杂志》（*American Journal of Human Genetics*）上。他的关键发现是两组中，女孩比男孩更容易发生与神经异常发育相关的突变。无论是拷贝数目变异（CNV，即DNA特定片段的染色体拷贝数目的变异）还是单核苷酸变异（SNV，即DNA信使中单个遗传字母的改变）都是如此。

从表面上看，这似乎是令人信服的女性保护模式证据。由于Jacquemont医生检验的所有孩子的数据都被诊断出有问题，如果比起男孩，女孩有更严重的突变，那就表明她们生理机能的其他方面掩盖了结果。因此，如果这个解释是正确的，女性比男性更容易受到保护，而不出现症状。而且，作为进一步的证实，Jacquemont医生的发现与一个三年前发布的研究成果相符，该研究发现CNV在自闭症女孩中横跨多个基因（也因此可能更具破坏性），比患自闭症男孩的多。

相反的观点认为，如果女孩能更好地隐藏她们的症状，那么在被诊断的小组中，只有更极端的女性案例会出现。如果这是正确的，那么在有症状的女孩身上可能有更大程度的突变。然而，Jacquemont医生和他的同事同样也发现，破坏性CNV遗传于母亲的可能性大于他或她的父亲。他们解释这进一步证明了女性保护机制。自闭症症状使男女双方都不太可能成为父母。如果母亲是孩子中大多数自闭症诱导基因的来源，那么这表明他们受这些基因的影响较小。

然而，这些都不能解释男孩比女孩更易受到影响的确切机制。在这个问题上，也有两个主要的理论。一个理论认为男性更敏感,因为他们只有一条 X 染色体。这使得他们更容易受该染色体突变的影响,因为任何受损基因都没有另一条"双胞胎"X 染色体来掩盖它。由于这个原因,一种认知障碍,即脆性 X 染色体综合征,在男性中确实更常见。然而,Jacquemont 医生的研究发现,X 染色体突变的作用有限。这表明该差异的遗传基础是分布在整个基因组中的。

另一个理论则是解剖学意义上的。它基于男性和女性大脑内部连接模式差异的脑成像研究。男性大脑比起女性大脑而言,有较强的本地连接、较弱的长范围连接。这与自闭症患者和非自闭症患者的大脑差异相似。这表明,男性类型连接模式在某种程度上更加易受引发自闭症和其他认知障碍因素的干扰。然而,为什么会这样,仍是未解之谜。

2. 中译英

(1) 段落翻译。

①The successful launching of China's first experimental communication satellite, which was propelled by a three-stage rocket and has been in operation ever since, indicates that our nation has entered a new stage in the development of carrier rockets and electronic technology.

②When the graphite crystal is crushed, it breaks into micro-scaled laminar powder. There are many dangling bonds on the edge of an isolated graphene sheet which elevates its energy and makes it unstable. When a graphene sheet curls into a carbon nanotube, the number of dangling bonds decreases and the system energy is reduced accordingly.

③Tongue diagnosis is an important diagnostic method in traditional Chinese medicine. However, one important problem in tongue diagnosis is that its practice is subjective and difficult to describe quantitatively. With the development of computer technology, it is a trend to utilize the image processing and pattern recognition technology in aid of tongue diagnosis.

(2) 篇章翻译。

5G is actually the abbreviation of the fifth generation of mobile communication technology. 1G was applied in the 1970s and 1980s, mainly represented by the cellular phone, with a single function—mainly voice communication. 2G had the function of voice communication and text, such as phone call and SMS. We can download pictures via 3G. We can download videos directly from the phone via 4G. Compared with previous generations of mobile communication technology, 5G has the characteristics and advantages of internet of everything, high speed, ubiquitous network, low delay, low power consumption, reconfiguration security, etc.

The development of 5G has a significant impact on various industries, and has made a huge chemical effect in the integration of various industries. 5G drives the development of the global economy, and 5G also drives the development of the industry chain: the upstream is mainly about the base station upgrade (including base station RF, baseband chip), the midstream is about the network construction (network planning and design companies, network optimization and maintenance companies), the downstream is about the product applications and terminal product application scenarios (cloud computing, Internet of Vehicles, Internet of Things, VR/AR). 5G will focus on upgrading its own network to meet the demand for services emerging from various related fields. 5G has a broad prospect for use in communications, smartphones, and the Internet of Things, and is a milestone breakthrough in the field of mobile communications technology. In the traditional Internet era, portal websites were generally not very relevant to traditional industries, while the mobile Internet has contributed to the reform and innovation in many industries, making the world more efficient and the organization flatter.

5G will not only change our life, but also will greatly change the social and economic situation. 5G era will witness significant changes in cultural production, dissemination and consumption, mainly in the following aspects. Firstly, the acceleration of Internet speed will promote cultural content producers to present more diversified characteristics, from creators of cultural enterprises and non-profit cultural units to cultural creation freelancers, and even everyone may become cultural content producers. Secondly, cultural production and dissemination will pay more attention to the individual cultural needs, showing personalized characteristics. And the matching between product supply and consumption will be greatly improved, as well as the cultural consumption experience and service quality. Thirdly, the trend of digitalization of the cultural industry will be strengthened. Fourthly, 5G will drive the emergence of more applied programs and application scenarios, which will be widely used in radio, film, television, video, theme parks and other tourism projects, animation and game industries. VR, AR and other technologies will become more developed and mature, and more vivid and interesting leisure industries and projects will be developed, so that the consumers' cultural and entertainment experience will be upgraded. Fifthly, 5G will make the products and technologies of 4G era more improved. For example, TV and video HD technology will make mobile newspapers, mobile TV, mobile games and other products more popular. Sixthly, the use of mobile phone network will cost less and it can cross the spatial boundaries of busy cities and remote mountainous areas, enabling ubiquity and low prices, thus promoting the growth of mobile phone-derived cultural consumption and cultural industries. Seventhly, the integration of 5G with AI (Atificial Intelligence) and IoT technology to form a super Internet will generate new cultural industry business models and products in this integration process, which will bring important applications and impacts to smart homes, smart cities, creative cities, and traditional cultural industries (such as the performing arts industry and the film industry). Eighthly, it can reduce the production and distribution costs of the cultural industry. Ninthly, it can make the large-scale dissemination of cultural products faster and wider, and weaken the barrier of international spatial distance for cultural experience and caltural consumption.

第十章　科技产品简介

(一) 略

(二) 翻译练习

1. 词组翻译

superior quality

finely processed

complete in specifications

attractive design

to have a long standing reputation

strong resistance to heat and hard wearing

packaging diagram

spare no cost/at all costs/at any cost

waterproof, shock – resistant and antimagnetic

crease – resistance

with traditional methods

酒精温度计

空气热交换器

模拟指示器

喷雾加湿器

热风供暖器

吸气式湿度计

自然通风冷却塔

风冷式空调器（制冷系统配有风冷式冷凝器的空调器）

绝热交换器

空气挡板

空气套冷凝器

空气制冷机（利用压缩机膨胀的制冷机器）

雷达测高仪

激光地形仪

电荷耦合器件

摄影测量仪器

电子经纬仪

电磁波测距仪

双频测深仪

微波测距仪

高效空气过滤器

2. 句子翻译

（1）与其他牌子相比，这种轮胎每英里损耗较少，也耐磨一些，因为它是用一流橡胶做成的。

（2）由于我们的产品具备了您所需要的所有特性，而且比日本产品便宜20%，所以我向您极力推荐这款产品。

（3）这种牌子的真空吸尘器在国际市场上颇具竞争力，是同类产品中最畅销的。

（4）我们生产的计算机的特点是质量好、体积小、节能，而且易学好用。

（5）一旦使用该机器，公司将会增产30%，而且一个人可以顶三个人使用。

3. 篇章翻译

（1）iPad Air 是保持 WiFi 和 LTE 连接的完美方式。无论身在何处，你都可发起 FaceTime 视频通话，加入视频会议，或者与好友、同学合作开展小组项目。而有了先进的摄像头和麦克风，面容看着很真切，声音听着也清晰。iPad Air 拥有超快的 WiFi 6 功能，而通过 LTE 连接，即使在没有 WiFi 的地方，你也能进行连线。

iPad Air 灵活多用，功能丰富，用它来工作格外得心应手。你可以配个键盘式智能双面夹，也可以配个妙控键盘，把灵敏的键盘和内置触控板同时用起来。无论是日常收发邮件，还是心心念念的短篇小说创作，这些事用 iPad Air 做起来特别适合。再加上满足一天所需的电池续航，它随时能与你并肩工作，撑你到底。

A14 仿生惊人的图形处理性能，配上 iPad Air 绚丽的显示屏，带给你完全沉浸其中的娱乐体验。你可以在支持 P3 广色域的 Liquid 视网膜屏上，看看大片；可以通过高品质横向立体声扬声器，倾听音乐；还可以尽享主机级游戏的震撼画面，甚至可以连上手柄来玩。

A14 仿生芯片的中央处理器提速 40%，图形处理器快了 30%，新一代神经网络引擎让机器学习性能提升到过去的两倍。真正强大的科技，应该是让每一个人都能使用的科技。因此，iPad Air 内置了一系列辅助功能，以满足对视力、听力、肢体活动能力与认知能力有不同需求的人士。

（2）一名工业设计专业的学生发明了一款高科技产品"第三只眼"，这只"眼睛"可以贴在人的额头上，当他们边走路边埋头看手机时，这只"眼睛"可以帮他们留意前方的障碍物。

无可否认，智能手机已经成为现代生活不可分割的一部分。多数人每天都花几个小时看手机，有些人甚至在走路或开车的时候也看手机。你应该还见过有人因为看手机掉进喷泉或洞里的搞笑视频，说不定你自己也经历过类似的事情。不过，有了设计师彭敏旭的"第三只眼"，你就可以一边走路一边发短信或刷社交媒体，不用担心发生事故。

彭敏旭发明的"第三只眼"包括一个半透明的塑料盒，用薄胶垫就可以直接贴在佩戴者的额头上。塑料眼睛内置了一个小扬声器、回转传感器和声波定位传感器。当回转传感器检测到用户低头时，就会开启第三只眼的塑料眼皮，声波定位传感器就会开始监测用户前方区域。当传感器检测到障碍物时，就会通过连接的扬声器警告佩戴者。

设计师表示："看起来像瞳孔的黑色组件是感知距离的超声传感器。当用户前方有障碍物时，超声传感器会检测到障碍物，并通过连接的蜂鸣器通知用户。"

彭敏旭告诉建筑设计杂志 *Dezeen* 说，他的"第三只眼"是首个试着想象未来"手机人类"面貌的项目。手机已经在改变我们的身体，而它们不过才出现了一二十年，想象一下手机会对未来几代人产生什么样的影响吧！

彭敏旭称："使用手机的不良姿势已经导致人们的颈椎前倾，造成龟颈综合征，托住手机的小指也逐渐变得弯曲。几代过后，使用手机带来的这些小改变将会积少成多，创造出一种截然不同的新人类。"

参考文献

［1］HAGGARD T R, KUNEY G W. Legal drafting: process, techniques, and exercises［M］. 2 nd ed. St. Paul, Minn.: Thomson West, 2007.

［2］NEWMARK P. A textbook of translation［M］. Upper Saddle River, N. J.: Prentice Hall, 1988.

［3］NORD, C. Text analysis in translation: theory, methodology, and didactic application of a model for translation – oriented text analysis［M］. 2nd ed. Amsterdam: Rodopi Bv Editions, 2005.

［4］ROHWER C D, SCHABER G D. Contracts in a nut shell［M］. St. Paul, Minn.: West Publishing Co., 1997.

［5］SHEVLIN C. Writing for business［M］. London: Penguin Books Ltd., 2005.

［6］曹怀军. 中国专利翻译研究：回顾与展望［J］. 上海翻译, 2022（1）: 21 – 26, 95.

［7］陈美莲. 谈科技英语中被动语态的翻译技巧［J］. 辽宁行政学院学报, 2006, 8（6）: 174 – 175, 177.

［8］本书编写组. 英语应用文写作大全［M］. 北京：社会科学文献出版社, 2003.

［9］戴玉霞, 石春让. 网络科技新闻标题英译汉常见问题及优化策略［J］. 中国科技翻译, 2016, 29（3）: 26, 33 – 35.

［10］邓军涛, 许明武. 科技论文摘要汉译英典型问题探究［J］. 中国科技翻译, 2013, 26（1）: 11 – 14.

［11］段薇. 论英文家电说明书的特点和翻译方法［J］. 科技信息, 2010（9）: 603, 633.

［12］樊才云, 钟含春. 科技术语翻译例析［J］. 中国翻译, 2003（1）: 59 – 61.

［13］费一楠. 浅谈专利摘要中有益效果的翻译技巧［J］. 中国发明与专利, 2013（9）: 72 – 74.

［14］方茜. 科技英语的文体特征与翻译技巧［J］. 长春教育学院学报, 2017, 33（11）: 46 – 48.

［15］韩琪, 朱宇, 郑雨欣, 等. 翻译目的论视角下的科技产品广告汉译——以苹果公司广告语为例［J］. 湖北经济学院学报（人文社会科学版）, 2019, 16（6）: 118 – 120.

［16］韩琴. 科技英语特点及其翻译［J］. 中国科技翻译, 2007, 20（3）: 5 – 9.

［17］何其莘, 仲伟合, 许钧. 科技翻译［M］. 北京：外语教学与研究出版社, 2012.

［18］侯宝晶. 科技新闻英汉翻译与文化语境［J］. 科技信息, 2009（28）: 136, 138.

［19］黄翀, 赛音托娅. 专利文献翻译模式和翻译技巧探析［J］. 中国翻译, 2016, 37（2）: 105 – 109.

［20］黄翀，张奇. 浅析专利文献翻译及其对技术创新的影响［J］. 中国翻译，2019，40（1）：167-173.

［21］胡庚申，王春晖，申云桢. 国际商务合同起草与翻译［M］，北京：外文出版社，2001.

［22］贾和平. 企业简介英译中的问题及应对策略［J］. 中国商论，2014（15）：181-182.

［23］贾慧，张雪娜. 网络科技新闻标题的文体特征及翻译［J］. 中国科技投资，2013（20）：149.

［24］靳静波. 科技英语文体特征及翻译技巧［J］. 科技信息，2009（7）：193，230.

［25］金其斌. 宣传资料的翻译策略初探［J］. 中国科技翻译，2003（4）：23-27，61.

［26］赖兴娟. 机械类产品使用说明书的翻译［J］. 现代交际，2019（6）：98-100.

［27］雷超. 国际商务合同英语的文体特色研究［J］. 黑龙江对外经贸，2009（2）：70-72.

［28］李创. 通信专利汉译英中无主语句的处理［J］. 中国科技翻译，2021，34（3）：5-7，46.

［29］李乐乐. 修辞劝说视角下企业外宣翻译策略的分析——以华为官网的公司简介为例［J］. 外文研究，2020，8（1）：67-72，108.

［30］李秀存，李耀先，张永强. 科技论文英文摘要的特点及写作［J］. 广西气象，2001（3）：58-60.

［31］林庆扬，石春让. 基于语料库的企业简介文体分析及英译启示［J］. 长春师范学院学报（人文社会科学版），2011，30（1）：107-111.

［32］刘明东. 英语被动语态的语用分析及其翻译［J］. 中国科技翻译，2001（1）：1-4.

［33］刘燕. 科技英语新能源产品的翻译策略［J］. 知识文库，2020（1）：5-6.

［34］刘源甫. 科技论文摘要英译技巧［J］. 中国科技翻译，2003（1）：6-9，26.

［35］龙志勇，徐长勇，尤璐. 商务英语翻译［M］. 3版. 北京：对外经济贸易大学出版社，2019：175-177.

［36］罗琼. 会展外宣文本汉译实践研究报告［D］. 西安：西北大学，2017.

［37］卢小军. 中美网站企业概况的文本对比与外宣英译［J］. 中国翻译，2012，33（1）：92-97.

［38］莫再树. 商务合同英语的文体特征［J］. 湖南大学学报（社会科学版），2003（3）：83-88.

［39］戚云方. 合同与合同英语［M］. 杭州：浙江大学出版社，2004.

［40］曲艺，李鹏. 科技产品介绍的文体特点［J］. 现代交际，2019（15）：90-91.

［41］任楚威. 英文专利文献的语言特点及其翻译研究［J］. 中国科技翻译，1994（1）：25-27，58.

［42］沈育英. 科技论文英文摘要的特点及写作［J］. 中国科技翻译，2001（2）：20-22.

[43] 宋德文. 国际贸易英文合同文体与翻译研究 [M]. 北京：北京大学出版社，2006.

[44] 苏士雄. 科技英语广告翻译技巧的探讨 [J]. 中国科技翻译，1990 (2)：25-28.

[45] 腾超. 会展翻译研究与实践 [M]. 杭州：浙江大学出版社，2012：160-175.

[46] 滕真如，谭万成. 英文摘要的时态、语态问题 [J]. 中国科技翻译，2004 (1)：5-7.

[47] 田传茂，许明武. 报刊科技英语的积极修辞及其翻译 [J]. 中国科技翻译，2001 (1)：26-29.

[48] 王国亮. 浅析科技新闻翻译的特点及翻译方法——以《参考消息》科学技术版的翻译为例 [J]. 长春理工大学学报（社会科学版），2010，23 (3)：84-86.

[49] 王留喜. 产品说明书的若干句型 [J]. 上海科技翻译，1988 (5)：17-18.

[50] 王璐. "功能加忠诚"视角下科技产品操作说明书翻译实践报告 [D]. 天津：天津大学，2017.

[51] 王平辉. 会展文案写作：规范与范例 [M]. 南宁：广西人民出版社，2008.

[52] 王身健. 被动句在科技英语翻译中的应用 [J]. 中国科技翻译，1997 (2)：59-60.

[53] 王思菡，韩荣. 从跨文化交际角度探究科技新闻翻译 [J]. 国网技术学院学报，2013，16 (6)：73-75.

[54] 王振南. 当前对外会展宣传翻译中的常见问题 [J]. 上海翻译，2009 (4)：25，34-37.

[55] 吴斐. 文化图式视域下科技新闻的翻译研究 [J]. 科技视界，2013 (34)：43，47.

[56] 邬金. 科技英语说明书的文体特征及翻译技巧初探 [J]. 辽宁广播电视大学学报，2013 (3)：91-93.

[57] 向国敏. 会展文案：写作与评改 [M]. 上海：华东师范大学出版社，2008.

[58] 向群. 浅谈专利翻译规范 [J]. 价值工程，2014，33 (23)：292-293.

[59] 相廷礼. 产品说明书的特点及翻译 [J]. 企业导报，2009 (2)：140-141.

[60] 熊玲林. 试论以对外宣传为目的的企业简介翻译——兼论文本的译前处理 [J]. 科技信息（学术研究），2007 (2)：29-30.

[61] 徐爱君. 部分进口产品英文说明书的表达特点及翻译实践 [J]. 沙洲职业工学院学报，2019，22 (1)：50-53.

[62] 许传桂，徐锡华. 科技论文英文摘要写作技巧 [J]. 中国科技翻译，1998 (3)：19-23.

[63] 徐芳芳，徐馨. "公司简介"英译的分析与探究 [J]. 浙江教育学院学报，2005 (1)：44-49.

[64] 徐沛文，冷冰冰. 专利文献术语英译常见困难与实用策略 [J]. 中国科技翻译，2019，32 (4)：28-31.

[65] 阎庆甲. 科技英语翻译方法 [M]. 北京：冶金工业出版社，1981.

［66］杨清平. 应用翻译的规律与原则应当如何表述——评林克难教授"看易写"原则［J］. 上海翻译，2007，(3).

［67］杨寿康. 科技文章英译的精练与简洁［J］. 中国科技翻译，1995 (1)：56-59.

［68］杨寿康. 科技文章英译的分句法［J］. 中国翻译，1996 (3)：28-29.

［69］杨一秋. 合同英语文体特点及翻译要点［J］. 中国科技翻译，2003 (4)：11，40-42.

［70］余高峰. 科技英语长句翻译技巧探析［J］. 中国科技翻译，2012，25 (3)：1-3.

［71］曾海帆. "切"也是科技翻译应该遵循的原则之一［J］. 中国科技翻译，1994 (4)：1-3，10.

［72］曾剑平. 机械设备使用说明书的文体特点及其翻译［J］. 中国翻译，2004 (6)：72-74.

［73］张超，邹轶，张振华，等. 农业科技论文英语摘要翻译的特点与技巧［J］. 农业科学研究，2014，35 (1)：73-76.

［74］张成智，王鹏. 专利翻译研究现状综述［J］. 才智，2016 (36)：217.

［75］张克亮，崔钦华. 科技英语文体特点的实证法分析［J］. 外语研究，2001 (1)：63-64.

［76］张立民，阎兴朋，孙泰霖，等. 英汉对照应用文大全［M］. 2版. 南京：江苏科学技术出版社，1984.

［77］张彤. 论科技品牌与产品名称翻译中的影响因素［D］. 北京：北京邮电大学，2017.

［78］张文英，张晔. 英语科技应用文翻译实践教程［M］. 北京：国防工业出版社，2015.

［79］战英民. 专利文摘翻译谈略［J］. 上海科技翻译，1989 (4)：20-21.

［80］郑琼京. 公司简介翻译的冗余度失衡研究［J］. 传播力研究，2019，3 (4)：185-186.

［81］周燕，廖瑛. 英文商务合同长句的语用分析及其翻译［J］. 中国科技翻译，2004，17 (4)：29-32.

［82］朱箐兰. 科技广告研究［D］. 长沙：湖南大学，2010.

［83］程妍. 中美企业简介体裁对比分析［J］. 湖北经济学院学报（人文社会科学版），2012，9 (5)：133-134，145.

［84］邹渝，顾明. 法学大辞典［M］. 北京：中国政法大学出版社，1991.

［85］刘军平. 西方翻译理论通史［M］. 武汉：武汉大学出版社，2009.

［86］赵萱，郑仰成. 科技英语翻译［M］. 北京：外语教学与研究出版社，2006.

反侵权盗版声明

  电子工业出版社依法对本作品享有专有出版权。任何未经权利人书面许可，复制、销售或通过信息网络传播本作品的行为；歪曲、篡改、剽窃本作品的行为，均违反《中华人民共和国著作权法》，其行为人应承担相应的民事责任和行政责任，构成犯罪的，将被依法追究刑事责任。

  为了维护市场秩序，保护权利人的合法权益，我社将依法查处和打击侵权盗版的单位和个人。欢迎社会各界人士积极举报侵权盗版行为，本社将奖励举报有功人员，并保证举报人的信息不被泄露。

举报电话：（010）88254396；（010）88258888

传 真：（010）88254397

E-mail：　dbqq@phei.com.cn

通信地址：北京市万寿路 173 信箱

     电子工业出版社总编办公室

邮 编：100036